THE NEW ECOLOGICAL HOME

Selected Titles by the Author

The Solar House: Passive Heating and Cooling
  (Chelsea Green)

Superbia! 31 Ways to Create Sustainable Neighborhoods
  (New Society Publishers)

*The Natural House: A Complete Guide to Healthy, Energy-Efficient,
  Environmental Homes* (Chelsea Green)

*The Natural Plaster Book: Earth, Lime, and Gypsum Plasters for
  Natural Homes* (New Society Publishers)

Lessons from Nature: Learning to Live Sustainably on the Earth
  (Island Press)

Beyond the Fray: Reshaping America's Environmental Response
  (Johnson Books)

*Voices for the Earth: Vital Ideas from America's Best Environmental
  Books* (Johnson Books)

Environmental Science: Creating a Sustainable Future, 6th ed.
  (Jones and Bartlett)

Natural Resource Conservation: Management for a Sustainable Future,
  9th ed. (Prentice Hall)

Human Body Systems: Organization and Structure
  (Jones and Bartlett)

Human Biology: Health, Homeostasis, and the Environment, 4th ed.
  (Jones and Bartlett)

Biology: The Web of Life (West)

Essential Study Skills (Brooks-Cole)

THE NEW ECOLOGICAL HOME

The Complete Guide to Green Building Options

DANIEL D. CHIRAS

Chelsea Green Publishing Company
White River Junction, Vermont

Chapters 10 and 11 of this book were adapted with permission
from articles published in *Mother Earth News.*

Designed by Peter Holm, Sterling Hill Productions
Printed in the United States of America
First printing, March 2004
10 9 8 7 6 5 4 3

Printed on acid-free, recycled paper

Library of Congress Cataloging-in-Publication Data
Chiras, Daniel D.
The new ecological home : a complete guide to green building options / Daniel D. Chiras.
p. cm.
Includes bibliographical references and index.
ISBN 1-931498-16-4 (alk. paper)
1. Ecological houses. 2. Architecture and society. 3. Green products.
4. Dwellings—Energy conservation. I. Title.
TH4860.C45 2004
690'.8—dc22

2004000103

Chelsea Green Publishing Company
Post Office Box 428
White River Junction, VT 05001
Editorial and Sales offices: (802) 295-6300
To place an order: (800) 639-4099

www.chelseagreen.com

To the pioneers of green building, who have given this movement life and continue to contribute tirelessly to the creation of a sustainable future, one building at a time . . .

CONTENTS

ACKNOWLEDGMENTS

I am deeply indebted to the numerous people who have aided me in the preparation of this book, answering questions, sending information, or simply steering me in the right direction. A great many thanks to Alex Wilson (BuildingGreen), David Johnston (What's Working), Randy Udall (CORE), David Adamson (EcoBuild), Ron Judkoff (NREL), Chuck Kutscher (NREL), Brian Parsons (NREL), Kristin Shewfelt (McStain Neighborhoods), Doug Schwartz (Grayrock Commons Cohousing), Debbie Behrens (Highline Crossing Cohousing), Jay Scafe (Terra-Dome), Brad Lancaster (Drylands Permaculture Institute), Cedar Rose Guelberth (Building for Health Materials Center), Greg Marsh (Gregory K. Marsh and Associates), Doug Hargrave (SBIC), John and Lynn Bower (The Healthy House Institute), James Plagmann (HumanNature), Doug Seiter (NREL), Jennie Fairchild (P.A.L. Foundation), Marcus von Skepsgardh (P.A.L. Foundation), Samantha McDonald (Spatial Alchemy), Charles Bolta (American Environmental Products), Bill Eckert (Friendly Fire), Niko Horster (Chelsea Green), Bruce Brownell (Adirondack Alternate Energy), Heinz Flurer (Biofire), Vashek Berka (Bohemia International), and Doni Kiffmeyer and Kaki Hunter (OK OK OK Productions). Thanks also to the many individuals and companies who unselfishly provided photos for this work.

I am also grateful to my colleagues at Chelsea Green: Jim Schley, who helped me early on to shape this book and focus it on the right audience; my copyeditor, Alan Berolzheimer, for helping refine the manuscript; and Collette Fugere, who ushered this book through production.

I owe a huge debt of gratitude to my two sons, Forrest and Skyler, for their love and affection and constant good humor. As always, you help me maintain my perspective on what's important. Finally, I owe many thanks to my partner, Linda, for her patience, understanding, kindness, and unwavering love and support.

Creating Sustainable Shelter

Shelter is one of the most basic of all human needs, rivaling water, food, and clothing. It safeguards us from foul weather and shields us from sweltering heat and oppressive humidity. Without it, human survival in the diverse and sometimes harsh climates we inhabit would be difficult, if not impossible.

But shelter offers more than protection. It serves as the nucleus of our family life. It is here that we and our loved ones share our lives, grow, and celebrate life's victories and mourn intermittent defeats. Although not always a respite from the hectic pace of modern life, our homes can provide peace, solitude, and a place for reflection and quiet contemplation.

Shelter is also a financial investment that can reap huge economic rewards. And in many countries, providing shelter is big business—very big business. According to the U.S. Department of Commerce Bureau of the Census, newly built single-family homes, duplexes, condos, townhouses, and apartments constructed in 2002 were worth an estimated $389 billion! Hundreds of thousands of jobs, from forest managers, lumberjacks, and truck drivers to framers, carpet installers, and painters, are created by the massive home-building industry.

However, shelter, like many other elements of human existence, comes at an extraordinary cost to the planet and its inhabitants. For example, construction of the 1.2 million new homes annually in the United States results in a massive drain on the Earth's natural resource base. Consider our forestlands. In the United States, 85 percent of all new homes are framed with wood. Wood is used elsewhere in construction, too, for instance, to manufacture exterior sheathing, doors, and floors. Today, nearly 60 percent of all timber cut in the United States is used to build houses (Figure 0-1 a & b).

According to the National Association of Home Builders, a typical 2,200-square-foot home requires 13,000 board feet of framing lumber: two by fours, two by sixes, and larger dimensional lumber. If laid end to end, the framing lumber required to build an average-sized new home would stretch two and a half miles. If all the dimensional lumber used to build these 1.2 million new homes constructed in the United States each year were laid end to end, it would extend 3 million miles—to the moon and back six and a half times.

Many other countries also rely heavily on forests for wood. Humankind's massive demand for timber results in extensive tree cutting. In many regions, timber harvesting is carried out with little concern for the environment. Trees are clear-cut and the land is laid bare, often on an enormous

Figure 0-1 a & b.
Modern home construction
requires mountains of wood,
harvested from the world's
diminishing forests. Each
new 2,000-square-foot
home requires about
an acre of clear-cut.

Source: Dan Chiras

An average-sized new
wood-frame home in the
United States results in a
clear-cut approximately
one acre in size. The larger
the home, the larger the
clear-cut.

scale. Massive deforestation, in turn, results in a long string of environmental problems: soil erosion, sediment pollution in nearby streams, loss of
wildlife habitat, and species extinction (see figure 0–2).

Although forests can be replanted and managed sustainably, deforestation continues to be a problem worldwide and will likely worsen as the
human population's insatiable demand for wood and wood products continues to grow.

Home construction is also a source of enormous waste. Building an average 2,200-square-foot house, for instance, generates three to seven tons of
waste. This small mountain of waste, which is typically hauled off to local

Figure 0-2.
Wood for home building often comes from huge clear-cuts like these in the Pacific Northwest. Clear-cutting leaves enormous scars on the land.

Source: Gerald and Buff Corsi; Visuals Unlimited

landfills, contains an assortment of materials, including scrap wood, cardboard, plastic, glass, and drywall. In the United States, home construction, remodeling, and demolition of dilapidated homes are responsible for approximately 25 to 30 percent of the nation's annual municipal solid waste stream.

Building homes clearly depletes natural resources and produces a huge amount of trash. However, the environmental problems created by shelter continue long after the construction of a home ends. Huge amounts of water, electricity, natural gas, food, and countless household products stream into our homes throughout their useful life span. Although our individual needs may not seem like much, they add up quickly. According to the U.S. Department of Energy, each year America's homes consume approximately one-fifth of the nation's fossil fuel energy, for heating, cooling, lighting, running appliances, and a host of other purposes. This fossil fuel consumption is responsible for one-fifth of the nation's annual carbon dioxide emissions, which, many scientists believe, is contributing to dramatic changes in the world's climate with severe economic and ecological repercussions worldwide. Bizarre and ever more violent weather, rising sea levels, warming seas, and record-breaking temperatures, with hundreds—sometimes thousands—perishing each year from the heat, are just a few of the documented consequences of the planet's unsettling change in climate (see figure 0-3).

The day-to-day operation of conventional homes is also a source of other types of pollution. In industrial nations, billions of gallons of waste water are released each day from our homes. This waste, containing human excrement, toxic cleaning agents, and other household chemicals, is routed to septic tanks in rural areas or municipal sewage systems in cities and towns.

Buildings damage our planet and our future both during construction and during their occupation.

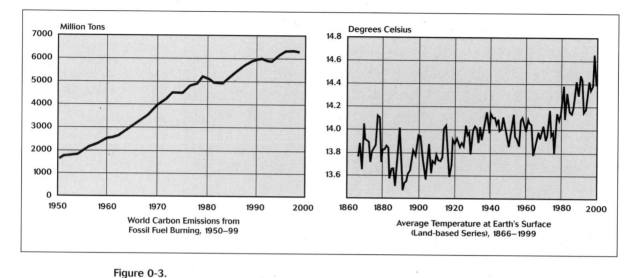

Figure 0-3.
Rising fossil fuel use over the past 100 years has increased carbon dioxide levels in the atmosphere by around 30 percent and resulted in a small but significant increase in average global temperature (global warming) that is having dramatic and costly impacts on our climate, our lives, our economy, and the lives of millions of species that share this planet with us.

Source: Worldwatch Institute

Our homes contribute significantly to the on-going deterioration of the air we breathe, the water we drink, the soil we depend on, and the wildlife that enrich our lives. And some homes are even poisoning the occupants they were designed to protect through indoor air pollution.

From septic tanks, the wastes seep into the ground, where they may trickle into and pollute underground water supplies. Although sewage treatment plants remove much of the waste, collectively they release millions of gallons of potentially harmful pollutants into surface waters each day—the lakes and streams from which we obtain our drinking water.

Unbeknownst to many, then, our homes are steadily poisoning the planet with acid rain, sewage, synthetic chemicals, radioactive wastes, and toxic sludge from power plants. Furthermore, researchers have found that our homes contain many building materials and products, including furniture, furnishings, appliances, paints, stains, and finishes, that release toxic fumes into the indoor air. Among this large list of toxicants is formaldehyde. Found in plywood, oriented strand board, cabinets, furniture, fabrics, and a host of other products in our homes, formaldehyde causes numerous health problems. In some people, it has triggered a debilitating disease known as *multiple chemical sensitivity* (MCS). Victims of MCS often become nearly incapacitated by exposure to formaldehyde or other chemicals. Some evidence suggests that formaldehyde may cause cancer. Moreover, formaldehyde is just one of a hundred or more pollutants that degrade the air inside our homes.

When we think of our homes, we don't tend to think about resource depletion, environmental pollution, habitat destruction, species extinction, or crippling exposure to indoor pollutants. We imagine our homes as comfortable refuges, not as sources of personal and environmental harm.

Maverick architect and builder Michael Reynolds, based in Taos, New Mexico, identifies another problem with our homes: the extreme dependency they foster. Our homes—and those of us living within them—are totally dependent on human-made life-support systems for food, energy, water, and waste disposal, he argues. As with a patient in an intensive care unit, cutting the lines to the outside world, even for a day, causes suffering.

The personal economic cost of our dependency, the thousands of dollars we spend each year on heating and cooling our homes, providing electricity, and supplying other amenities, is also significant.

Fortunately, many of these basic needs can be provided at a fraction of the economic and ecologic cost.

The Rise of Green Building

Today, a new generation of architects and builders is emerging, intent on creating homes that meet human needs without depleting the planet's resources, polluting the environment, poisoning their occupants, and driving innocent species to extinction. Relying on new building principles, practices, and technologies, this bold cadre of green builders is seeking to create shelter while protecting the life-support systems of the planet that sustain people and economies.

Environmentally responsible homes, often called *green homes,* come in many forms and use a wide variety of building materials—some conventional, some unfamiliar and innovative—from engineered lumber to natural earthen materials. They all have one thing in common: They provide shelter at a fraction of the environmental impact of conventional housing. They are built to promote human health and to emancipate us from fossil fuels and other costly inputs. As Dave Johnston, green building consultant, puts it in his book *Building Green in a Black and White World,* "Green building is about creating the future with intention: the intention to build a better world for our children and our children's children." It also contributes to the long-term welfare of the millions of species that share this planet with us. Ultimately, green-built shelter is good for people, the planet, and our economy.

Although green homes come in many forms and use a wide variety of building materials, they all have one thing in common: They provide shelter at a fraction of the environmental impact of conventional housing.

Figure 0-4.
Straw bale homes like this one, with thick, insulated walls, provide extraordinary comfort, reduce wood use, and can be exquisitely beautiful. Part of a movement called natural building, straw bale construction is just one of many options for creating more economic, people-friendly, and environmentally sensitive shelter.

Source: Catherine Wanek

Green building is beginning to attract a lot of attention. In 1999, for instance, the National Association of Home Builders—the trade organization for the entire home-building industry—sponsored its first green building conference. The conference now occurs annually, growing each year as more and more contractors, architects, and homeowners turn to what many people see as an inevitable evolution in building practices. Today, many other organizations are sponsoring similar events.

Another encouraging sign of change is the presence of a growing number of specialty retailers such as Planetary Solutions in Boulder, Colorado; Eco-Wise in Austin, Texas; and Environmental Construction Outfitters in New York City. These and other similar companies (listed in the resource guide) supply a wide range of green building products, such as carpeting made from recycled plastic pop bottles and nontoxic paints, stains, and finishes. They provide materials to local builders but also ship nationwide.

Perhaps even more encouraging, green building products are also beginning to appear in mainstream building supply stores. Home Depot, which reportedly supplies one-tenth of America's lumber, has pledged to offer wood and wood products from sustainably harvested forests. This supplier also offers green building products such as energy-efficient compact fluorescent light bulbs and, in some locations, solar electric systems.

> Green builders are creating shelter that is good for people, the planet, and our economy, satisfying the triple bottom line.

Green building programs are yet another heartening sign of change. At this writing, nearly twenty U.S. cities, counties, and states have established green building programs that certify homes built to meet standards for efficiency and the use of healthy, low-environmental-impact building materials. Austin's Green Building Program, the first of its kind and a spark plug for the movement, supplies a wealth of information on products that homeowners and builders can use to make homes healthier, safer, more economical, and more environmentally friendly.

Many home builders are jumping on the bandwagon, from small contractors who build a few homes each year to massive production builders who construct hundreds, even thousands, of new houses. Centex and U.S. Homes, two of the nation's largest commercial home builders, are now turning to green building practices. As Ron Jones of Sierra Custom Builders in Placitas, New Mexico, notes, "Green building is no longer the province of starry-eyed visionaries in Birkenstocks."

Even the U.S. Navy's engineers, who are responsible for all building and infrastructure carried out by this branch of the armed services, have adopted an official green building policy. It calls on them to incorporate affordable sustainability principles and concepts in the design of all facilities and infrastructure projects to the fullest extent possible. In 1995, the Naval Facilities Engineering Command began conducting sustainability workshops for its personnel and has incorporated green building principles and practices into many projects.

Further propelling this exciting movement forward, both the Department of Energy and the U.S. Environmental Protection Agency have launched green building programs. Even lending institutions are getting into the act, offering special loans for people who build or buy energy-efficient homes (see chapter 6).

Clearly, a powerful "greening" of the building sector is underway in America. But don't think that the movement is restricted to home construction. Far from it. Many new office buildings, universities, schools, and government buildings are incorporating long lists of green building materials and ideas.

And don't think that Americans hold the corner on the market. Hardly. Europeans have been building green much longer than we have, and they frequently produce buildings whose energy- and resource-conserving performance exceeds that of some of our best-built structures.

What Is *The New Ecological Home* About?

This book is written for a broad audience, from home buyers to owner-builders to architects, contractors, and any other individuals who are interested in building a sustainable future. It contains a wealth of up-to-date, practical, and reliable information on green building techniques, materials, products, and technologies that can be applied in part to make a home—or any building—more environmentally sound, or in whole to create truly sustainable shelter.

My main objective has been to provide an overview of the green building techniques, materials, products, and technologies that are currently available and some that are coming out in the near future. My intention is not to make readers green building experts but to provide sufficient detail to enable you to be conversant in all aspects of green building and design. When finished with the book, you will have a solid grasp of the components of truly sustainable shelter.

Whether you are a professional involved in designing and building homes or a citizen who is planning on buying or building a new home or simply remodeling an existing house, this book should help you meet your dreams of safe, comfortable, affordable shelter that is environmentally friendly, healthful, attractive, comfortable, independent, and economical to operate. The advice and guidance offered here, gleaned from my own experience and the experience of countless others, could save you thousands of dollars by helping you avoid many common problems when building green.

Organization of the Book

This book is divided into three parts. Part 1 explores the principles of green building and explains why you should strongly consider building or buying a green home. It also explains why it makes sense to build as green as possible. Part 1 ends with a discussion of one of the most fundamental aspects of green building and design: site selection and protection.

In part 2, we turn our attention to other crucial aspects of green building, notably indoor air quality, the use of green building materials, the efficient use of wood, energy conservation, and ergonomic design. We will then examine a variety of building options, looking at ways to use conventional and nonconventional materials, such as straw bales or earthen materials, to build homes or additions that meet our needs yet protect people and the planet and provide a greater measure of independence.

Part 3 focuses on renewable and environmentally benign systems that provide the necessities of life: electricity, heating and cooling, water, waste treatment, and food. It also explores ways to landscape a home for energy efficiency and environmental protection.

At the end of the book, you will find a comprehensive resource guide that lists other books, videos, articles, suppliers, and more, that offer more in-depth information. Three appendices are also included. Appendix A provides a comprehensive list of suggestions from this book, which you can use when buying or building a green home. Appendices B and C are useful checklists from two exemplary green building programs. You'll also find a glossary starting on page 313 that defines some of the building terminology.

PART ONE

SETTING THE FRAMEWORK

CHAPTER 1

Green Building and Beyond:
Where Do You Begin?

Imagine a home that stays cool in the summer and warm in the winter, yet requires only a small fraction of the energy a standard house of similar size uses. Imagine, too, that the home is made from an assortment of green building materials and products that significantly reduce global demand for natural resources, thus minimizing environmental damage. Imagine, furthermore, that a large portion of the waste generated during the construction of the house is recycled into an assortment of useful products, such as paper, garden mulch, and particle board, rather than being tossed into a landfill.

This description is not a dream of things that might be but, rather, a profile of a home built by one of America's newest green builders, a high-volume, national home builder, Centex Homes, headquartered in Concord, California.

The 21st Century Performance Home by Centex is "beautiful on the outside and works better on the inside, making it comfortable, economical, and friendly to the environment," says the company's brochure.

Working in partnership with the U.S. Department of Energy, the National Renewable Energy Laboratory, and a number of other organizations, Centex hopes to build many other similar homes in the Bay Area. These homes will offer all of the amenities of modern life, comfort, and beauty. They will also incorporate a wide range of green building products, techniques, and technologies. For example, each home will feature solar panels that generate electricity from sunlight, known as *solar electric* or *photovoltaic modules* (PVs). Another technology, solar hot-water panels, provides hot water for showers, baths, and dishwashing. The rest of the hot water comes from an energy-efficient instantaneous (or tankless) water heater that produces hot water on demand—as soon as you turn on the faucet. Together, the solar hot-water panels and high-efficiency water heater cut fossil fuel use in each home, providing significant savings on monthly fuel bills.

Energy savings in Centex's 21st Century Performance Home also derive from highly efficient windows, designed to prevent heat from radiating into the house during the summer, and a special type of roof sheathing that incorporates a layer of thin aluminum foil that blocks 97 percent of the sun's heat during the summer, reducing attic temperatures by as much as 30°F on the hottest days. As you will see in chapter 12, reducing attic temperatures reduces temperatures in the living space and increases comfort levels.

To further reduce external heat gain, that is, heat entering the building

> "What's the use of a house if you haven't got a tolerable planet to put it on?"
>
> **Henry David Thoreau**

Each year Centex donates $35 from each home it builds and sells nationwide to The Nature Conservancy to help it save America's endangered habitats. The total annual contribution comes to a whopping $1.3 million.

from the outside, the designers have included porches, roof overhangs, and plant-festooned trellises, features that add curb appeal as well. Combined with effective ceiling and wall insulation and energy-efficient windows, these measures cut summer cooling requirements by around 70 percent.

Centex's 21st Century Performance Home incorporates an assortment of environmentally friendly building products as well, among them foundation concrete made from 15 percent recycled fly ash, a waste product from the pollution control devices on coal-fired power plants. This practice reduces the amount of cement required to build homes, decreases waste, and greatly lessens the energy required by and the environmental impact resulting from cement production, for reasons explained in chapter 4.

Special long-lasting siding made from waste wood fiber and cement protects the home and comes with a much-welcomed fifty-year warranty. The siding will withstand hurricane-force winds, won't rot, and is virtually immune to water damage. Adding to its benefits, the siding is fire and termite resistant. It also holds paint longer than traditional wood siding does, which cuts down on maintenance. The net effect of reduced exterior maintenance: more free time to enjoy life.

Exterior walls of the 21st Century Performance Home are insulated with cellulose made from recycled newsprint. This material is resistant to mold and pests and fills wall cavities tightly, reducing air filtration into and out of the house. Its insulative properties and its ability to reduce air infiltration result in year-round energy savings. Foam- and caulk-sealing of cracks also contribute to energy efficiency by blocking air movement into and out of the house through the building envelope: its walls, roofs, and windows.

Further contributing to its environmental benefits, the 21st Century Performance Home is built from engineered lumber, such as particle board. Engineered lumber is manufactured from wood fibers that are glued together to produce sheathing (such as oriented strand board) and framing lumber (headers, beams, and floor joists). As you will see in chapter 5, engineered lumber uses a higher percentage of the wood fiber from trees, meaning fewer trees need to be cut down to provide wood. It also typically is made from smaller-diameter trees, reducing the need to harvest large trees in old-growth forests, and thereby protecting the temperate rain forests of the Pacific Northwest and southeastern Canada, an ecosystem nearly eliminated by clear-cutting.

As good as they are, engineered wood products are typically manufactured using resins that contain formaldehyde. In Centex's home, those engineered wood products in direct contact with indoor air are sealed to prevent this potentially toxic chemical from contaminating indoor air.

Further contributing to high-quality indoor air, paints and adhesives used in the home contain no formaldehyde and much lower levels of volatile organic compounds than are found in commonly used finish products. In addition, a ventilation system provides replacement air, eliminating pollutants from daily activities such as cooking and cleaning, while ensuring a steady supply of fresh air.

The already impressive list of green building features in Centex's home does not stop here. The company installed energy-efficient compact fluorescent lights, ceiling fans to cool the home during the summer, and water-efficient faucets and toilets. Insulation along the outside edge of the foundation results in a warmer floor in the winter and less air conditioner use in the summer. Thicker drywall on walls and tiled surfaces on floors in the 21st Century Performance Home help absorb heat and thus reduce temperature fluctuations year-round, for reasons explained in chapters 11 and 12. Tile floors do not provide a breeding ground for dust mites like carpeted floors, and they do not release or harbor potentially toxic chemicals.

Although even more can be done to create environmentally friendly buildings, Centex's home is truly inspirational. Serving as a model of the home of the future, it offers many direct benefits: Not only is this home cheaper to operate and maintain, it is built from healthy materials and it promotes energy self-reliance.

As we examine the principles and practices of green building, it is important to remember that green building lies on a continuum, ranging from modest changes that help promote environmental protection, human health, and our long-term economic welfare to dramatic changes designed to make a home as socially, economically, and environmentally sustainable as humanly possible.

Principles of Green Building

For much of my adult life, I've been searching for principles of sustainability that could be used to help human society steer back onto a sustainable course. My search began during the early 1980s, after I "retired" from the University of Colorado at age 30 to pursue a full-time writing career.

Over the next decade or so it became clear that human society was on an unsustainable course. The signs were everywhere. In the early 1990s, I undertook a detailed study of over two dozen environmental indicators, examining trends over a period of three decades. This study showed that despite improvements, the vast majority of these trends were unsustainable.

Over time, it also became clear that the problems we face—such as global air pollution and species extinction—are really symptoms of another, more troubling problem, notably that the systems we rely on for providing food, housing, water, energy, and other goods and services are fundamentally unsustainable. Designed in a time of abundance and relative ignorance of the problems we were creating, these systems may provide us with what we want, but they are rapidly eroding the Earth's ability to support the ever growing human population. But what made them—and us—so unsustainable?

Shades of Green

"There are as many ways to build green as there are builders in this country. There are also many shades of green. The darker green it is, the more environmentally friendly or sustainable the house is."

David Johnston
Building Green in a Black and White World

The answer came to me a few summers after I had written an environmental science textbook. At the heart of the problem is unrestrained population growth, accompanied by unrestrained economic activity. Our lack of restraint in these areas, in turn, stems in part from our lack of insight into the broadcast changes we are causing in the planet's ecosystems, the life-support systems of the planet, and a lack of understanding of the importance of these systems to our economic and personal well-being.

Spurred by a sense of limitless resources and faith in technology to solve all problems large and small, we in the industrial nations of the world have created a society that, while extraordinary in many ways, is extremely wasteful and inefficient.

It seemed to me that our problems also stem from a linear view of the world we embrace and use to design systems. Companies extract resources, hammer them together to make their products, and send them to market, where they are dutifully gobbled up. When their useful life is over, or the products fall out of fashion, they often end up in the trash bin. From here it is a short trip to the landfill. Very little waste is recycled and very few products are made from recycled waste. As former World Bank economist Hermany Daly put it, "Most corporations are treating the Earth as if it were a corporation in liquidation."

But there is more to explain our unsustainable saga.

My studies suggested that our problems also arise from the fact that we have become heavily dependent on—perhaps even addicted to—fossil fuels. Today, fossil fuels are the lifeblood of modern society. Our heavy reliance on fossil fuels has spawned a host of problems, from the blight of urban air pollution to the harmful downpours of acid rain to the ever widening and increasingly more evident threat of global climate change. Although renewables could free us from these problems, we barely give lip service to such promising energy sources as wind and solar energy.

Furthermore, in our quest for comfort and convenience we humans have demonstrated a tireless proclivity for unsustainable resource management. We take what we need, deplete the resource, then move on to begin the cycle of exploitation again . . . and again. As I studied human society, it seemed to me that this lack of sustainable management stems largely from an effervescent belief in the abundance of nature and our supremacy over all other species when it comes to meeting resource needs. The motto of our times is "There's always more, and it is all for us."

In nature, though, I found a different model of existence. In natural systems, for example, resource conservation is generally the rule. Organisms typically use what they need and use it efficiently. Unsustainable harvest is a rarity in wild species. You won't find a robin's nest in your backyard with another built alongside just to hold stuff. Although there are some notable exceptions, waste does not remain for long. In nature, one organism's waste is another's food. Nothing is really wasted. In natural systems, then, resource conserva-

---

Spurred by a sense of limitless resources and faith in technology to solve all problems large and small, we in the industrial nations of the world have created a society that, while extraordinary in many ways, is extremely wasteful and inefficient.

---

tion—using what an organism needs and using it efficiently—is a key to survival and, more important, sustainability. But should this fail, there is an efficient recycling system to ensure that some organism benefits.

It also occurred to me that nature sustains itself because natural systems rely primarily on renewable resources: soil, water, air, plants, and sunlight. These infinitely replenishable resources are in essence the raw materials of ecosystems and the organisms in them. They are renewed constantly to ensure the continuation of life. Clearly, if what you rely on can be continually replenished and it is harvested at sustainable rates, survival is pretty much assured.

It also occurred to me that, unlike humans, natural populations are kept in check by natural mechanisms. Natural checks and balances help create natural restraints on resource demand, and populations live within limits.

As I pondered the contrast between humans and nature, I found that nature provided many mechanisms to repair the damage created by the diverse array of living organisms on Earth. Nature persists, in part, because of this ability to restore. Humans head toward oblivion because we practice restoration as a second thought, if we practice it at all!

Trying not to idealize nature, it nonetheless appeared to me that nature holds many secrets for human success. It was a surprise to me at the time, but as a wise student once wrote in a research paper, "After all, nature is the master of sustainability." She's persisted for millions upon millions of years and has worked out her strategies through trial and error, as many of us are attempting to do now.

In natural systems, resource conservation is generally the rule. Organisms typically use what they need and use it efficiently. Unsustainable harvest is a rarity in wild species.

All this theorizing got me to thinking: Could the principles of ecological sustainability—conservation, recycling, renewable resource use, sustainable harvest, restoration, and population control—be used to rethink and restructure human society? Could we apply these principles to human systems to create a sustainable future?

Those questions led me on a long and ongoing search and the publication of a book on the subject, *Lessons from Nature: Learning to Live Sustainably on the Earth*. Published in 1992, the book was largely ignored by just about everyone. Undaunted, I produced several much more scholarly publications in which I outlined ways in which the principles of ecological sustainability could be applied to community development and human systems. This work fell on equally deaf ears.

In 1995, though, I got a chance to apply the principles in the construction of a new home. It was time to test my theory: Could principles of conservation, recycling, renewable resource use, and restoration really be applied to something as common as a home? Would the application of this simple set of guidelines result in a sustainable home?

As the design of the home began, I found myself using these guidelines as something of a filter through which each decision on design, materials, technologies, and techniques was passed.

To my delight, the guidelines worked.

As in any endeavor, though, as I proceeded I began to learn more. I discovered, for instance, that green building requires more than strategies designed to achieve resource conservation and prevent pollution. There is a human element to green building. One factor that I hadn't given much thought to was the need to create a healthy indoor environment for people. Others reminded me that green building also requires measures that ensure serviceable, accessible, and affordable shelter.

As my understanding of green building deepened, I learned that there was still more to this challenge than producing homes that meet criteria for planetary protection and human comfort and health, affordability, and accessibility. Green building, I discovered, requires efforts that ensure the development of a rich community life, where opportunities to interact with neighbors exist, where friendship and cooperative bonds can form, all the while ensuring some degree of privacy.

These discoveries and realizations have led to the following set of guiding principles. I encourage you to use them, even tailor them to your own project, and add additional ideas you think are necessary.

> Conservation is applied common sense. The more we use, the less there is for future generations and other species.
>
> **David Johnston**
> *Green Building in a Black and White World*

Principle 1: Conservation—Use What You Need and Use It Efficiently

Conservation is applied common sense, to paraphrase David Johnston, author of *Green Building in a Black and White World,* who operates a consulting company known as What's Working. "The more we use," he explains, "the less there is for future generations and other species." Conversely, the more efficiently we use resources, the more is available for future generations and the species that live alongside us.

Conservation is the key to meeting our needs, and it ensures future generations the resources they require to satisfy their needs. The conservation principle entails two fundamental ideas: the frugality principle, that is, using what you need, and the efficiency principle, that is, using resources optimally to minimize waste. In home construction, there are a great many ways to put these ideas into practice.

Renovate Older Buildings. Renovating existing homes to achieve green building goals represents the epitome of conservation and is arguably one of the most sustainable forms of construction. Although renovation can be costly, it uses existing resources such as land, foundations, and walls. No new land must be bulldozed or cleared to make room for a new home; trees do not need to be cut down. Further benefits can be achieved if wastes generated from the project are recycled. Framing lumber, if in good shape, can be used to reframe parts of the building. If you are interested in green building, you may want to consider remodeling an existing structure rather than starting anew.

Build Small. One of the most overlooked means of applying the conservation principles is building small. Building a small, compact home greatly

Figure 1-1.
Bigger is not always better. Small homes cost less, use fewer resources, and disturb less land.

Source: Dan Chiras

reduces our consumption of resources and our impact on the environment. Smaller homes require less framing lumber, concrete, copper pipe, tile, furniture . . . less of everything.

Besides requiring fewer resources to build, smaller homes create less waste during construction. They also require fewer resources to furnish, maintain, and operate. We'll examine smaller houses in more detail in chapter 5.

Reduce Wood Use through Design. Resource demand can also be reduced by careful design and engineering. Today, most homes are overdesigned—that is, built with more materials, especially wood, than are required to provide sufficient strength. In addition, because most building materials come in increments of two feet, designs based on two-foot multiples also help reduce wood consumption. These approaches are discussed in chapter 5.

Design and Build an Energy-Efficient Home. In addition to building efficiently to conserve materials such as wood and concrete, green builders pay close attention to energy conservation. And for good reasons. Energy costs money, and producing energy from fossil fuels is at the root of many major environmental problems.

Adequate insulation of walls and ceilings and the installation of energy-efficient windows are the keys to saving energy. These simple, cost-effective measures help retain heat during the winter, making a home considerably more comfortable. Thus, they reduce the amount of oil, natural gas, electricity, or wood needed to heat a house in the winter and can save homeowners a considerable amount of money each year. And, as explained later, effective insulation makes it easier to heat a home naturally, via solar energy. Lest we forget, insulating also helps cool a home in the summer and reduces

Figure 1-2.
Air enters homes from
cracks in the building
envelope, mostly around
doors and windows and the
juncture of walls and
foundations and roofs.

Source: David Smith
(from *The Solar House*, by D.Chiras)

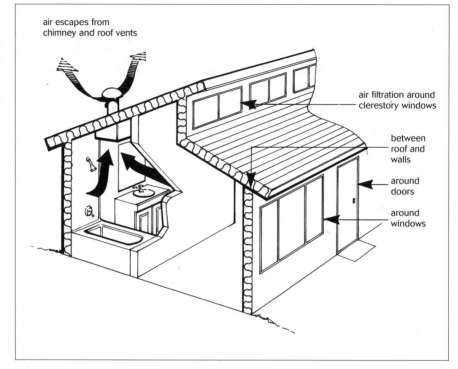

air escapes from
chimney and roof vents

air filtration around
clerestory windows

between
roof and
walls

around
doors

around
windows

summertime energy bills, which often can exceed wintertime heating costs, especially in hotter climates.

Caulking and weatherstripping also help reduce fuel consumption and increase comfort in a home, as they reduce air movement (see figure 1-2) into and out of a home, known as *infiltration* and *exfiltration,* respectively.

In this era of overlighting—some call it "headache-level lighting"—considerable amounts of energy can be saved through efficient lighting strategies. Resisting the temptation to "overlight" a room helps enormously. Supplying more wall switches, so lighting can be more carefully controlled to meet needs, also assists in cutting electrical demand. Task lighting, preferentially supplying light in areas where certain tasks are performed, for example, over a desk in a study, also decreases electrical demand, as do high-efficiency light bulbs, such as the compact fluorescent bulbs. Compact fluorescent light bulbs use 75 percent less energy to produce the same amount of light of a standard incandescent bulb. In the process, they also produce less waste heat. This, in turn, can lower cooling requirements.

Design a Water-Efficient Home. In many parts of North America, water supplies are hard-pressed to meet demand and water rationing is becoming a more common occurrence. Because of global warming, many climatologists predict even more severe water shortages in years to come.

Water shortages dictate conservation measures in many areas, but that is not the only reason for judicious use of this precious resource. Water is a valuable resource to a host of other species that make their homes in or

With water shortages looming ever large on the horizon and with populations continuing to expand, water efficiency ranks high on the list of conservation priorities.

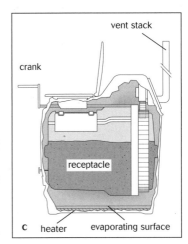

Figure 1-3. Designed to minimize odors, these waterless toilets turn human waste into a rich, organic material called humus. (a) Some units, like this Sun-Mar Excel NE made in Canada, are compact and self-contained. Waste drops into a drum, undergoes rapid decomposition, then falls into a removal tray. (b) Centralized systems like the Sun-Mar Centrex 2000 consist of a throne and a much larger receptacle usually located in the basement below. (c) Cross section through a self-contained composting toilet.

around lakes, rivers, and streams. Water conservation, both inside and outside a home, is essential to ensuring these species' survival. Besides protecting water supplies, water conservation cuts down on energy use—in this instance, energy required to run pumps that draw water from wells or lakes. It also reduces the amount of chemicals such as chlorine used to treat domestic water to make it "safe" for human use.

For these and other reasons, water-efficient showerheads, low-flush toilets, and water-miserly dishwashers and clothes washers are essential furnishings of a green-built home. Even waterless fixtures such as the composting toilet will become more popular as shortages worsen (figure 1-3).

Outdoor water conservation measures are equally important. In Phoenix, Tucson, and Denver, for instance, lawn watering consumes the bulk of a family's annual water usage. In dry climates where lawn irrigation is commonly practiced, planting low-water grass and other water-miserly (xeric) plants greatly reduces demand for irrigation water while also producing beautiful landscaping. We'll explore these and other water conservation measures in more detail in chapters 14 and 15.

Use Low-Embodied-Energy Materials. The energy that goes into a home does not all come through pipes or electrical wires. A considerable amount of energy "arrives" in the materials used to build and furnish a building. It is called *embodied energy*.

The term *embodied energy* is a used to describe all of the energy required to make a product, including the energy needed to harvest or mine raw materials, process them, and manufacture products. It also includes the energy required to transport raw materials to production facilities as well as to ship finished products to retail outlets and end users.

Consider Life-Cycle Costs of All Materials and Products

In addition to embodied energy, builders should consider the life-cycle costs of all materials and products that go into a home. The life-cycle costs are the social, economic, and environmental costs of a material or product from cradle to grave—that is, from the extraction of the raw ore needed to make it through the manufacturing to the end use to disposal or recycling.

Table 1-1

Embodied Energy of Common Building Materials and Products

Material	Embodied Energy in MJ/kg (million joules per kilogram)
Baled straw	0.24
Adobe block (traditional—mud and straw)	0.47
Concrete block	0.94
Concrete (poured on site)	1.0–1.6
Concrete (precast)	2.0
Hardwood timber, kiln dried, rough sawn	2.0
Softwood timber, kiln dried, finished	2.5
Cellulose insulation	3.3
Plasterboard	6.1
Cement	7–8
Plywood	10.4
Fiberglass insulation	30.3
Steel (virgin)	32
Carpet (nylon)	148

From: Andrew Alcom. *Embodied Energy Coefficients of Building Materials.* Wellington, New Zealand: Centre for Building Performance Research, 1998.

Building materials vary markedly in their embodied energy (table 1-1). Steel and aluminum, for example, have a much higher embodied energy than wood, because the mining and processing of the mineral ores requires much more energy than cutting trees and processing them into lumber. A wood-frame house, therefore, has a lower embodied energy than a steel-frame house.

Buying locally produced building materials is a simple way of lowering the embodied energy of the raw materials that go into making a home, because delivering a local product to the consumer requires less energy than delivering a product from a distant supplier. If you live in Salt Lake City, Utah, and have the choice of a recycled tile shipped from Georgia or a tile made from clay harvested nearby, the locally produced tile is probably a better bet from an energy standpoint.

Design for Human Efficiency. A consideration worth a great deal of thought in the design phase of any house is the efficiency of the design from a human perspective. How efficient will the house be to occupants as they move about? Will traffic flow nicely? How convenient will it be to move from one location to another? Will access to areas of high use be blocked by furniture? Will you have to run down two flights of stairs every time you want a bite to eat? Are outlets and cabinets efficiently located? Will you have to get on your hands and knees to plug in an appliance? Will you have to climb onto a stepladder to reach food in cupboards? Will countertops be situated conveniently for all members of the family to participate in food preparation?

Although designs that address these concerns are not often considered in green building, they are another important element in achieving comfort

Besides decreasing the embodied energy of the building materials in your new home, renovation, or remodel, purchasing locally available materials also helps support local economies. Moreover, buying from local producers reduces pollution generated from transportation.

and utility. Efficient design makes life easier and promotes physical health by reducing unnecessary strain on the body. Chapter 7 offers a more thorough discussion of the subject.

Compact Development. "Buy land," Will Rogers once advised, "they ain't making any more." And he's right. As cities expand and land prices skyrocket, cities, towns, and developers are looking for ways to conserve what we have, for example, by encouraging more compact development. Compact development may involve in-filling, that is, putting unoccupied, often abandoned or even degraded land within existing cities and towns to good use. Compact development also seeks to make newer development more concentrated.

By compacting new housing, planners seek to put more people on the limited land base. Using creative designs, though, they can prevent people from feeling cramped. In fact, with careful design a more dense development may actually feel more open or generously apportioned than a less dense development planned less carefully. Compact development in and around cities and towns also makes mass transit more feasible, reduces automobile dependence, and lowers the cost of providing services, such as fire protection, water, and electricity, thus conserving vital natural resources.

Four other innovative ideas—cohousing, ecovillages, traditional villages, and cluster development—also promise advantageous strategies for comfortably accommodating more people on a limited supply of land, while helping to preserve farmland, open space, and wildlife habitat from development.

Cohousing, ecovillages, and new towns represent a compact form of development that reduces the amount of land plowed under to make room for housing. But they also serve to build community—close-knit neighborhoods where people share their lives and support one another, creating a healthy social and psychological environment for children and adults. (See Appendix C for ideas.)

Principle 2: Recycle, Recycle, Recycle . . . and Compost

Since the late 1960s, Americans have been encouraged by governments and environmental groups to recycle, and by and large we've embraced this simple, cost-effective means of contributing to a brighter future.

In the 1980s, however, the recycling tune changed, as advocates discovered that recycling isn't enough. To make the system work, they learned, we need to manufacture goods made from recycled waste. And we need viable markets, that is, customers who are willing to buy products made from recycled materials.

In keeping with this advice, numerous state and federal government procurement programs were launched to encourage the purchase of products made from recycled materials. Many individuals and businesses also responded, increasing their acquisition of recycled-content goods. In the late 1980s and early 1990s, architects, builders, and designers began to ask if

> By compacting new housing and services, planners seek to put more people on the limited land base. Using creative designs, they can prevent people from feeling cramped. In fact, with careful design a more dense development may actually feel more open or generously apportioned than a less dense development planned less carefully.

they too could join in the recycling revolution by building houses from materials made from recycled waste.

Over the years, building material manufacturers have produced a plethora of products made from "waste" that would otherwise have ended up in a landfill or an incinerator. From doormats made from used automobile tires to cellulose wall and ceiling insulation manufactured from recycled newsprint to backyard decks made from recycled plastic "lumber" to the kitchen tile made from recycled automobile glass, virtually every component of a house can be made from a recycled material. Don't worry: recycled-content building materials won't compromise structural strength or quality. Most perform as well, if not better, than their conventional counterparts—and they are often priced competitively.

Salvaged materials are also useful when building or remodeling. Salvaged lumber, trim, certain plumbing fixtures, and hardware may be available from friends and neighbors or at local used building material outlets. I used several salvaged doors and a barn ladder for my children's loft. After a little sanding and application of some reclaimed paint, they were as good as new. We also used salvaged barn wood for cabinets and vanities (figure 1-4). It needed to be remilled and stained, but it turned out to be quite attractive.

In the green building movement, waste from industrial nations is turning out to be a huge asset. The list of waste materials now being used to produce high-quality building materials is mind-boggling: used automobile tires, straw, pop bottles, drywall, clothing scraps, newspapers, cardboard, plastic milk jugs, scrubber sludge, and a host of other common waste materials.

Some people are even building homes entirely out of waste. The walls of my home, for instance, are built from eight hundred used automobile tires gathered from local tire shops. The tires were packed with dirt from the building site, then coated with cement stucco. We also installed carpeting made from recycled plastic pop bottles. In Fraser, Colorado, Rich Messer built a house out

So many recycled-content building materials are now available that builders can create an entire house from waste.

BUILDING NOTE

When using salvaged materials, be sure to test their safety and energy performance. Old windows that are leaky or whose trim is covered in lead paint are not worth salvaging.

Figure 1-4. Cabinets and vanities in the author's home were made from salvaged barn wood, taken from a 150-year-old barn slated for demolition. The gray, weathered wood was remilled, sanded, and refinished with environmentally friendly stains and finishes. The results were stunning.

Source: Dan Chiras

of bales of waste paper and plastic (see chapter 4). In Spokane, Washington, a ban on burning leftover straw in the fields after wheat is harvested, enacted to reduce air pollution, has resulted in the availability of nearly 30,000 tons of bales, spawning an ambitious project to build "spec" straw bale houses priced competitively with other new homes in the area.

Chapter 4 describes numerous green building products manufactured from recycled waste and other environmentally friendly materials. The resource guide provides information on where you can find these products. Chapter 9 describes natural building techniques, such as straw bale and rammed-earth tire construction.

Building from recycled materials represents an important advance in green building, but it is only part of the equation. Home construction and remodeling create mountains of waste—approximately 25 to 30 percent of the trash in our landfills. Reusing waste from building projects is another important step along the path to a sustainable building industry. Not only does it reduce the burden on our landfills and conserve natural resources, reusing job waste also supports a vital new industry and can lower construction costs.

Architects and builders can also select building materials based on their ease of recyclability. Choosing steel roofing, which is readily recyclable, over asphalt shingles, which are not, is a good example. When steel roofing needs to be replaced, it can be shipped to recyclers. Who knows, it may even find its way back onto the roof of another home.

While you're at it, make sure your design incorporates a recycling center in your home. It should be adequately sized to accommodate the waste your family produces and convenient to clean. In addition, don't forget to incorporate outdoor composting bins to recycle organic wastes. Numerous manufacturers produce composting bins made from plastic in which organic wastes from kitchens and yards are deposited. Mixed with dirt and watered from time to time during dry spells, these wastes decompose with little, if any, odor, producing a rich organic matter that can be added to vegetable gardens, flower beds, and lawns to improve soil quality. Composting kitchen wastes lowers the amount of organic matter dumped down the drain each day and reduces the strain on septic tanks and sewage treatment plants. In addition, composting yard wastes reduces trash pickup and landfill waste and the energy required to transport wastes to landfills.

Finally, homes can be plumbed to recycle graywater—waste water from sinks, showers, and washing machines. Safe and effective technologies are also available to recycle blackwater from toilets. We will look at these options in chapter 14.

Principle 3: Use Renewable Resources

The economy of nature depends on renewable resources: the sun, soil, water, and plants. In sharp contrast, the human economy is largely maintained by a steady input of nonrenewable resources, such as coal, oil, natural gas, steel,

Putting Waste to Good Use

According to one source, 25 to 30 percent of the garbage in U.S. landfills is construction and demolition waste. Unless you and your builder are environmentally minded, most of this perfectly usable material will end up in a dumpster and will be hauled off to a local landfill.

Recycling Centers in Our Homes

As cities and towns the world over divert more and more materials from their waste streams into growing recycling networks, recycling of household trash will increase dramatically. Builders can make this task easier and more pleasant by incorporating a recycling center into every home.

aluminum, and many other metals. The differences between these types of resources are many and profound. Most obviously, as their name implies, nonrenewables cannot be regenerated by natural processes and are, therefore, finite in supply. Reliance on them is therefore a rather chancy proposition, especially given the fact that our population and economy continue to grow.

In contrast, renewable resources such as trees and soil are regenerated by natural processes. If managed carefully, these biological renewables could provide a steady stream of resources, including building materials, to human society *ad infinitum*. If not managed with care, they too can be depleted, much like nonrenewable resources.

In contrast, renewable energy, such as wind and sunlight, cannot be depleted. In most parts of the world, we don't have to worry about supplies, either. Moreover, the potential of renewable energy resources far outstrips the potential of nonrenewable fuel supplies. In a study for the U.S. Department of Energy, Robert L. San Martin compared the two, and to the surprise of many, he found that nearly ten times more energy is available *annually* from renewable resources using current technologies than is available from *all* of the fossil fuel remaining in the United States. That's right: *The annual renewable potential is nearly ten times greater than the total potential of the nation's finite fossil fuel supplies.*

Building green relies on techniques and technologies that tap into enormous supplies of renewable energy, notably solar energy to provide heat and electricity and wind to supply electricity. Fortunately, homeowners can take advantage of renewable energy in many ways. They can purchase green power from their local utility or install a set of solar electric panels on the roof or a wind generator in the backyard. With rebates and other incentives for renewable-energy installations available in some parts of the country, costs are becoming less of a barrier. Heating and cooling green-built homes with the sun and natural forces is an option in many locales and adds little, if anything, to the cost of the home while saving tens of thousands of dollars in energy costs over the lifetime of the house (figure 1-5). Chapters 11, 12, and 13 will have more to say about renewable energy.

Principle 4: Promote Environmental Restoration and Sustainable Resource Management

In nature, organisms and ecosystems possess many mechanisms that permit them to repair damage. These natural restorative mechanisms are vital to the sustainability of the rich web of life. To create a prosperous and sustainable future, we humans must take an active role in restoring the Earth's ecosystems—for example, by reviving farm fields and forests that have been damaged over the years by human mismanagement. Such actions are important, indeed essential, to meet the needs of the world's growing population and its steadily increasing demand for resources, especially building materials. Besides ensuring a sustainable supply of resources, restoring previously dev-

Infinite growth in a finite system is an impossibility.

E. F. Schumacher, *Small Is Beautiful*

Renewable resources, such as trees and soil, are regenerated by natural processes. If managed carefully, these biological renewables could provide a steady stream of resources, including building materials, to human society *ad infinitum*. If not managed with care, they too can be depleted, much like non-renewable resources.

Figure 1-5.
A passive solar home like this one relies on south-facing windows to emit the low-angled winter sun. Sunlight warms the interior and is stored inside by thermal mass such as tile and concrete slabs. Heat stored in the mass is released into the interior during the evening and during long, cloudy spells, creating wintertime comfort and potentially huge economic savings.

Source: Dan Chiras

astated landscapes helps reinstate important ecological services that ecosystems once supplied, such as flood control, air purification, water pollution control, oxygen production, and pest control.

Whether you're a builder or a homeowner, you can make a positive contribution to the restoration of natural systems. For example, revegetating a building site with native species, plants indigenous to your area, increases the chance of a speedy recovery and encourages and supports native wildlife. Native species are better adapted to local soil conditions, temperature, and rainfall than exotic species, that is, plants that originated in other climate zones. Among other things, planting native species reduces the demand for the irrigation water and fertilizer that are required for exotic species. Revegetation is discussed more thoroughly in chapter 15.

You might even consider earth sheltering your house (berming and/or burying large parts while maintaining a bright, comfortable interior) or installing a living roof—a roof covered by soil and vegetation. If properly constructed and well sealed to prevent leaks, a living roof will provide a lifetime of protection while returning a large portion of the building site to vegetation. Earth sheltering and living roofs are described in more detail in chapter 10.

Another way to promote sustainable resource use is to support suppliers of wood from companies that have adopted sound timber management and restoration policies. Such wood is known as *certified lumber.* The number of such suppliers is growing dramatically in the United States. and only promises to grow larger now that Home Depot, which sells 10 percent of all processed lumber worldwide, and its major competitor, Lowes, have pledged to supply their stores with wood from sustainably managed forests. See chapter 5 for details.

Restoring ecologically impoverished landscapes is not just an environmental do-gooders project; it is essential to our long-term survival. Planet care is the ultimate form of self-care.

Principle 5: Create Homes That Are Good for People, Too

So far, most of the discussion has been focused on improving the environmental performance of a home, that is, reducing pollution and habitat destruction. Although planet care is the ultimate form of self-care, green builders also strive to create safe, healthful living spaces that have tangible benefits for people. Today, many green builders use materials and technologies that ensure toxicant-free, healthy homes. A healthy home is also warm and free of drafts in the winter and cool in the summer. It is shielded from outside noise, well lighted, and pleasing to the spirit and the eye, both inside and out.

Build and Furnish a Home That Is Free of Toxic Chemicals. As you will see in chapter 3, many commercially available paints, stains, and finishes contain potentially toxic substances, such as formaldehyde, mercury, and volatile organic compounds (VOCs). These compounds are released into room air when these materials are applied, but also long after the painters have left in a process called *outgassing.* Plywood, oriented strand board, fiberglass insulation, glues, adhesives, carpeting, curtains, and furniture also contain chemicals that outgas, contaminating the air in a home for many months, even years, after a family moves in. In addition, indoor combustion sources, such as furnaces and water heaters, may release a number of potentially harmful chemical substances including carbon monoxide and nitrogen dioxide.

Today, many companies are producing nontoxic building products and pollution-free appliances. Chapter 4 discusses many of the building products needed for a healthy home; chapter 6 discusses indoor-air-quality-friendly appliances; and chapter 11 describes heating options that are good for the planet and contribute to healthy indoor air.

Make Your Home Easy to Operate, Service, and Maintain. Another feature of green homes is that they should be easy to operate, service, and maintain—and should require as little servicing and maintenance as possible. Passive solar homes are a good example of this principle.

Passive solar heating is a means of providing interior warmth for homes, offices, stores, and a host of other buildings. I frequently describe it as a heating system with only one moving part, the sun. Passive solar heating relies on south-facing windows (in the northern hemisphere) to let sunlight stream into a home; thermal mass to store solar energy in the interior of the house; and insulation and curtains or shades to help retain heat. Because there are no motors, fans, boilers, or furnaces, passive solar is not only easy to operate, it is reliable. Repairs and maintenance are minimal. The system will last as long as the house stands and the sun shines. To my way of thinking, passive solar energy is the lowest maintenance and easiest to operate heating system ever invented. (We will discuss passive solar heating and its summertime cohort, passive cooling, in chapters 11 and 12.)

Plaster and stucco exteriors also reduce maintenance. As discussed in chapter 9, many natural building materials, such as stone and adobe, can be derived from the earth. They provide durability and low maintenance. Native vegetation saves time by reducing landscape "maintenance" but also reduces, or even eliminates, the need for toxic chemicals (pesticides) typically used to protect nonnatives.

Another way to reduce maintenance and operation concerns is by eliminating pumps and other electrical equipment used to move water. Gravity-fed rain catchment systems, which capture water from roofs to water lawns and vegetation, for example, require far less energy than conventional water systems, such as wells, and are described in more detail in chapter 14.

Design Your Home to Be Accessible and Adaptable. Creating a safe, healthy home also requires measures that ensure easy access and use by healthy adults and children and by those in wheelchairs or on crutches. Don't forget that we all age and many of us will become injured during our lifetimes from automobile accidents and such. Making our homes easy to navigate and operate under those circumstances is important. If such measures are designed into the house in the first place, homeowners can save an enormous amount of resources, time, and money.

Homes can be designed to be adaptable to changes in our families—for example, when children grow up and leave home. Being adaptable means that a house can be easily and inexpensively modified, for example, by converting an upper-level room to an apartment suitable for a renter. Making a home adaptable also means designing in ways to make parts of a home, such as a kitchen counter, adjustable—for example, installing an adjustable countertop in a kitchen makes it useful for children as they grow. We'll explore all of these topics and more in chapter 7.

Create High-Quality, Durable Homes. Virtually any product designed to outlast its cheaply made competitors is a boon for the environment. A well-made wheelbarrow that lasts for twenty-five years, for instance, requires one-fifth of the energy and resources of a cheaply made counterpart that gives out in five years. The same is true for houses. A well-made house that lasts two hundred years requires one-fourth of the resources of the four cheaply made homes that fall apart in fifty years. Durability also reduces maintenance costs, minizes the use of resources to repair buildings, and saves money. I'll discuss this strategy more in chapter 5.

Create Community through Cohousing and Ecovillages. Meaningful human interaction is as vital to our long-term future as environmental protection. For this reason, many people have been turning to cohousing, ecovillages, and new towns.

A cohousing community typically consists of single-family homes or condominium-like structures on a commonly owned piece of ground.

Cohousing communities offer many social, economic, and environmental benefits. Many cohousing developments, for example, are built around a common house where residents often share meals or meet to discuss common decisions or just to chat. The common house may be used to accommodate visitors and may contain washing machines that everyone can use.

Guest rooms and other resources available to members of a cohousing community benefit the residents by reducing the floor space of their homes and the number of appliances they require. A large kitchen in a common house, for example, means that individual homes can be built with smaller kitchens that suffice for the vast majority of a family's needs. If you are throwing a party that will require a larger kitchen, you can reserve the common house and kitchen for the evening.

Many cohousing communities have community gardens and many set aside some of their land—often a substantial portion—as open space for children and wildlife. Forward-thinking communities often try to share equipment. Harmony Village in Golden, Colorado, for example, shares one lawn mower among the entire community—twenty-seven units in all!

An ecovillage typically consists of two to five cohousing communities organized around a small village with services within easy reach by foot or bicycle. Urban areas have also organized themselves to be mini ecovillages. Each ecovillage is a self-contained living environment that seeks to create community while it protects the environment.

A new town, or neotraditional village, seeks to mimic towns of earlier times. Seaside, Florida, is an example. Built in the early 1980s along the Gulf coast of Florida, Seaside is a mix of homes and walking paths. It is ideal for retired folks and people like myself who work at home and want to be able to access restaurants, grocery stores, and other services without having to drive to a local strip mall.

Green Building and You

By now, the benefits of green buildings—and green communities—should be pretty clear. They are healthier for the planet and healthier—both physically and mentally—for occupants. They offer quality construction and durability. And green building, although possibly more costly up front, offers huge economic savings over the lifetime of the building.

For those who care about their health and their own future, care about the health of the planet and the future of our children and their children, and care about the future of the planet and the millions of species that share this lovely place with us, green building is the way to go. But will you be going out on a financial limb?

CAN YOU AFFORD A GREEN-BUILT HOME?

The future of green building depends, in large part, on green homes being affordable to build, furnish, and operate. Contrary to the myth that building an environmental home is a costly venture reserved only for the well-to-do, this option can be quite cost-competitive. Even with some unfortunate cost overruns, my home cost $5 to $15 per square foot less than most other new spec homes in my area. And its operating costs are much lower. But that is not to say that every aspect of building and furnishing the house was cheaper. A few environmentally friendly products, like my super-efficient SunFrost refrigerator and compact fluorescent light bulbs, cost more, sometimes a lot more, than their less efficient counterparts. However, many other green building products, such as recycled tile, carpeting, carpet pad, and insulation, cost the same or less than their counterparts.

Green builders I've talked to say that green building costs from zero to 3 percent more. But remember that investments in some green products and materials may end up providing additional comfort and saving money right away. Adding extra insulation, for example, keeps a house warmer in the winter and cooler in the summer. This not only makes living space more comfortable, it dramatically lowers fuel bills, potentially saving tens of thousands of dollars over the lifetime of a house. Furthermore, because a well-insulated house requires less heat, you'll be able to install a smaller and less expensive furnace, saving money right from the start. Smaller air-conditioning units are also needed, and in some cases measures to make a home more energy efficient make air-conditioning unnecessary. Installing water-efficient toilets and showerheads reduces the demand for water, decreases electricity used for pumping water from wells, and could reduce the size of the leach field required for homes on septic systems. Because these options are no more expensive than standard fixtures, you end up saving money in the short and the long run.

Resale Value: Am I going out on a financial limb if I build a green home?
Buying, building, or remodeling an existing home is a big decision, very likely the single largest financial decision you will make in your life, often involving the expenditure of several hundred thousand dollars (especially when the interest from a thirty-year mortgage is factored into the cost equation). Although our purchases are designed to suit our personal needs and tastes, one big factor lingers in the background, ever vying for attention: *resale*.

Two questions about resale often crop up in buyers' and builders' minds. First, will I be able to sell my house down the road? And, second, will I be able to sell it at a good price? In other words, will others find my home desirable and be willing to pay a fair price for it?

Fortunately, the news on this front is good.

Very good, actually.

According to David Johnston, who advises commercial builders and municipalities on green building, environment- and people-friendly homes reflect what many purchasers want out of life: health, comfort, and an opportunity to take an active role in the preservation of the natural world. "Commercial builders who have responded to the changing market have prospered and claim they will never build the (old) way again," says

Johnston. If builders are prospering in the new green-home building market, chances are you will be able to find a willing buyer and command a decent price for your home—if you can bear to let it go.

"People care deeply about the health of their families, they always want to save money where they can, and they want to contribute to something greater than themselves," says Johnston. And market research, he adds, bears out the contention that green building is not only a good idea but that demand for green-built homes will very likely increase in the years to come.

Buying is a practical decision, based on location, quality, price, and many other factors. However, it is also a decision that is driven by values. There is no better way to say "I care about the environment" and to show your willingness to manifest your convictions than by purchasing a green-built home or one that is remodeled according to the principles and practices of green building.

A huge group of potential buyers of your green home is the Cultural Creatives, says Johnston. Cultural Creatives is a term used to describe a group of people, approximately one-fourth of the adult population in the United States, who believe that our "society faces significant problems and needs to reinvent its culture, institutions, and practices to solve problems and to provide a positive future for its children." Cultural Creatives "are integrating their values into their everyday lives and are taking action on a wide range of social, environmental, and spiritual concerns," he goes on to say. They want their purchases to reflect their core values, one of which is ecological sustainability. Green building could not be a better match.

To reach the Cultural Creatives, though, you may have to work with a real estate agent, training him or her on the value of your home beyond standard concerns such as the number of bedrooms and the size of the kitchen. You might even run a few ads yourself and prepare a flyer that points out the beneficial features of your home.

Passing Fad or Wave of the Future?

Sure you say, there's a market for green-built homes now, but is green building a trend or just a short-term fad likely to fade into oblivion?

While no one can predict the future of green building, it is likely here to stay because all of the forces that dictate the need for green building—expanding populations, continuing economic growth, environmental pollution, and resource depletion—are bound to continue well into the future. Concern for the environment and interest in taking an active role in improving the world, preserving our health, and protecting other species from the damaging effects of human progress are also likely to increase.

In short, then, trends dictate more green building in the immediate and long-term future. Green building is very likely the wave of the future.

There is no better way to say "I care about the environment" and to show your willingness to manifest your convictions than by purchasing a green-built home or one that is remodeled according to the principles and practices of green building.

"Purchasing a green home makes homebuyers feel like they are a part of the whole environmental movement. Not only are they saving money, they are making a decision for the long term. Not only the original buyer but three homeowners down the road may also receive the benefit."

Tom Hoyt
owner, McStain Enterprises
Boulder, CO

CHAPTER 2

Site Matters: Site Selection and Protection

I arrived at the Real Goods Solar Living Center in Hopland, California, two hours north of San Francisco, on a hot August day in 2000. I was in town to present a workshop on natural building and to participate in a panel discussion on sustainable design at SolFest, the Institute for Solar Living's annual renewable energy expo.

As I stepped out of my rental car in a nearby parking lot, the oppressive heat from the early afternoon hit me like blast furnace. The air temperature was in the high nineties and sweat beads immediately began to form on my body.

With perspiration dripping down my forehead, I strode to the Real Goods main entrance imagining that I was going to die from heat prostration during the weekend. But my concerns quickly evaporated when I passed through the front gate. There before me was an oasis, twelve acres of tall trees, lush grass, gardens, and cool, inviting ponds (figure 2-1). Songbirds gave music to the trees, providing a cheerful refrain to the blissful landscape.

Not only was the air temperature considerably lower than that of the surrounding landscape, but I soon found out that eight years earlier the verdant grounds had been a California Department of Transportation dump site. With hazardous chemicals seeping into the nearby surface waters, the land had once been an eyesore, something developers call a *brownfield,* devoid of soils and prone to flash floods.

My visit to SolFest that year got me to thinking about siting, one of the most important aspects of green building. Siting plays a pivotal role in how a home, or any building, performs.

In this chapter, we will explore criteria that should be considered when buying land and selecting a building site, focusing primarily on those that contribute to the energy performance and comfort of a home and those that help advance the environmental goals of green building. We'll end with a brief discussion of ways to protect land and the environment during construction.

Building within City Limits

Before we begin to explore criteria for selecting a building site, let's pause for a moment to consider where you want to live. If you're like many Americans, you may be entertaining ideas of living in the country, out of the hustle and bustle, and the crime and pollution, of our cities. Escaping to the country, even

Figure 2-1.
Who would have thought that this site—home of Gaim Real Goods and The Solar Living Institute—was, only a few years ago, a Department of Transportation waste site?

to an outlying suburb, does have its advantages. But those who escape often lament the loss of many of the amenities of living closer in. Living close in, for instance, often dramatically reduces commute time, giving us more free time to spend with our families. It may even permit one to walk, bicycle, or take mass transit to work, saving the daily nightmare of rush-hour traffic. Such alternatives help reduce local air pollution. In fact, seeking refuge from pollution by moving to the country may make urban pollution worse.

In addition, the countryside is under increasing pressure from development. According to the Farmland Trust, each year approximately 1.25 million acres fall to development, mostly new roads, subdivisions, and shopping centers. Much of this land is prime farmland. Although some naysayers contend that we North Americans are blessed with excess farmland, others take exception, calling for much more stringent measures to protect these lands. Their importance in domestic and global food production should not be overlooked. Weighing in on the debate, I like to remind the naysayers that we mustn't forget that the human population is not stagnant: Approximately 84 million new people are added every year—or about 250,000 people per day! Each one needs food. As a result, any loss of farmland represents a serious threat to future generations. And lest we forget, the loss of outlying lands also robs us of valuable wildlife habitat and open space.

As David Pearson points out in his book, *The New Natural House,* "It is much more ecologically sensitive . . . to improve the environment of the cities we have and live better in them" than to "lose more countryside to

Seeking refuge from pollution by moving to the country may make urban pollution worse.

yet more development." So, give strong consideration to purchasing land within city limits or within metropolitan areas to build your new home. You may be surprised to find a large number of vacant lots within the zone of development, including land once occupied by buildings that fell into deterioration as people fled the inner parts of our cities for the promise of the suburbs. They're typically known as *grayfields*.

What to Look for in Land

No matter where you decide to build your home, you will need to look carefully at each site, assessing each option by a variety of criteria. The most common criteria people use are the proximity factors: how close a site is to work, schools, stores, recreation, and services such as hospitals, police, fire protection, electrical service, gas lines, and mass transit.

Although at first glance proximity factors may seem entirely anthropocentric, serving people only, they do have significant environmental implications. Living close to work, schools, and grocery stores, for instance, can dramatically reduce time spent behind the steering wheel, thereby reducing gasoline consumption and air pollution. Walking and bicycling provide opportunities for the exercise needed for a long and healthy life.

Other very practical factors must also be taken into consideration when shopping for land, for example, easements, zoning restrictions, covenants, and future land development plans. Shortly after I moved into my last home, the local cable company strung a bright silver line across my property, bisecting an otherwise unobstructed view of meadow and forest. We complained, but there wasn't a thing we could do. They had an easement, a legal right to cross my property.

Be certain to check into zoning restrictions, as well. Zoning laws may prohibit homeowners from installing a wind generator or raising chickens. In subdivisions, covenants (legal restrictions on property use) may restrict the type of house one can build. My newest home, for instance, lies in a rural mountain subdivision with a long string of covenants. Here, the Architectural Review Board must approve all building projects, even the color of paint one chooses. (You can imagine their consternation when I proposed to build a home out of straw bales and recycled automobile tires!) Watch out for restrictions that could limit your use of solar energy.

Information on easements, zoning restrictions, and covenants should be supplied to you well before you sign a contract to purchase a home or a piece of property. Insist on it, and be sure to receive this information in written form. Review it carefully before signing a contract to purchase the land, so you have no surprises later.

Also, be certain to check with a local real estate agent, the county or city building department, or other knowledgeable folks about future development plans in the area. You don't want to purchase a piece of property only to find that a convenience store or a Wal-Mart is slated for construction next door.

When looking for a building site, whether in the country or within the already developed urban/suburban landscape, remember that a good site provides many "free services," such as natural drainage or access to the sun for heating your home in the winter. These features may save a homeowner thousands of dollars during construction.

Explore the Community

What do you know about the community you're choosing to live in? Does it contain many like-minded people? Is it an area that welcomes newcomers? Will you and your family be considered outsiders and treated accordingly? For most of us, belonging to a community of people is essential to our long-term psychological health.

Those readers who are guided by strong environmental values should consider "environmental amenities" such as recycling and composting facilities, bike paths for commuting to work or exercising, and mass transit. And what about open space, parks, and recreation? Are these amenities available and readily accessible?

Be ruthless and demanding when it comes to buying land. As the late Paul G. McHenry Jr. writes in his book, *Adobe: Build It Yourself.* "A cold-blooded analysis should be made before purchasing any particular parcel of real estate."

Selecting Your House Site

In addition to deciding upon where you want to live, you need to figure out exactly where on the property to site your home. Several factors contribute to a site with optimal green features.

Choose a Site with Good Solar Access

As I noted in chapter 1, solar energy is a vital component of green building. It can provide a substantial portion of a home's annual heating requirement (see sidebar). Solar energy can also be used to produce electricity and hot water. In fact, if a home is sited properly and designed efficiently, solar energy can provide all of the heat, hot water, and electricity a family will need (see chapters 11 and 13).

If you live in the Northern Hemisphere, proper siting for solar design requires a plot that is open to the south, providing access to the sun for as many hours as possible, but especially between 9 A.M. and 3 P.M. (In the Southern Hemisphere, of course, you need land that gives access to sun in the north.) Pay close attention to obstructions such as trees, cliffs, mountains, and buildings that could block the low-angled winter sun. Chapter 11 will explain solar siting and help you choose a good solar site.

Be certain to check with a local real estate agent, the county or city building department, or other knowledgeable folks about future development plans in the area. You don't want to purchase a piece of property only to find that a convenience store or a Wal-Mart is slated for construction next door.

Solar Savings

Orienting a house to the south reduces the annual heating bill by 10 percent. Shifting a few of the windows to the south side increases the solar gain and can decrease heating bills by up to 30 percent. Concentrating windows on the south side and other measures such as installing better insulation can boost your solar heating even more, cutting fuel bills by 50 to 80 percent.

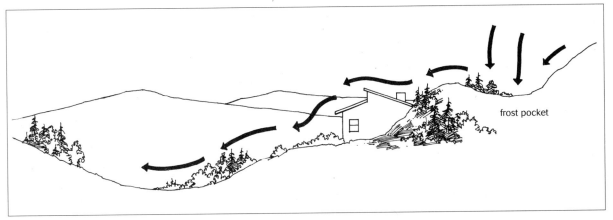

frost pocket

Figure 2-2.
Cold air collects in depressions, especially along valley
floors, and can dramatically
increase heating costs. Avoid
frost pockets when siting a
home.

Consider Wind Currents and Air Drainage

If you are thinking about producing some of your own electricity with a wind generator, select a site that experiences strong, reliable winds. Remember, however, that winds can also rob heat from a house during the heating season. If you are smart, you can have your cake and eat it too. That is, you can install a wind generator to provide electricity, yet protect the house from the cold winter winds, for example, by earth sheltering your home or planting a windbreak upwind from it. Situate the wind generator upwind from the windbreak.

Cold air pockets are another important consideration when assessing the suitability of a site. Cold air is more dense than warm air. As a result, at night cold air tends to flow downslope, collecting in valleys or other low points, creating cold air pockets, also known as "frost pockets" because they are the first places to freeze (figure 2-2). Avoid frost pockets, if you can. They will certainly increase your home's heating requirements and could also make gardening more difficult.

Choose a Sloping Site for Earth Sheltering

Earth sheltering a home—pushing dirt against water-proofed walls (berming) or the nearly complete burial of a home—keeps a house warm in the winter and cool in the summer, especially when coupled with passive solar design. Combined, earth sheltering and passive solar design reduce fuel consumption and dramatically decrease energy bills and environmental pollution. Earth sheltering also reduces external maintenance, as much of the house's exterior is underground.

Earth sheltering on sloped land is generally easier than on flat land, because less dirt must be bulldozed against the structure. If you are planning to build an earth-sheltered passive solar home, the land should slope toward the sun.

Earth sheltering offers other benefits as well. Earth-covered roofs, for instance, can be planted with an assortment of native grasses and wild flowers, creating a beautiful living roof. The living roof reduces the amount of

"There simply isn't anything like earth cover to endow a building with a sense of eternal appropriateness."

Malcolm Wells
The Earth-Sheltered House

land taken out of commission by home building. So, earth sheltering not only promotes energy conservation but also helps restore the Earth.

Seek Favorable Microclimates

Over the years, I have found that climate varies, sometimes rather dramatically, within a region. These small climatically distinct areas are called *microclimates*. South-facing slopes for instance, are generally warmer than north-facing slopes and typically create their own microclimate. In colder regions, south-facing slopes make an excellent building site.

Larger microclimates are also common. For example, I live in the foothills of the Rocky Mountains, almost due east of the 14,000-foot Mt. Evans. Besides providing locals with a beautiful view, this gentle slump-shouldered giant seems to affect our weather, causing storms to move north or south of us. The valley I live in is therefore often bathed in sunshine, especially in the winter when I need the sun the most. Ten miles north, south, and west of me, big burly storm clouds bunch together, blocking out the sun, while my little slice of heaven—which I happened upon by dumb luck, not any advance planning or brilliant site analysis, I'm sorry to say—is bright and sunny, permitting me to produce all of my electricity from solar panels (PVs) and to heat my home with solar energy as well.

An intimate knowledge of a region can help you locate suitable micro-climates. Ask around to find out about these special places. If you are successful in finding one, you'll be rewarded many times over.

Select a Dry, Well-Drained Site

Although every aspect of siting is important, one of the most important is to find a site that's well-drained. Well-drained sites—ones that drain on their own accord—require little, if any, grading. This, in turn, minimizes land disturbance, habitat loss, and the energy required to build a home, and it saves money. (We'll explore natural drainage in more detail in chapter 15.)

Natural drainage relies on slope and/or porous soils that permit water from rain or melting snow to escape, either flowing aboveground or percolating into the earth to become groundwater. Watch out for clay-rich soils. They can prove to be a nightmare, as they trap water and convert the ground around a home into a muddy mess. In Vermont, for instance, where the soil often has a high clay content, locals often joke about winter ending not in spring but in the "mud season." I experienced the same phenomenon in the farm country of western New York where I grew up.

A dry, well-drained site is also essential because it prevents water from accumulating in the soil around a foundation and seeping into the basement. Mild seepage can create a damp, musty interior. More serious seepage can result in flooding.

In colder climates, water in the soil around a home may freeze in winter. The freeze/thaw expansion and contraction of the soils under the foundation of a home can lead to cracking. Cracks in the foundation can cause cracks in the walls and, over time, can destroy a home. Repairing damage, if possible, will require considerable time, energy, natural resources, and money.

Soil moisture around a home also sucks heat out of the building through the foundation like a wet T-shirt on a cold winter day. Generally, the drier the soil surrounding the foundation, the less heat loss.

If the soils in a site are not well drained, be sure to grade the land so that it slopes away from the house. Providing special drainage around the foundation, for example, by the French drain shown in figure 2-3, is helpful, especially in wetter climates or when building an earth-sheltered home or a home with a basement. You may even want to consider running water from gutters and downspouts into pipes to transport it away from the house.

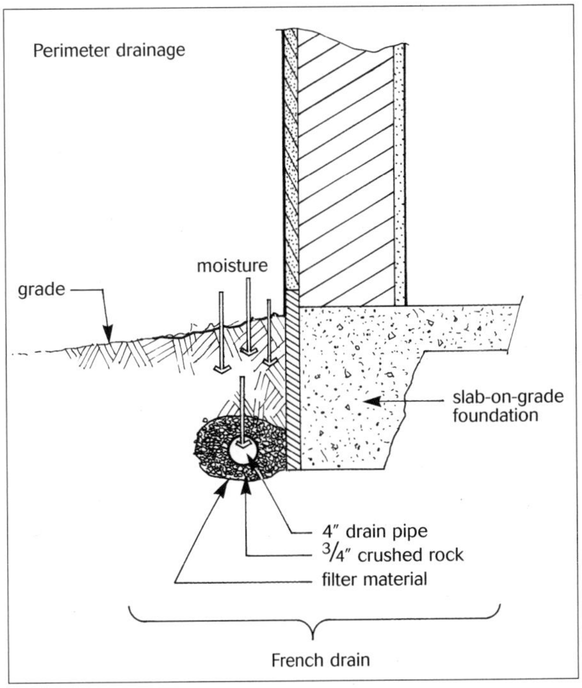

Figure 2-3.
Perimeter foundation drainage (shown here) helps prevent frost heave and reduces moisture penetration into basements. By keeping the soil drier, this French drain also reduces heat loss from the foundation.

Source: Michael Middleton (from *The Natural House* by D. Chiras)

Select a Site with Stable Subsoils

A house is only as good as its foundation. And a foundation is only as good as the earth upon which it rests. In most cases, this isn't a problem. Houses can be built on bedrock or a stable compact subsoil. In some locations, however, subsoils can be quite problematic. For example, subsoils that contain a lot of bentonite clay undergo considerable expansion and contraction in response to changing moisture levels.

Because bentonite clay can expand as much as nineteen times its dry volume when wet, subsoils containing this particular type of clay should be avoided or, if necessary, built on with great care. Stable subsoils, like well-drained soils, minimize the risk of foundation and wall cracking, thereby helping to ensure the longevity of a home.

To learn about the subsoil in your area, you may want to talk with local builders or, better yet, hire a soil scientist to examine the site. A local building department may also be of some assistance.

While you're exploring soils, you should also consider the earthquake potential of any building site. Earthquakes can wreak havoc on a home. At the very least, you should know what the earthquake hazard is, then design accordingly. Local building department officials can provide information and maps of the relative earthquake danger in your area. Architects and structural engineers can provide advice on building in earthquake zones.

Avoid Natural Hazards

Another very important, though frequently ignored, rule is to avoid building in the path of natural hazards. One of the most common hazards is the

Although people have been living in floodplains for centuries, nature truly did not intend them for anything but temporary human habitation.

floodplain, a low-lying area along the banks of a river. Floodplains belong to rivers, as even the best-behaved tributaries periodically spill over their banks. In heavy rains or after a sudden snow melt, the mildest stream can transform into a raging torrent.

Although people have been living in floodplains for centuries, nature truly did not intend them for anything but temporary human habitation. Given the rise in the occurence of hurricanes and other violent storms resulting from the climatic mess we're creating by deforestation and the release of greenhouse gases, flooding is likely to get worse. You can protect yourself by staying out of harm's way.

In arid regions, building in or near dry gullies, known as *arroyos,* is equally problematic. Even though an arroyo may have been bone-dry for decades, heavy rainstorms, often many miles away, can transform the parched gully into an ugly, raging deluge that will wreck a home in a flash . . . flood that is. Building in the path of landslides, mud slides, or avalanches is equally dangerous. Don't do it.

Avoid Marshy Areas

When searching for a building site, you should also avoid marshy areas or depressions in which water naturally accumulates during rainstorms or after snows begin to melt. Marshes and wet areas may serve as a breeding site for mosquitoes. Mosquitoes make rotten neighbors and now may carry the potentially lethal West Nile virus. Other marsh creatures, such as the strikingly handsome red-winged blackbird, while vital to the life of the swamp, can be quite noisy.

Building near wetlands can damage them and, in some cases, destroy them entirely. As conservationists unceasingly remind us, marshes serve as valuable wildlife habitat and offer many free services to humans, such as water purification. Marshes also help replenish groundwater supplies and prevent flooding in large part by their ability to absorb flood waters.

Despite their importance, in most parts of the world wetlands have been severely depleted. In the United States, for instance, wetlands once covered an area twice the size of California—about 220 million acres. Today, half of these wetlands are gone, with the greatest losses occurring in California (91 percent), Ohio (90 percent), and Iowa (89 percent). Although wetland destruction has greatly decelerated in recent times, thanks to state and federal laws, the losses are still estimated to be about 33,000 acres per year.

Despite these laws, many swampy areas still fall to development. Wetlands continue to decline because legal loopholes allow developers to fill in or drain them, so long as they replace them with artificial substitutes, known as *constructed wetlands.* Developers can also legally destroy wetlands that stand in their way if they take measures to protect threatened wetlands in other areas.

Vanishing Wetlands

In total, U.S. wetlands once covered an area twice the size of California, about 220 million acres. Today, half of these wetlands are gone, with the greatest losses occurring in California (91 percent), Ohio (90 percent), and Iowa (89 percent). Although wetland destruction has greatly decelerated in recent times, the losses are still estimated to be about 33,000 acres per year.

While laudable, artificial wetlands are a feeble replacement for the real thing. The result is much like replacing a city's symphonic orchestra with a sixth-grade band. Protecting endangered wetlands, the second strategy, while commendable, also results in a net loss of these biologically rich ecosystems. Because wetlands are so vital to the health of our planet, please think long and hard about building on one or near one.

Select a Site Suitable for Growing Food

Homegrown vegetables and fruits nourish the body. Flowers nourish the soul. Trees provide comforting shade, refuge for birds, and fuel for the evening fire. For those who want to benefit from homegrown food and fiber, perhaps to achieve a level of independence, good soil is a necessity.

Experienced gardeners can determine the quality of the soil by its appearance, smell, and feel. They can also judge soil quality by the plants that grow on it. If land is overrun with weeds, it is an indication that the previous owner has abused it—overgrazing it or depleting the soil of its nutrients.

Take a good look at the land you are considering. Is it riddled with gullies from erosion? Is it overrun with weeds? Or does it support rich, lush grass or a forest? Take a shovel along with you so you can examine the depth of the topsoil. Take a sample or two of the topsoil and send it to a soil testing lab, usually located at a state agricultural college. (Check with them first to determine how they prefer samples to be extracted.) The lab will provide data on the pH (acidity or alkalinity) of the soil, as well as nutrient levels and organic content, all at a reasonable price.

When considering a green-built home, be sure to take a soil sample from prospective garden spaces—and be sure the garden spot will receive plenty of sunshine.

Select a Site That Offers Building Resources

Most homes are built with materials shipped from manufacturers many miles from the building site. This practice, while popular and affordable thanks to economies of scale and cheap fossil fuels, increases the embodied energy of the materials that go into a home. And because more energy is needed to transport materials from producers to retailers to end users, more pollution is produced.

Those interested in building as sustainably as possible may want to consider their site as a potential source of building materials. One of my editors built his home in the lush hills of Vermont out of locally harvested timber, which he and friends cut and milled on site, creating all of the framing lumber required to build the solar home. You can't get much more sustainable than that—especially if you harvest trees in ways that minimize damage to other healthy trees. Clay-rich soils can also be used to build an assortment of natural homes, including adobe and rammed earth buildings, discussed in chapter 9.

Choose a Site with a Good Water Supply

Water is vital to our survival. Without it, few, if any, households could manage for very long. When buying a home that is already built or looking for land to build on, be sure to check out the water source before you sign on the dotted line. In most cases, water is supplied by wells or local municipal systems.

If you are considering drilling a well, ask locals and experts about the state of the aquifers, the underground water supplies, in your area. In some parts of the country, aquifers are on a steep decline. Be sure to ask how deep typical water wells are in your area. In some locations close to my home, I've known people who had to drill as deep as 700 feet to find water. In one case, the homeowner got only half a gallon a minute. With wells costing $10 to $15 per foot, drilling this deep becomes a pretty expensive venture with little reward. And don't forget: A deep well will require lots of electricity to pump water to the surface. A catchwater system (see chapter 14) might be cheaper and more reliable.

Another approach is to tap into nearby streams, lakes, ponds, or springs—so long as they can be developed without damaging fish and wildlife that depend on the water or violating other people's water rights (as may be the case in the West). Be sure there are no potential sources of contamination, such as factories, abandoned hazardous waste dumps, or farmyards containing stockpiles of manure that may run off into nearby surface waters. And ascertain that the stream doesn't dry up during the summer, leaving you and your family without water. Four gallons a minute is generally sufficient for most homes, although less will work if you have a way to store it for high-use periods.

When planning to use groundwater, be sure to taste the water, if possible. You should also obtain a sample for testing. State and county health departments will usually test drinking water for a small fee. Check for radon or natural gas in your water. Both can present significant hazards.

Tapping into a municipal water system relegates you and your family to the consumption of water laced with chlorine and other contaminants such as lead. Although the scientific studies I've reviewed suggest that chlorinated drinking water is relatively harmless, it may not taste too good and certainly smells bad. If you want to avoid the taste and smell of chlorine and any possible health impacts, consider installing chlorine filters.

You may want to consider an alternative water system, notably a catchwater system. A catchwater system captures rain and snowmelt from the roof. Water is then stored in a tank, an underground or aboveground cistern. This water can be used to irrigate lawns and gardens or to wash cars. It can also be used inside a home to flush toilets or for bathing, showering, cooking, and drinking when properly filtered and purified. A location with ample rainfall is important; however, even in dry climates, a household can collect tens of thousands of gallons off a decent-sized roof. In the high desert around Taos, New Mexico, many families live off rainwater collected from their roofs. Even though very little rain and snow falls on the area most of the year, spring and summer rains can be quite intense, providing sufficient water to fill the cistern with a year's supply of water. By using water-efficient showerheads, front-loading washing machines, and low-flush toilets, these hardy souls can meet 100 percent of their demand for fresh water from the generous gift from the sky.

Be sure to send a sample from your potential drinking water supply to your local health department for testing. Results of their tests will indicate how much filtering you will have to do to make the water potable—and may alert you of any potential problems.

Minimize Bulldozing for Roads and Driveways

In many parts of the country, land for building houses is becoming scarce. As a result, many new homes are built on marginal sites. In the foothills of the Rockies, where I live, most of the good building sites are gone, so contractors frequently build new homes on steep hillsides. Driveways often plunge down precipitous slopes or rise steeply from the roadway like the hair-raising inclines of a roller coaster. Driveways can be pure hell to navigate during the winter when covered with ice. Moreover, many gravel driveways erode severely in heavy spring rains, creating deep gullies and polluting nearby creeks. In some instances, accessing a building site means cutting through pristine meadows, destroying beauty.

When looking for a site, think practically. The more bulldozer work that is needed, the less desirable the site. My advice is not to bulldoze any more land than you have to. Roadwork not only damages the environment, it can be quite costly. Driveways can also increase erosion, causing siltation in nearby streams and lakes, damaging the habitat of fish and other aquatic organisms.

Avoid Noise and Pollution

Noise is a pollutant that pervades our society and is so prevalent that most of us have tuned it out. Even though noise in our environment slowly deafens us and adds to our daily stress, few people stop to consider this ubiquitous pollutant when looking for a building site. When I bought my first home, for instance, I never gave noise a moment's thought. The house was a few blocks from a major bus route, which made commuting to the university where I taught convenient.

The house had a small backyard with many trees, and I was looking forward to sleeping out on a hammock on hot summer nights. Shortly after I moved in, I set up my sleeping pad, then crawled into my sleeping bag, ready for a good night's rest. That was when the distant roar of cars became evident. At about 2 A.M., after trying to sleep through the din of trucks and cars, I hauled my sleeping bag and my weary body into the house and collapsed on the bed. So much for my dream.

Unfortunately, noise is not the only pollutant we need to be aware of when searching for a new home or a building site. Air pollution can be a

When looking for a site, think practically. The more bulldozer work that is needed, the less desirable the site. My advice is not to bulldoze any more land than you have to. Roadwork not only damages the environment, it can be quite costly.

THE PRICE YOU PAY FOR BEAUTY

Orienting a home to capture a view often destroys or greatly reduces natural heating and cooling potential. Situating the long axis of a home so that you can look to the west to capture a breathtaking view, for example, drastically reduces solar gain during the winter. As you will learn in chapter 11, in the winter the sun cuts a low arc across the sky. A house whose long axis is oriented to the south will capture the sun's energy through south-facing windows. Orienting a house to the west instead to capture a magnificent view won't provide much surface for solar gain. The view won't provide much solace as you stand there shivering in your pajamas. In the summer and fall, though, when heat is not needed, west-facing windows often permit huge amounts of sunlight to enter as the sun descends in the sky in the afternoon, causing overheating. You'll have to draw the shades, hiding your view, to keep from baking.

BUILDING NOTE

To view the EPA's radon zone map, log on to www.epa.gov/iaq/radon/zonemap.html.

problem, as can radon gas. Radon is a radioactive gas given off from naturally occurring materials in the soil and rocks. To find out if you are in an area where radon is a problem, check out EPA's nationwide map (see the accompanying building note). It's best to test for radon anyway, even if the map indicates you're in a relatively safe area.

When selecting a site, take note of the location of power plants, factories, and other potential sources of pollution. Will you be locating downwind from them?

Balance View with Vital Needs

View is a major consideration when selecting property or selecting a home site. If you purchase land with a good view, it is only natural to want to orient your home toward the visual amenity: the mountain range, river valley, meadow, forest, pond, lake, park, or other natural feature for which you've paid so dearly.

Unfortunately, aligning a house to capture a stunning view often means orienting in a way that minimizes its capacity to capture solar energy for heat, electricity, and hot water. The result is a lifetime of high energy bills and winter discomfort, and possibly significant overheating in the summer and fall. (For an explanation, see the sidebar above.) Fortunately, there are ways to capture views and heat and cool your home naturally when the sun and views fail to align. I'll show you a few tricks in chapter 11.

Don't Destroy Beauty in Your Search for It

Over the past thirty years, I've seen one beautiful mountain valley after another "uglify" as builders cram them with buildings. Some of my favorite western towns have slowly transformed from picturesque villages into hodge-podge conglomerations of architecturally dissimilar homes, fast-food restaurants, convenience stores, and highways.

CLUSTER DEVELOPMENT

Clustering homes enables builders to preserve parks, open space, wildlife habitat, and farmland. According to the Sustainable Buildings Industry Council, developers have found that they can cluster homes to create a slightly more densely populated community and preserve open space at a greater profit than if they had developed the area in a more traditional fashion. In addition, they note that most home buyers are attracted to—and some are willing to pay a premium for—homes that are adjacent to open spaces, recreational areas, or other amenities. Clustering can also foster an increased sense of community.

Building a sustainable future depends on building homes in a way that protects all of our natural resources, that is, not just forests from timber cutting but also the beautiful areas that are responsible for a community's unique character. Developers can contribute to this goal by clustering homes and placing the clusters out of sight so that the visual landscape remains intact.

Outside Chicago, for example, Bigelow Homes—one of the nation's leading green builders—clustered 1,100 homes on 150 acres, a density three times greater than that of average single-family-home subdivisions. By doing so, they were able to protect 300 acres of farmland and forest. Avoiding building homes in open fields or on hilltops also helps protect views.

Individuals can protect beauty as well by purchasing homes that have been sensitively nestled in the landscape and by building their own homes out of sight in the least obtrusive manner. Earth sheltering often helps meet this requirement (see chapter 10).

When siting a home, do not place it in the most beautiful spot on the property. "Leave those areas that are the most precious, beautiful, comfortable, and healthy as they are," writes Christopher Alexander and his coauthors in *A Pattern Language,* "and build new structures in those parts of the site which are least pleasant now."

Malcom Wells agrees. He advises us to leave the untouched sites to nature. Buy ugly spots, abandoned lots if necessary, and build homes that are adorned with a jungle of vegetation. Create beauty; don't be an agent of its destruction.

Make Systematic Comparisons

One of the secrets of successful selection and siting is spending a lot of time on the property, then carefully and objectively analyzing the facts. To simplify the task, I strongly recommend using the list of criteria in table 2-1 to assess each and every piece of property you visit. You can then eliminate the least suitable sites and compare the more desirable ones objectively.

Once you have selected the best property, take time for a second and third look. In a fast-moving market, a small down payment (called *earnest money*)

Table 2-1

Assessing and Comparing Potential Building Sites

Features	Site 1	Site 2	Site 3	Site 4
Location *list address and* *other information*				
Proximity to work				
Proximity to stores				
Availability of mass transit				
Access to bike paths or walkways				
Proximity to recreation				
Access to hospitals				
Availability of fire protection				
Availability of police protection				
Access to power and other utilities				
Future development plans				
Crime rates				
Quality of schools				
Solar access				
Wind resources				
Desirable temperature				
Sloping site for earth sheltering				
Favorable microclimate				
Dry, well-drained soils				
Stable subsoils				
Natural hazards				
Marshy areas				
Soils suitable for growing				
Building resources on-site				
Adequate water supply				
Driveway access				
View				
Beauty				
Noise or other pollution				
Community				
Environmental services *(recycling centers,* *composting facilities, etc.)*				
Other				

may be required to hold the land while you perform a more thorough inspection. Earnest money is nonrefundable if you renege on the deal for reasons not stipulated in the contract you sign, for example, if you find a piece of property you like more. If you find something legitimately wrong with the lot—for example, there's something wrong with the title or the land flunks the perc tests needed for a septic tank—you can usually get your money back. Be certain you or your real estate agent list all potential reasons for refund of earnest money on the contract offer you make to purchase the land so there won't be any misunderstandings.

Protecting a Site during Construction

When you've located a suitable building site, it is time to begin designing your home. Although most people carry images of their dream homes around in their heads or have photos and sketches of them filed away, I recommend designing a home *after* the site has been determined for a number of reasons.

First, this approach allows you to design a home that suits the site aesthetically—that is, it fits into the surrounding landscape, blends in beautifully with the trees and other physical features of the land, and is compatible with the existing architecture (figure 2-4b). The size of the home, its architectural style, and its exterior siding and roofing all influence how well a home blends in with the landscape. Although your dream home might look great in Florida, it might be an eyesore in the mountains of North Carolina (figure 2-4a).

Second, designing a home to the site—site-specific design—allows one to create a building that works best with the view, topography, and climate. Site-specific design, in other words, helps us create homes that perform better and last longer on a given site.

Once the design has been finalized, it is time to start building. Most home construction involves a considerable amount of excavation for foundations, driveways, utilities, and drainage. If you are going to build a house out of

Houses should blend with the landscape architecturally, complementing the character of the land and the architectural style of buildings in the region.

Figure 2-4a & b.
Many homeowners build their dream homes without considering the natural terrain, vegetation, and surrounding architecture. The results can be quite jarring (a). Whether you're building in the city, town, or country, design your home for visual integration (b).

Source: Dan Chiras

adobe, cob, or some other natural material (described in chapter 9), you may have to disturb the land a bit more to acquire earthen building materials—although in many instances excavation for foundations may supply you with an ample supply of material.

Minimizing Land Disturbance

Before construction begins, you and your builder should devise a plan to minimize damage during construction. The plan should begin with the designation of an access route and an area for parking vehicles. Restricting access to one route and limiting parking to one location, or asking workers to park along the road, will help protect vegetation on the site from damage. These measures also minimize soil compaction, which reduces the porosity of soils, making it harder for plants, especially trees, to obtain the oxygen and water they need. Posting a sign showing where parking is allowed is a good idea, especially if you or the builder won't be on-site all of the time, as many subcontractors and building inspectors will visit a site during construction.

A large volume of building materials will be delivered to your work site. Truckloads of sand, framing lumber, and sheathing in the case of a wood-frame house or straw bales and plaster components in the case of a straw bale home, for example, will be trucked in during construction. Designating one area for deliveries that is close to the building site and readily accessible will greatly minimize damage.

You may also want to cordon or fence off areas to protect them from earth-moving equipment or other heavy machinery. Overzealous bulldozer or backhoe operators may park on sensitive vegetation or run their machinery over areas that are better left undisturbed. By restricting their operations, you can greatly cut down on damage that will need to be repaired after construction is completed.

Handling Topsoil

On most building sites, topsoil is first scraped from the site, then stockpiled for reapplication. Be sure to stockpile it close to the site, but not so close that it interferes with operations. Try to cover as little land as possible and don't pile topsoil—or any soil for that matter—around the bases of trees. This can cut off the oxygen to the trees' roots, stressing and even killing them.

Topsoil that is stockpiled on site should be protected from wind and water erosion. It should also be isolated from subsoil to facilitate reapplication.

While we are on the topic of topsoil, be certain that topsoils are not accidentally used for backfill. Backfilling is best reserved for subsoils. Topsoil takes hundreds, sometimes thousands of years to form and should be reserved for gardens and lawns. Moreover, topsoil is a poor backfill material because it contains a substantial amount of organic material, that may decay over time, causing the soil to settle.

BUILDING NOTE

Subsoils are usually used for backfill, while topsoil is used for lawns, gardens, and orchards.

Stockpiling and Recycling Waste

It is important to make room to organize construction waste for maximum recycling. Designate areas for wood, cardboard, metals, and glass. If you're using dumpsters, be sure they are conveniently placed and well marked to enhance worker participation. Be sure workers are encouraged to recycle and are properly trained. Because you don't want to spend a lot of time sorting through cardboard to remove scraps of wood and metal that should have been thrown into their own bins, you or your builder may have to designate someone to manage the recycling of construction waste.

While you are at it, be sure to set aside a place to stockpile potentially usable wood, for example, framing lumber cut-offs that might be used for other purposes. Wood that can't be reused for building projects can always be used for firewood in somebody's wood stove. Some companies, notably high-volume builders, divert their waste wood to recyclers that turn it into mulch, wood chips for animal bedding, particle board, or other useful products.

Also, be sure to provide convenient receptacles for recycling aluminum cans, bottles, and plastic bottles from workers' lunches. You may also want to set aside excess building materials, such as extra shingles or siding or sheathing, rather than tossing them out. Such materials can be donated to Habitat for Humanity for resale in their stores (called Habitat Restores) or sold or donated to construction recycling outfits, other contractors, or owner-builders with a penchant for scrounging.

To protect a site, it is also beneficial to prohibit activities such as cleaning cement mixers or to confine them to a designated area. The disposal of any hazardous materials, such as solvents or fuels, should not be permitted on-site.

Protecting Trees

Shade trees increase the beauty and value of residential property and also make passive cooling possible. This, in turn, helps reduce or eliminate cooling costs. As a consequence, many builders attempt to protect trees during construction. Because they can be preserved with very little effort or expense, it makes no sense to cut down a mature shade tree only to replace it a few weeks later with a sapling that will take fifty years to provide the shade and beauty of the one just toppled with a chain saw. Trees and shrubs can sometimes be removed and transplanted to another location or another lot, although this can be costly. You may want to hire or consult with an expert—a city forester, a licensed arborist, or a certified tree appraiser—who will evaluate the trees on the property. He or she can gauge the health of each tree and then help you decide which ones should be saved. Some large shade trees that appear healthy may suffer from heart rot, a fungus that destroys the heartwood and eventually causes trees to die.

Arborists note that older trees are less adaptable to change than younger trees. Protecting a 1- to 8-inch-diameter tree, therefore, may make more sense than protecting a larger-diameter mature tree. In addition, some species

BUILDING NOTE

Large trees standing within 10 feet of houses and driveways are likely to be damaged and will therefore need to be cut down. Build a safe distance from large trees whenever possible.

For more information on protecting trees during construction, contact the National Arbor Day Foundation. Their Building With Trees program, cosponsored by the National Association of Home Builders and Firewise Communities, offers information on tree protection practices for builders. Builders and developers who plan and design projects to protect trees and sign a pledge to continue their commitment during and after construction are eligible to receive recognition.

adapt better to change and soil disturbance from construction. A tree special-ist can tell you which ones will tolerate root damage and which ones won't.

After assessing the health and adaptability of the trees on your property, the arborist will mark trees that are to be removed with colored tape. Those that are to be felled should be cut down carefully so as not to damage those that are to remain. And those that are left in place should be physically pro-tected. You may want to wrap their trunks with cardboard to protect them from injury by heavy equipment. Damage to bark makes trees more suscep-tible to insects and disease.

Efforts also need to be made to protect the roots of trees. Large, woody roots usually grow out horizontally, approximately 6 to 24 inches below the ground's surface. These roots give off many smaller roots that grow close to the ground surface, absorbing water and nutrients. As a general rule, roots extend horizontally a distance approximately equal to the height of the tree. At least half of the root system needs to be preserved to maintain a tree's vigor and health.

Because people forget, it may be advisable to erect barricades around trees and post signs that remind workers that they are nearing a protected-tree zone. Barricades should be made from wood or wire fencing and should be placed according to the tree size and species. For newly planted trees, one to four years of age, the barricade should be erected at the dripline, the periphery of the trees' branches (figure 2-5). Barricades for larger trees are generally placed using the 1-foot-from-the-trunk-per-1-inch-of-trunk-diameter rule of thumb. For a tree with a 12-inch base, the barrier should be 12 feet from the trunk. A 4- to 6-inch-deep layer of mulch should then extend to the dripline. Obviously, the further the barricade is from the trunk, the greater the protection. Be sure that any trenches are situated as far from trees as possible to keep from damaging their roots. If underground power or water lines are installed, ask contractors not to cut major roots, but rather to tunnel or auger underneath them.

Trees should be fertilized before construction begins and should be watered every two weeks during construction. Fertilizer and periodic watering help reduce stress and increase the vigor.

Don't forget that you can minimize the number of trees that must be cleared by selecting a site very carefully—choosing the one with the fewest trees possible. In the City of Austin's Green Building Program, points are given to home builders who put felled trees to good use—for example, they're mulched or cut up for firewood,

BUILDING NOTE

Even though you may talk with the general contractor and the subcontractors about protecting trees, you may need to convene on-site meeting with the work-ers themselves to go over the guidelines. It is wise to include stipulations in the contract arranging for com-pensation in the event that specified trees are dam-aged, based on the species, the age of the affected trees, and extent of the damage.

Figure 2-5. Protect trees along the dripline, as shown here. Trees add beauty and value to a home site, and replacing them can be costly.

Source: Lineworks

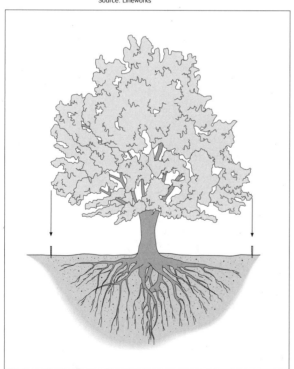

rather than trucked to a landfill. It is a good idea for any home construction project.

In some areas, you will find companies that take the trees, cut them into usable lumber, and sell the wood products to local builders. In Oakland, California, for example, the tree recycling yard operated by the Protect All Life (P.A.L.) Foundation, a nonprofit organization run by Marcus von Skepsgardh, salvages redwood, pine, and eucalyptus cut down in the city. They charge a nominal fee to pick up usable trees, that is less than or comparable to disposal fees at the local landfill. The trees are then milled and the wood is sold to builders for making countertops, bar tops, table tops, construction lumber, flooring, and decking. Besides rescuing trees from the landfill, this program helps reduce pressure on our nation's forests.

Figure 2-6.
The P.A.L. Foundation in Oakland, California, collects wood from urban sites to produce usable lumber, reducing strain on our nation's forests and landfills.

Source: P.A.L. Foundation

Minimizing Erosion

Erosion control is essential on many building sites. Erosion control takes several forms. Minimizing the surface area that is disturbed—that is, the land area that is bulldozed, cleared of vegetation, or driven on—greatly reduces soil erosion. But some disturbance is inevitable. In such cases, plastic fences may be installed across the fall line to prevent erosion. Straw bales can be strategically placed in drainage ditches or gullies to reduce water flow and erosion.

Be sure to design driveways to minimize erosion. Sloping of the driveway very slightly so that water runs off laterally, rather than running down its full length, prevents erosion, as do switchbacks, which snake the road up a hillside to minimize the grade. I used a series of water bars—4-inch-diameter logs buried in the road surface—to slow surface water flowing down a steep driveway in my last home. This simple, inexpensive measure virtually stopped the erosion and saved me tons of work each year transporting sand back up to the top of the driveway to fill the gullies that used to form when the rains came.

Preparing for Revegetation

Protecting a job site during construction makes revegetating the site when construction is complete far easier and can save a considerable amount of time and money. But there are other steps you can take to facilitate revegetation. For example, you can reseed your land, or part of it, with native grasses and wildflowers. To do so, you can gather seeds from plants on your property before construction begins. Seeds gathered in the fall should be stored in a cool, dry location until used.

"Every person, environmentalist or not, knows about recycled paper, but what about recycled trees? If we recycle paper to protect the forests, why not recycle trees? It only makes sense. But believe it or not, millions of tons of urban wood ends up in landfills each year."

Marcus von Skepsgardh
P.A.L. Foundation

Protecting a job site during construction makes revegetating the site when construction is complete far easier and can save a considerable amount of time and money.

Another trick that you may find useful is to use sod from the topsoil stockpiled on your building site for revegetation. I used this technique to plant the living roof on my house with great success. It took only a year for a lush vegetative covering to grow on my living roof, while the rest of the property, planted with seeds I gathered the previous year, took five full years to grow. (For more on revegetation, see chapter 15.)

Selecting a good site to build a home requires diligence, patience, knowledge, and, of course, a bit of money. Take your time and choose wisely. If you're interested in further information about site selection issues, see the more extensive discussion in my book *The Natural House.*

Once you select a site, be sure to protect it from unnecessary harm during construction. Remember, protecting a job site is to building what preventive medicine is to health care. It makes the task of bringing land back to life much easier and far less costly. It does require vigilance, however.

When building is all done, the topsoil is spread back over the site and the land is usually graded to shed water away from the house. It is then revegetated. If you've been smart, you'll soon be living in a lush garden of vegetation.

PART TWO

GREEN BUILDING AND REMODELING

CHAPTER 3

The Healthy House

Years ago, author and healthy house advocate Lynn Bower began having trouble sleeping. Her muscles and joints ached. She experienced difficulty breathing and digesting food and often had trouble thinking clearly. Adding to her misery, she suffered from constant sinus pain caused by a chronic inflammation. Chemical smells, even the odor of printing ink and new clothing, began to bother her.

Like millions of other people in the industrial nations, Lynn suffers from a disorder known as *multiple chemical sensitivity,* or MCS for short. This immune system disorder is brought on by exposure to chemicals in the environment, especially the home environment.

For Lynn, symptoms began to appear after she and her husband John remodeled their 1850s farmhouse. As best as they can determine, exposure to chemicals in the building materials they used to refurbish their home was responsible for Lynn's deteriorating health. "Frankly, we were shocked," write the Bowers in their book, *The Healthy House Book.* "We had used the same plywood, cabinets, paints, and carpeting that anyone could buy at lumber yards and building supply centers."

Scientists believe that the reason Lynn Bower and others like her feel ill is that some chemicals in modern building materials and furnishings, such as formaldehyde, bind to body proteins. In the process, they create foreign substances that the immune system attacks, making people feel terribly ill. (This reaction, in turn, is believed to be the cause of MCS.)

Exposure to chemicals in carpeting, stains, paints, finishes, plywood, sofas, and other furniture makes millions of people sick. Unfortunately, many of us have no idea why we feel so rotten. Troublesome as these sources of indoor air pollutants are, they are not the only ones to be concerned about. The air inside our homes—and many other buildings, too—can be contaminated by a whole host of potentially toxic substances from a variety of sources. Health problems attributable to indoor air pollution range from annoying maladies such as headaches, chronic

---

Symptoms of Multiple Chemical Sensitivity

• Difficulty breathing
• Insomnia
• Difficulty concentrating
• Memory loss
• Migraine headaches
• Nausea
• Abdominal pain
• Chronic fatigue
• Aching joints and muscles
• Irritated eyes, nose, ears, throat, and/or skin

Some with MCS show impaired balance and increased sensitivity not just to odors but also to loud noises, bright lights, touch, extremes of heat and cold, and electromagnetic fields.

Source: MCS Referral and Resources at www.mcsrr.org

How Do You Know If You're Chemically Sensitive?

Chemical sensitivity is a relatively new diagnosis. Most doctors don't know about it and therefore are likely to think that your symptoms are "all in your head." Or, if they've heard of MCS, they may not know much about it or think it is a "fringe" diagnosis. If you suspect you have a chemical sensitivity, contact the American Academy of Environmental Medicine or the American Academy of Otolaryngologic Allergists for a recommendation of physicians in your area who can diagnose and treat you. These groups are listed in the resource guide.

No one is born with allergies, asthma, or multiple chemical sensitivity. They acquire these conditions for a variety of reasons; some reasons are genetic, many are environmental. So just because you are healthy today doesn't mean you will be tomorrow. It is wise for everyone to avoid as many environmental pollutants as possible.

John and Lynn Bower
personal interview

bronchitis, asthma, and allergies to debilitating diseases such as multiple chemical sensitivity to life-threatening conditions such as cancer.

Indoor air pollution is an issue of great importance. Whether you are building new, remodeling a home, or simply purchasing new shelving, painting a room, or adding new carpet, you have the potential to dramatically alter the air quality in a home—and the health of those who live in it.

But are the concerns over indoor air quality and health well founded? Is this really a large a problem?

The short answer is yes. Indoor air pollution is a problem in part because most people spend the bulk of their lives indoors, either at home or at work. If you work in an office or teach in a school, for example, you easily spend 80 to 90 percent of your life inside. Even construction workers are indoors about 50 percent of the time. So, the potential for exposure is significant.

Not only do we spend most of our lives indoors, but most homes are affected by indoor air pollution. When asked how prevalent the problem is, John and Lynn Bower note that "most houses have some degree of indoor air quality problem. The degree varies considerably from quite mildly contaminated to uninhabitable."

Moreover, evidence indicates that large numbers of people are affected by indoor air pollutants. The list of complaints includes minor skin rashes, joint pain, allergies, and respiratory difficulties. Some indoor air pollutants such as mold cause sneezing, watery eyes, shortness of breath, dizziness, lack of energy, fever, digestive problems, and even flulike symptoms.

When asked for an estimate of the number of people who are adversely affected by unhealthy room air at home and in the office, the Bowers noted that "50 percent of all illness is caused by indoor air pollution." They quickly added that "this sounds unbelievable unless you consider all possible symptoms—everything ranging from the sniffles to headaches, all the way up to the ultimate symptom, death. It has been estimated that 17 million Americans have asthma and 30 percent of us have allergies, and these conditions are often caused or aggravated by poor indoor air quality."

Making matters worse, many people function under extraordinary stress. In our pressure-cooker society in which "immediate" is frequently "not soon enough," stress wears down our immune systems. Under stress, the adrenal glands release a hormone called *cortisol*. Although cortisol helps us cope with stress, elevated cortisol levels over long periods tend to suppress the immune system, making us more vulnerable to disease and very likely to indoor air pollutants. To protect ourselves, we can of course reduce stress. We can also make our homes more healthful by reducing indoor air pollutants.

This chapter provides guidance on the subject of healthy home building, focused primarily on ways to ensure healthy indoor air—a vital goal of green building. I begin by outlining the major sources of indoor air pollution so you know what to avoid, then discuss a three-part strategy that helps ensure healthy indoor air. Healthy buildings not only promote human health, they help create a cleaner, healthier world. Green paints and stains, for instance, are safer not just for homeowners and the painters who apply them. They're safer for the people who make them. And they result in fewer pollutants released into the atmosphere, helping us all breathe a little easier.

Our hyperstressed lifestyles render us more likely to be adversely affected by indoor air pollutants.

Sources of Indoor Air Pollution

According to *The Inside Story: A Guide to Indoor Air Quality,* published by the United States Environmental Protection Agency and the Consumer Product Safety Commission, indoor air pollutants include gases and particles from five major sources: (1) combustion appliances, such as woodstoves and furnaces; (2) building materials and furnishings; (3) household chemicals, such as cleaning products, personal care products, and paints and solvents used by hobbyists; (4) central heating and cooling systems and humidification devices, and (5) outdoor sources. Electrical wires and various electrical devices, such as waterbed heaters, are also added to the list for reasons that will become clear shortly. Let's take a brief look at each one.

Healthy building materials are good for occupants of a home, the workers who make them, and the workers who apply or install them. They also result in fewer pollutants released into the air and, thus, help all living things live healthier lives.

Combustion Sources

Combustion within the confines of our homes takes places in an assortment of appliances and devices designed to make our lives more convenient and comfortable. In the kitchen is the gas-powered range and oven. In the utility room is the water heater. In the basement is the furnace or boiler that delivers heat to the house. In the living room is the wall-mounted gas heater, a fireplace, or a woodstove. These devices burn an assortment of fuels, including home heating oil, natural gas, propane, wood, kerosene, and, occasionally, coal. Combustion sources also includes candles and tobacco products—cigarettes, pipes, and cigars.

Combustion of organic fuels within our homes generates a number of potentially harmful particulates and gases, including carbon monoxide and nitrogen dioxide. Each one can adversely affect our health. Carbon monoxide in the air we breathe, for example, binds tightly to oxygen-carrying hemoglobin in red blood cells, reducing the blood's ability to transport oxygen to cells. Although carbon monoxide is believed to cause no problems at low levels, when concentrations are high, problems may occur. In healthy individuals, an elevated level of carbon monoxide in the blood may cause headaches. In elderly individuals and people suffering from cardiovascular disease, carbon monoxide deprives the heart muscle of oxygen and can cause chest pain (angina) and heart attacks.

Nitrogen dioxide in the air we breathe is converted to a strong acid (nitric acid) in the lungs when it combines with water. Nitric acid, in turn, erodes the lining of the tiny air sacs

Sources of Indoor Air Pollution

- Combustion appliances
- Building materials and furnishings
- Household chemicals
- Central heating and cooling systems
- Humidification systems
- Outdoor air
- Electrical wires and electrical devices

in the lung. As the air sacs break down, breathing becomes difficult. The result is a chronic, debilitating condition known as *emphysema*. Particulates can penetrate deeply into the lungs and cause a number of problems, including lung cancer.

Building Materials and Furnishings

Building materials and furnishings, such as carpets and furniture, are another major source of indoor air pollution. Building materials often are made with engineered wood such as particle board, plywood, oriented strand board, and laminated beams. Unfortunately, these products are manufactured with resins or glues that contain urea formaldehyde or phenol formaldehyde. Formaldehyde is an irritant in indoor air when present at high concentrations. At low concentrations, it sensitizes people to other chemicals and, as noted earlier, is a leading cause of multiple chemical sensitivity.

Formaldehyde resins are also found in particle boards used to manufacture cabinets and shelving (figure 3-1). For years, fiberglass insulation was made with a binding agent containing formaldehyde, although at least one major manufacturer has now replaced it with a nontoxic latex resin. Even new carpeting and new furniture may contain formaldehyde resins.

Building materials, furniture, and furnishings release formaldehyde into our homes. This process, known as *outgassing,* may continue for many months, sometimes several years, after a home is built or remodeled or after new furniture and furnishings are installed. In addition, carpeting and furniture may also house bacteria, mold, mildew, and dust mites, which can cause allergies and other symptoms.

Making matters worse, many home builders use glues and other adhesives when installing tile, carpeting, linoleum, and other products in our homes. These adhesives can be a source of irritating, even harmful, chemicals.

Household Chemicals

Most American households contain an arsenal of utility chemicals, including toilet bowl cleaners, degreasers, scouring powders, disinfectants, bleaches, and the like. These products frequently contain toxic substances that are released into the air we breathe when used.

Even personal care and hobby products can be a source of indoor air pollution. Hair spray, nail polish, and polish remover, for example, all release toxic chemicals into indoor air. Spray paints contain methylene chloride. Solvents release volatile organic chemicals, as do glues and other chemicals used in the pursuit of arts and crafts.

Central Heating and Cooling Systems

Central heating and cooling systems are also a potential source of indoor air pollution. Besides releasing carbon monoxide into our homes, furnace duct systems may transport mold, mildew, and bacteria that build up inside them.

IAQ in Apartments

"Apartments can have the same indoor air problems as single-family homes because many of the pollution sources, such as the interior building materials, furnishings, and household products, are similar. Indoor air problems . . . are caused by such sources as contaminated ventilation systems, improperly placed outdoor air intakes, or maintenance activities."

U.S. Enviornmental Protection Agency
The Inside Story: A Guide to Indoor Air Quality

Figure 3-1.
Cabinets in most new homes are typically made from engineered lumber (particle board) manufactured with a formaldehyde-containing resin that can outgas for months, even years, into our homes.

Source: Dan Chiras

These irritants can be dispersed throughout the house by fans that blow air through the ducts.

Outside Air Pollution

Indoor air pollution may arise from outside our homes. Air pollutants from cars on a nearby highway, for instance, can seep into nearby homes through cracks in the building envelope when a house is closed up or through open windows and doors. In many parts of the country, radon seeps into homes from the underlying soil. As noted earlier, radon is a radioactive gas that arises from naturally occurring uranium in the rocks and soil. Although it is colorless and odorless and doesn't cause any immediate health problems, radon can cause lung cancer. The EPA estimates that radon causes approximately 14,000 cases of lung cancer each year in the United States. Unfortunately, the majority of people who contract lung cancer die from it.

In rural areas, agricultural chemicals can pose an additional problem. Large-scale or widespread spraying on farm fields can contaminate the air inside homes in the area.

Magnetic Fields from Wires and Electronic Equipment

Electric lines and an assortment of electric devices emit magnetic fields that some health experts think may be hazardous to human health, causing cancer, birth defects, and miscarriages. However, an exhaustive review of the more than 500 studies on the subject and interviews with dozens of researchers in the field, convened by a panel of scientists belonging to the National Research Council, concluded that the bulk of the evidence suggests that exposure to magnetic fields does not cause cancer, neurological problems, or behavioral problems or damage reproductive cells or the developing fetus.

Our Indoor Air Is Polluted

Just how badly depends on the circumstances. But studies show that in many homes the air is 5 to 10 times worse inside than it is outside! "It doesn't matter where you live," write John and Lynn Bower, "in a major city or a rural area— the air is almost always worse indoors."

Source: *The Healthy House Book*

Figure 3-2.
Indoor air pollution may arise
from outside sources, such
as nearby roads.

Source: Dan Chiras

But don't close the book on the subject. Studies do indicate that these electromagnetic fields may stimulate the growth of cancer cells. In other words, even if they don't cause cancer, there's some evidence to suggest that once a cancer forms, exposure to magnetic fields may accelerate tumor growth. As a precaution, healthy home experts generally recommend that homeowners and builders take measures to limit exposure to them. It's better to be safe than sorry. I'll present some ideas shortly.

Airtight Design

It's not just building materials, appliances, furnishings, and household chemicals that are responsible for indoor air pollution. Part of the problem lies in the fact that many new homes are designed and built to be very airtight. Airtight design and construction reduces air infiltration, saving energy for heating and cooling, and greatly increases comfort levels. To create more airtight homes, builders apply caulk and weatherstripping to seal cracks in the building envelope and install vapor barriers and house wraps to further reduce air movement into and out of their structure.

Although energy efficiency is vital to our efforts to conserve natural resources, making a home more airtight does come at a cost: It traps pollutants, released from various sources just described, often allowing them to build up to dangerous levels. Fortunately, there are ways to create an airtight, energy-efficient home without poisoning its inhabitants, as described below.

Creating a Healthy Home

A healthy building provides the best possible environment for people. The indoor air is free of toxicants, irritants, and allergens (substances such as dander that cause allergic reactions). "A healthy home is not just a place free of hazards and toxins," write David Rousseau and James Wasley in their book, *Healthy by Design,* "rather it is a place that provides positive life-affirming conditions in which its occupants can live and thrive." The temperature is stable and comfortable. There are no cold spots or hot spots. Lighting is ample and pleasing. The building itself is easy to maneuver in and conveniently laid out. Spaces are interesting and inviting. The structure operates quietly and efficiently. It prevents the sounds of outside traffic and barking dogs from intruding on our peace and quiet. Indoor noise is held to a minimum as well.

To create a truly healthy and sustainable shelter, green designers and builders strive to protect the larger environment. They realize that environmental responsibility and personal health are reciprocal goals. However, it is not always easy to achieve both goals simultaneously. Consider the electric stove. When it comes to protecting personal health, electric stoves offer substantial benefits over gas ranges. Unlike a gas range, electric stoves produce no indoor air pollution. However, electric stoves are a less than optimal choice from an environmental perspective. Electricity to run them is generated primarily by coal-fired and nuclear power plants, both with a long list of serious social, economic, and environmental impacts.

Sometimes the goals of planet care and self-care are coincident, however. For example, a passive solar home heats living space and does so cleanly for both people and the planet. Passive solar is an especially healthy option for people suffering from multiple chemical sensitivity, who are sensitive to dust blown in forced-air heating systems or even burning dust on baseboard heaters. Floor and wall tile made from recycled materials is another product that is beneficial to both people and the environment. It is manufactured from a waste material and requires less energy to make than standard tile, so manufacturing produces less air pollution. And tile doesn't outgas harmful chemicals as carpeting and vinyl flooring do.

"We believe in homes that are healthy and energy efficient, healthy and resource conserving, healthy and affordable, healthy and beautiful."

**David Rousseau and
James Wasley**
Healthy by Design

Keys to Building a Healthy House

Creating a healthy home isn't that difficult. It requires common sense and a little knowledge. To build a healthy home, many builders employ a three-pronged approach: eliminate, separate, and ventilate. Lynn and John Bower have dubbed these the "three healthy house principles."

Eliminate Sources of Harmful Pollutants. To eliminate means just that: to avoid unhealthy products and technologies in the design and specification stages of a project. Conventional paints, stains, and finishes containing volatile organic chemicals (VOCs), for instance, can be replaced by low- or no-VOC products to avoid chemical outgassing. Ceramic tile, plaster, hardwood, and natural fiber cotton and wool are generally the healthiest materials for interiors. Modern energy-efficient furnaces and water heaters are a good choice. Many new models draw air into the combustion chamber from outside the house, then vent exhaust gases to the outside as well, so combustion gases won't leak into the house. (They're discussed in chapter 6.) Even better options for both you and the environment are a passive solar heating system for space heat (see chapter 11) and a solar hot water system for domestic hot water .

In existing homes, elimination is not always easy. It is, for instance, very expensive to rip up a floor to get rid of formaldehyde-outgassing subfloor-ing. But you can get rid of a water heater or furnace that leaks combustion gases, replacing it with a newer, more efficient, cleaner model. You may not have the money or desire to replace a gas stove, but you can adjust it so that it burns more cleanly.

Healthy building experts generally identify four product categories to pay close attention to: (1) engineered or manufactured wood products, such as oriented strand board; (2) carpeting; (3) combustion appliances; and (4) paints, stains, and finishes.

Manufactured wood products are made from wood chips, wood fibers, sawdust, or other similar materials (figure 3-3). As noted earlier, these mate-rials are glued together with a resin that contains formaldehyde. Engineered wood is used to make cabinets, dressers, end tables, couches, chairs, and wood panels for interior walls. It is also widely used for floor and roof deck-ing as well as for exterior sheathing. And many new homes contain posts, beams, floor joists, and rafters made from manufactured wood. While these products offer some substantial benefits, (see the sidebar) virtually all of them outgas formaldehyde.

Fortunately, there are ways to lower the emissions of some engineered wood products—for example, by storing oriented strand board in a well-aerated, dry location for a couple of weeks before installing it. (You need to stack the wood so that air can freely circulate around the pieces to permit the harmful chemicals to escape.) Some manufacturers are now producing alternatives to engineered lumber products, including low-VOC oriented strand board and sheathing made from straw or recycled paper, both of which are manufactured with nontoxic binding agents (see chapter 5).

Another product that concerns healthy home builders is carpeting. "Carpeting is a problem," say the Bowers, because "carpet fibers, padding, chemical treatments, and cleaning products can outgas dozens of harmful chemicals . . . [and carpeting] harbors vast quantities of dirt, dust mites, and other allergy-provoking particles."

Combustion appliances, described earlier in the chapter, are the third "problem product." Woodstoves, fireplaces, gas ranges, and furnaces can all pollute interior air with carbon monoxide and nitrogen dioxide.

Engineered Wood

Engineered wood is gener-ally considered a more sus-tainable option than con-ventional wood products such as standard 2 x 4s and 2 x 6s. The reason is that engineered wood products are made from more readily available, smaller-diameter trees, rather than vanishing old-growth trees. Manufactured wood products also use more of the tree, so there's less waste. Some products are even stronger than the conventional materials they are replacing. Unfortunately, virtually all manufactured wood products contain formaldehyde.

Paints, stains, and finishes are also major offenders. When used on interior surfaces, most standard paints, stains, and finishes release harmful VOCs into the indoor air. Release of these chemicals begins the minute the can is opened, proceeds throughout the application, and continues as the products dry, which may take weeks or several months. Fortunately, there are a number of healthier alternatives, described in chapter 4.

Figure 3-3.
Recognizing that traditional practices require endangered old-growth trees and waste considerable amounts of wood, many lumber companies are now producing engineered lumber (shown here) for exterior sheathing and framing.

Source: Truss Joist Weyerhauser

Separate Yourself from Potentially Harmful Substances. Eliminating harmful materials and products from a house is the most effective means of ensuring healthy indoor air. It's the first line of defense. However, if all potentially hazardous products can't be eliminated from a house, there are strategies to prevent exposure. You can separate yourself from them. This constitutes the second line of defense.

Separating yourself from potential toxicants requires the creation of barriers between you and the potentially harmful materials or products. For example, in new construction using oriented strand board, say as subflooring, a builder can seal the sheathing (after airing it out) with a nontoxic sealant to prevent formaldehyde from entering the interior of a home. When oriented strand board is used for exterior sheathing or roof decking, be sure to install a vapor barrier (plastic sheathing) to prevent formaldehyde from entering the house. Vapor barriers are also valuable because they prevent moisture from entering walls and building up inside wall cavities, which reduces the effectiveness of insulation and promotes mold growth (see sidebar).

As another example, radon seeping into a house from a crawl space can be reduced or eliminated by laying a polyethylene plastic sheet on the ground in the crawl space.

If you want to block magnetic fields, you can, but the materials can be very costly. Electric lines can be shielded by iron, nickel, or cobalt, but such shields are effective only in certain situations. Hire a knowledgeable consultant or contractor. Take special precautions in rooms that are occupied more frequently, for example, bedrooms, kitchens, and living rooms. Position yourself away from appliances or electronic devices such as microwave ovens and televisions. Clock radios apparently give off strong fields. Position them away from the bed. A wind-up clock would be better choice. (And it saves on electricity!)

If you are concerned about the potential effect of magnetic fields, be careful where you locate your house. Avoid building or buying a home near high-tension wires, power substations, and transformers. Stay at least a half mile away from them, as well as from cell phone towers, radar towers, microwave transmission towers, and television and radio towers.

BUILDING NOTE

Vapor barriers are plastic sheathing designed to prevent water vapor from entering wall cavities and damaging insulation. They are applied beneath the exterior sheathing in warm climates to prevent moisture from entering a house from the outside air (which tends to be more moist than indoor air). In colder climates, vapor barriers are applied on the inside of the wall, just beneath the drywall or paneling, to prevent moisture from escaping through the wall from the interior (which tends to be more moist than outdoor air).

Figure 3-4.
Exhaust fans in bathrooms
and other locations such as
utility rooms can be used to
ventilate a house, ensuring
healthier indoor air.

Source: David Smith

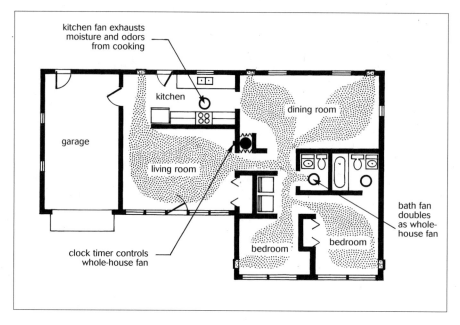

kitchen fan exhausts
moisture and odors
from cooking

kitchen

dining room

garage

living room

bath fan
doubles
as whole-
house fan

clock timer controls
whole-house fan

bedroom

bedroom

According to the EPA,
source controls such as
those discussed under the
categories of "eliminate"
and "separate" are usually
the most effective means
of improving indoor air
quality. They are also generally more cost-effective
than ventilation.

BUILDING NOTE

Every kitchen, bath, and
laundry room should have
a quiet, reliable exhaust
fan, which should be used
often, even placed on a
timer, if no other forms of
ventilation are available.
Some builders also like to
install exhaust fans in
attached garages to pre-
vent fumes from vehicles
entering a home; these
fumes can be a significant
source of indoor air
pollution.

Ventilate Your Home. If you can't eliminate all toxicants or block their release
into room air, you will very likely need to employ the third and most costly
approach: ventilation, providing fresh, clean air to replace the stale, polluted
room air.

Ventilation can be achieved by opening windows. Window fans can be
used to facilitate the process. This simple approach allows fresh air to enter
and pushes stale, polluted indoor air outside, although this strategy may not
be possible year-round in many locations. Exhaust fans in bathrooms,
kitchens, and utility rooms also provide assistance in purging polluted
indoor air. Even better is a whole-house fan (figure 3-4). Mounted in a central location, it exhausts room air through the attic. Fresh air typically enters
through open windows.

Best of all is a whole-house ventilation system, as shown in figure 3-5. In
whole-house ventilation systems, fresh air is generally drawn into buildings
through special openings in walls. Powered by a small fan, these systems provide much-needed fresh air and use very little electricity. However, during
the winter whole-house ventilation systems can waste a significant amount
of heat as they exhaust indoor air. To save heat in winter months, many
builders install a device known as an *air-to-air heat exchanger* or, more commonly, a *heat-recovery ventilator* (HRV).

A heat-recovery ventilator draws fresh air into a house while expelling
stale, polluted room air through a separate pipe, using an electric-pow-
ered fan. However, heat from outgoing air is transferred to incoming air
via a heat exchanger. Heat exchangers reduce heat loss in the winter by
around 60 to 80 percent. When operated during the summer with the air-
conditioning running, heat exchangers reduce heat gain, again saving
energy and creating greater comfort at a lower cost.

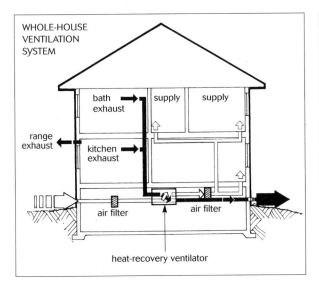

WHOLE-HOUSE VENTILATION SYSTEM

bath exhaust
supply
supply
range exhaust
kitchen exhaust
air filter
air filter
heat-recovery ventilator

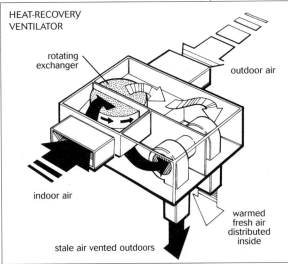

HEAT-RECOVERY VENTILATOR

rotating exchanger
outdoor air
indoor air
warmed fresh air distributed inside
stale air vented outdoors

Figure 3-5.
Ventilation systems equipped with heat-recovery ventilators not only provide fresh air but conserve energy.

Source: David Smith

Why Ventilate: Why Not Just Build a Home with Plenty of Natural Ventilation?

It may seem silly to make a home airtight, then run fans to move air in and out of the house to ensure adequate fresh air. Why not just build a home that's leaky—as we've done for many decades—and forgo the installation of a costly ventilation system?

The reasons for this seemingly contradictory strategy are numerous, and all involve control and comfort.

Leaky homes provide no means of controlling the ventilation rate. As a result, on days when the wind is blowing, which occur most often in the winter, too much fresh air enters the structure, gusting through cracks in the building envelope—for example, around doors, windows, and electrical outlets. As cold air pours in through cracks on the windy side of the house, heat is blown out through cracks on the downwind side. The combination of infiltration and exfiltration costs us dearly. Houses are uncomfortable and heating bills are sky high. On windless days, however, there may not be enough ventilation to ensure healthy indoor air.

Unregulated ventilation through cracks and crevices also has the potential to permit the transport of unwanted chemicals from exterior sheathing and insulation into a house. In other words, leaky home construction reduces your ability to physically isolate inhabitants from potential toxicants.

Leaky building envelopes may also permit moisture to build up in wall insulation. Most forms of insulation lose their ability to restrict heat loss when wet. Poor performance causes unnecessary discomfort and high energy bills. Moreover, mold may grow in the moistened insulation and then may be transported into the interior of the home.

When all factors are taken into account, it is generally less expensive and healthier to build a tightly sealed home with a mechanical ventilation system than to build a leaky home that is vulnerable to the whims of the weather.

To create a healthy home, you need to be able to control the amount of air entering or leaving the structure, as precisely as you can.

CATERING TO THE NEEDS OF THE CHEMICALLY SENSITIVE

If you are chemically sensitive or are building a home for someone who is, you need to know that air filters that successfully remove particulates and gases and reduce one's overall exposure to contaminants may also generate new pollutants that cause some people to react. Fiberglass or polyester fibers in particulate filters, for example, contain a synthetic resin, and some are sprayed with a fine layer of oil to enhance their ability to trap particulates. Some filters are treated with chemical substances that kill mold and other microbes. Although these filters outgas rather tiny amounts of these chemicals at levels that cause no adverse reaction in most of us, they are sufficient to cause reactions in some. One solution is to bake such a filter in the oven at 200°F for a couple hours, although this may cause some filters to deteriorate. Check with the manufacturer first. You may want to install a gas adsorption filter "downwind" from the filter, although this is not always advisable because some people are also sensitive to these media.

Some filters, notably the electrostatic precipitators and the negative ion generators (two types of particulate filters), also produce tiny amounts of ozone. Although most of us aren't bothered by low concentrations of ozone, some people may react badly to it. Additionally, electrostatic filters are made of plastic that outgases pollutants that bother some folks.

Chemically sensitive individuals may also react to adsorption media in gas filters. Carbon filters made from oxidized coconut husks may be tolerable, while carbon filters made from coal are not.

My advice on the subject of filters is to research options carefully. Talk to suppliers who are aware of the special needs of chemically sensitive people. They can usually set you up with filters that are tolerable to this special group. In addition, you will find that some manufacturers allow customers to try a filter in their home and return it if there's a problem. Buy from them.

Air Filters: The Last Resort

Implement Healthy Building Principles

The first thing to remember about air filters is that they're the last resort in new construction. If you implement the principles of healthy house design first, you shouldn't need to filter the air, except in instances where toxic chemicals are used inside the house. Air filters may be required if one or more of the occupants of a home suffers allergies to dust mites or dander from pets, has asthma, or is chemically sensitive.

By employing the three healthy building principles—eliminate, separate, and ventilate—you can create a home with interior air that is clean and healthful for almost everyone. (The only exception might be someone suffering from multiple chemical sensitivity, a subject addressed in the box above.) In existing homes or homes not built entirely in accordance with healthy building principles, air filtration may be required to improve indoor air quality. What do you need to know about air filtration systems?

Portable vs. Whole-House Filtration Systems

First of all, air filtration systems fall into two very broad groups: portable and whole-house. A portable filter is used to filter and purify the air in a single room, while a whole-house filter is designed to purify the air in the entire home.

Portable filters have limited usefulness. They work optimally when rooms are closed off from the rest of the living space—for example, in home offices, bedrooms, and workshops. Portable filters work well in bedrooms at night to filter air for people suffering from asthma or allergies.

Because portable room filters work best in closed rooms, however, they are of limited utility in houses equipped with central air-conditioning or forced-air heating systems. These systems circulate air throughout a home through ducts, so that isolating a room from the rest of the house is impossible, unless you close all of the registers. (Then, of course, you get no heat or cooling!) Whatever you do, don't waste your money on a desktop model. They are much too small to make any significant impact on room air quality.

Whole-house filters are usually integrated into forced-air heating or central air-conditioning systems. Heated or cooled air circulating through ductwork typically passes through a special air filter—not an ordinary furnace filter placed inside the ductwork. It removes pollutants. However, because heating and air-conditioning systems do not operate 24 hours a day, 365 days per year, a circulator fan may need to be run continuously year-round to achieve adequate air filtration. That, of course, can consume a lot of electricity.

Houses with mechanical ventilation systems can also be equipped with whole-house filters. However, the fan in most mechanical ventilation systems usually is not powerful enough to push air through a whole-house filter. A stronger, more energy-intensive fan must be installed.

In houses in which a ventilation system works in conjunction with a forced-air heating/central air-conditioning system, a single air filter can do double duty. This is, it can filter the air that is drawn into the house through the ventilation system *and* purify air that circulates through the heating and/or air-conditioning system.

Particulate vs. Gas Filters

Now that you understand the two basic types of air filtration systems, let's examine the types of filters used in them. Filters used in portable and whole-house systems come in two basic varieties: particulate and gas filters. Particulate filters, as their name implies, are generally designed to remove particulates such as dust, mold, dander, pollen, or smoke. Gas filters remove gases such as carbon monoxide and formaldehyde.

Particulate filters protect people with allergies and asthma by removing fine particles suspended in the air. Gas filters remove toxic gases that might trigger multiple chemical sensitivity or other health problems. To rid a house of both types of pollutants, particulates and gases, you will very likely need to install both types of filters or a dual-purpose filter that removes both offenders.

Particulate and gas filters come in many varieties with varying effectiveness. I suggest that you read product literature carefully and shop shrewdly. Some particulate filters, for instance, work well on large particles but do little to remove fine particulates from combustion sources such as tobacco smoke. Fine particulates can penetrate deeply into the lungs and can lead to lung cancer.

Asking Too Much of an Air Filter

Unfortunately, writes John Bower in *The Healthy House,* "People often ask more of an air filter than it's capable of doing. For example, they may (unsuccessfully) try to use an air filter to clean up the air in a very polluted house without first implementing the healthy house design principles." If you want to use a filter to clean up the air in a problem house, say the Bowers, "you'll need a very powerful system that will filter the air several times an hour." Not only will this cost a lot, it may result in excess air movement that decreases comfort levels, and it will surely be noisy. The Bowers go on to say, "Filters work best when combined with other strategies for improving indoor air quality."

To determine the potential performance of an air filter that removes particulates, you will need to consider its efficiency rating, how much pollution it traps or removes from air passing through it. Be careful when comparing models. Although manufacturers commonly list efficiency, they don't always provide enough information to make a sound buying decision. Fortunately, the industry is becoming standardized, so choosing a particulate filter is getting easier.

Another crucial indicator is the amount of air that moves through a particulate filter in a given period. This is usually measured in cubic feet or cubic meters per second. To achieve maximum performance, you want a filter that has a medium to high efficiency rating while handling a large amount of air, to ensure that much of the air in a room will pass through the filter within a short period.

Yet another consideration is the decline in performance that occurs as the filter begins to accumulate pollutants scoured from room air. Some models decrease in efficiency rapidly as they begin to fill up with pollutants. Others remain fairly efficient with use.

For most homes, all that is needed is a mid-range particulate filter rated at 25 to 45 percent efficiency (based on the spot-dust test). This type of filter is available from Carrier, General Filters, Honeywell, and Research Products Corp. More efficient HEPA (high-efficiency particulate accumulator) filters are also available, some with efficiencies over 99 percent. HEPA filters last one to five years and are by far the most efficient particulate filters on the market. However, for most homes they're unnecessary. In fact, they're overkill. The particulates that most bother the folks who suffer from allergies are removed by medium-efficiency filters.

To learn about air filter ratings, see the sidebar at right, and for more information on the subject, you may want to consult John Bower's book, *The Healthy House*. He does an excellent job of describing the major types of particulate and gas filters, the pros and cons of each one, and the ways efficiency is measured.

Houseplants as Air Cleaners

What about purifying the air in your home with houseplants? In the 1980s, the National Aeronautics and Space Administration (NASA) performed some experiments to assess the capacity of houseplants, such as spider plants and golden pothos (aka devil's ivy), to remove pollutants from indoor air. The researchers found that these plants were effective in removing several harmful gases, including formaldehyde, carbon monoxide, and nitrogen dioxide. Since that time, numerous magazine articles and television programs have boosted the image of plant purifiers.

Unfortunately, further studies showed that the responsible party was not the plants but the microorganisms in the soil the plants were growing in. They metabolize the pollutants, removing them from air. However, much to

> "There is a great deal of hype in advertising about air filters. As a result, many people are being sold more filtration capacity than they need, while others are sold gimmicks that do very little to clean the air."
>
> **John Bower**
> *The Healthy House*

RATING AIR FILTERS

When buying an air filter, one important criterion is efficiency. Unfortunately, there are no government standards for reporting air filter efficiency.

According to the Asthma and Allergy Foundation of America, "The Food and Drug Administration has twice asked groups of experts to recommend national standards, but neither effort succeeded. Both groups concluded that there isn't enough research data on the relationship between air filtration and actual health improvement to recommend national standards."

As a result, buyers must rely on one of two sources of information: (1) manufacturers' claims, which can be confusing or misleading, or (2) efficiency ratings by one of two professional organizations, the American Society of Heating, Refrigeration, and Air Conditioning Engineers (ASHRAE) or the Association of Home Appliance Manufacturers (AHAM). Look for their labels on products to ensure that air filtration systems have been tested. You can also use their ratings to compare competing models. Bear in mind that no rating system is based upon health impacts.

In February 2000, ASHRAE announced a new standard for testing and rating particulate filters. The test now measures efficiency for a variety of particle sizes, so there can be no misleading claims—for example, a manufacturer's claim that its filter is 90 percent efficient fails to tell the consumer that this value refers only to very large particles, which are less of a concern than medium-sized and small particles!

ASHRAE's rating is a minimum efficiency reporting value (or MERV). According to Charles Rose, a member of ASHRAE's technical committee on Particulate Air Contaminants and Particulate Contaminant Removal Equipment, "The MERV is a number from 1 to 16 [that allows a buyer] to compare air filters on the basis of the percent of dust they remove from the air. The higher the number, the higher the percent."

Portable air filters are often rated by the Association of Home Appliance Manufacturers. Its label on a filter indicates the clean air delivery rate (CADR). According to AHAM, CADR is a measure of the amount of air filtered by a unit, in cubic feet per minute, for particulates such as tobacco smoke, dust, and pollen. If an air cleaner has a CADR of 380 for tobacco smoke, for example, it reduces pollutant levels to the same concentration as would be achieved by adding 380 cubic feet of smoke-free air every minute! Obviously, the higher the CADR, the better the filter.

ASHRAE and AHAM ratings apply only to the filtering of particulates, which is by far the most complicated area and the one most rife with faulty or misleading claims. Removal efficiencies for gaseous pollutants are often reported in a more straightforward manner (there's no range of particulate size to deal with), with less room for misrepresentation.

"Although the FDA has no health-related standards," notes the Asthma and Allergy Foundation of America (AAFA), "it does consider some portable air filtration systems to be Class II medical devices." To obtain this rating, a manufacturer must show that the device is safe and that it has a medical benefit. "Look for both the UL [Underwriter's Laboratory] seal and a statement of the FDA's Class II approval [on the product]. If no FDA statement is available with the device, check the FDA's medical device listing before buying."

When shopping for an air filter system, the AAFA recommends that you buy a model that can recirculate eight or ten room volumes per hour. "This doesn't guarantee completely clean air," it says, "but it will be much cleaner than with systems that recirculate less." Asthma sufferers should purchase systems "that remove more than 90 percent of all particles larger than 0.3 microns in diameter. Most indoor allergens are larger than this, so this efficiency standard will handle them easily." This goal can be obtained by purchasing a unit with a HEPA filter.

the chagrin of plant lovers, current research indicates that potted plants remove far fewer pollutants than originally thought.

As is sometimes the case with scientific studies, it turns out that NASA's experiments were flawed. All of their studies were performed in sealed chambers to which a given amount of air pollution was introduced. Although the potted plants did indeed gobble up air pollution in the experiments, a sealed chamber containing a certain amount of introduced air pollution is not at all like a house. In a home, pollutants are continually replenished as they outgas from building materials and furnishings for years or are generated from additional sources.

For a list of links to numerous organizations involved in indoor air quality, log on to: www.epa.gov.iaq/more-info/html.

To test the effectiveness of plants in a real home, researchers at Ball State University performed an experiment that simulated real-life conditions. Their studies showed that houseplants (or the soil in their pots) do not reduce indoor air pollution to any significant degree. They apparently can't keep up with the continuous outgassing. This is currently the official position of the U.S. Environmental Protection Agency. Furthermore, as John Bower points out, "Having plants indoors tends to result in a higher relative humidity—and formaldehyde emissions increase as the relative humidity rises." He goes on to say, "So, in a real-life situation, the formaldehyde absorbed by the plants is replaced by an increased rate of outgassing." In addition, houseplants may release pollen spores that cause allergic reactions in some individuals. Elevated levels of humidity may also stimulate mold growth in the house, which can cause health problems as well.

Although there is a lot more to learn about filters, this introduction should help you start off on the right foot. Remember, a good filter is not a substitute for eliminating or reducing pollutants. And it certainly is no substitute for good ventilation.

Some Parting Thoughts

Healthy building is good for people and the planet. Although it may cost a little more to build a healthy home, the benefits are worth it. They will be felt immediately and throughout your life and the life of your home. For those who care about their health, and the health of their loved ones, employing healthy building principles is the only way to go.

CHAPTER 4

Green Building Materials

When John and Judy Matson (not their real names) set out to buy a new home, like millions of other couples they were seeking a wide assortment of amenities. They wanted a home in quiet neighborhood, close to work, with a big kitchen and a spacious bedroom. They were looking for a well-constructed home that would last, not a shoddily built structure designed to sell cheap. Like other energy-conscious home buyers, they wanted a house that was energy efficient, too.

John and Judy started their search in the newspaper. Here they found an overwhelming number of ads for new homes in new subdivisions in the rapidly expanding Denver metro area. The first-time home buyers were pleasantly surprised to find several houses that seemed to meet their needs. However, one stood out among the crowd, a green home built by McStain Neighborhoods in Boulder.

McStain is one of the nation's oldest and most successful green builders. This company began building environmentally friendly homes in 1996. For the Matsons, McStain's homes, which look like any other new home on the market, satisfied a deep commitment to environmental conservation.

Figure 4-1.
Tom and Caroline Hoyt of McStain Neighborhoods in Boulder, Colorado, are two of the nation's leaders in green building. Those wishing to join the movement can learn a great deal from their experiences.

Source: McStain Neighborhoods

Characteristics of Green Building Materials

• Produced by socially and environmentally responsible companies

• Produced sustainably–harvested, extracted, processed, and transported efficiently and cleanly

• Low embodied energy

• Locally produced

• Made from recycled waste

• Made from natural or renewable materials

• Durable

• Recyclable

• Nontoxic

• Efficient in their use of resources

• Reliant on renewable resources

• Nonpolluting

Run by Tom and Caroline Hoyt, McStain Neighborhoods builds homes that provide a safe, healthy interior for families. Low-VOC paints and finishes and mechanical ventilation systems help achieve this important goal and make its homes people-friendly. But McStain also incorporates many environmentally responsible building materials—called *green building materials*—that reduce pollution, habitat destruction, and resource consumption.

In this chapter, we'll focus specifically on green building materials used in homes like those built by McStain. Although the list is extensive, all green building materials support the environmental goals of green building outlined in chapter 1: (1) conservation, (2) recycling, (3) renewable resource use, and (4) restoration and sustainable management of resources. And many of them also help promote healthier interiors.

What Are Green Building Materials?

The term *green building materials* refers to a growing list of products used to build and furnish homes that are good for people and the planet. To make the list, these materials must meet at least one of the characteristics described in the next few paragraphs and summarized in the box above. Obviously, the more criteria each product meets, the greener it is.

One characteristic of a green building material is that it is produced and sold by companies that are socially and environmentally responsible. These are companies that treat their employees and customers well, exert a positive influence in the community, and engage in sustainable environmental practices in all aspects of the business: in the office, the factory, and the field.

A good example of social responsibility is workplace democracy, a form of organization in which workers have a say in company operations. Not only does it empower workers, it helps increase productivity.

Social responsibility also means that companies treat employees well—with respect, compassion, and dignity—and offer important worker benefits, such as paid maternity leave and on-site day-care facilities for employees' children. Equitable pay distribution among employees ranging from the CEO to the maintenance crew is yet another sign of social responsibility.

Good environmental practices include sustainable management of natural resources, such as forests; use of renewable resources, such as solar and wind energy; company-wide recycling of wastes, even office wastes; pollution prevention policies and practices; and energy conservation.

Although it is not always easy to determine a company's social and environmental policies, direct inquiries via phone, letter, or e-mail may turn up

One characteristic of a green building material is that it is produced and sold by companies that are socially and environmentally responsible. These are companies that treat their employees and customers well, exert a positive influence in the community, and engage in sustainable environmental practices in all aspects of the business: in the office, the factory, and the field.

valuable information. Articles in newspapers and magazines may also prove helpful in your quest.

Next on the list of essential criteria for green building materials is embodied energy. Embodied energy is the amount of energy required to extract or harvest the raw materials required to make building products, process them into finished products, and transport both raw materials and finished materials during the various stages of the production-consumption cycle. The lower the embodied energy, the better.

As noted in chapter 1, some materials, such as steel and aluminum, have much higher embodied energy than others, such as wood or earthen materials. As a rule, materials made from recycled waste have a lower embodied energy than materials manufactured from virgin resources—often much lower. In addition to requiring less energy to make, products made from recycled material also put valuable waste to good use, reducing the harvest or extraction of virgin materials, landfilling, and energy consumption.

Materials produced from locally available resources also have a much lower embodied energy than those manufactured from virgin materials at facilities thousands of miles from a building site. The ultimate in locally produced building materials are natural materials. Adobe and straw bales, for instance, are often locally abundant, requiring minimal transportation. Adobe dirt often can be acquired from foundation excavation. Adobe and straw offer other benefits as well, described in chapter 9.

Durability is another important criterion to consider when shopping for green building materials. The more durable a product is, the longer it will remain in service, and the lower the total environmental impact will be. Recycled aluminum or steel roofing, for instance, outlasts less durable competitors such as asphalt shingles, saving energy, materials, labor, and money. One durable, environmentally friendly roofing product, Ecoshake shingles, is made from recycled vinyl and wood fiber and comes with a fifty-year warranty! In addition, because this product resists fire and hail, many insurance companies offer substantial discounts—up to 28 percent—for homes on which it's been installed. Using durable products also reduces pollution and landfill waste. The net effect of durable materials is a higher-quality home that requires less maintenance than a less-well-built structure and has considerably less impact on the environment.

Many building products are reusable, too, or can be recycled after their useful life is over. Steel roofing and adobe blocks, for instance, can both be recycled, adding to their effective life span. Recycling such products reduces resource extraction, pollution, and energy use. However, to be truly recyclable or reusable, a product must have an outlet: a willing buyer, such as a local steel recycler. It must also be as pure as possible, that is, not mixed with other materials that make recycling difficult.

Green building materials and products must also be nontoxic to those who manufacture and install them as well as those who live with them.

As a rule, materials made from recycled waste have a much lower embodied energy than materials manufactured from virgin resources.

Durability Pays

The net effect of durable materials is a higher-quality home that requires less maintenance than a less-well-built structure and has considerably less impact on the environment.

Watch for volatile chemicals and other toxic substances that outgas into room air.

While our concern here is primarily for building materials and furnishings, many appliances such as stoves, water heaters, and heating systems are installed in our homes. To promote environmental sustainability, such devices should be as efficient as possible and, if reliant on fossil fuels, should burn as cleanly as possible. They must be designed and installed in ways that prevent indoor air pollution as well.

Life-Cycle Cost

One of the principal goals in green building is to use materials that promote both the health of the people who live in homes and the health of the planet—everyone's home. To do so, we must select those materials—as well those techniques, technologies, and designs—that have the lowest life-cycle costs. *Life-cycle cost* refers to all of the costs of a product over its life cycle, from the extraction to the manufacture to the sale and use of a product to its ultimate disposal. To achieve full-cost accounting, we look not just at economic costs but at social and environmental costs as well. The lower the life-cycle cost, the more sustainable a product is.

Although it is difficult to determine life-cycle costs, attempts are being made. One notable example is *The Environmental Resource Guide* by the American Institute of Architects. This publication provides fairly detailed life-cycle assessments of a number of building materials. I've listed the now out-of-print book in the resource guide at the end of this book. For more on the subject, including on-line assistance, see the sidebar at left and Nadiv Malin's article on life-cycle analysis in the March 2002 issue of *Environmental Building News*.

In your search, don't be dismayed if exact costs cannot be computed. Even a crude estimate of product costs can help steer a designer or builder in the correct direction—and it's far better than total ignorance of the costs.

Determining Life-Cycle Costs

For assistance, you can log on to the National Institute for Standards and Technology's web site to check out their Building for Economic and Environmental Sustainability software. It helps builders identify green building products using a set of predefined life-cycle criteria. Builders can simply choose the criteria they wish to apply. To log on, go to: www.fire.nist.gob/bfrlpubs/build01/art081.html.

From the Foundation to the Roof

When I began construction on my house in 1995, green building materials were few and far between. I couldn't swing by the local lumberyard and pick up sustainably harvested lumber or no-VOC oriented strand board. Nor could I call the carpet retailers in my area to purchase recycled-content carpeting or tile. Fortunately, I live in a state blessed with green building suppliers (at the time, EcoProducts and Planetary Solutions) that provide such items. With their help, I was able to purchase a wide assortment of green building materials—so many, in fact, that nearly my entire house is built from these environmentally friendly products!

Since I built my house, the number of green building materials and products has expanded dramatically—and, even more exciting, so has their availability! Most of them are still available only through green building suppliers, which I've listed in the resource guide. However, many conventional building outlets, such as Home Depot, now carry a number of green building products, from low-VOC paints and finishes to energy-efficient windows to landscaping lumber and decking materials made from recycled plastic milk bottles. Some local lumber retailers are happy to special-order products, such as low-VOC oriented strand board, for larger projects.

You may be encouraged to learn that there isn't a single product that goes into a house—from the foundation to the roof—that doesn't have at least one green alternative. You can even buy nails made from recycled steel, if you like!

Owner-builders, contractors, and others who want to learn more about green building materials may also be pleasantly surprised to learn that there are a half dozen books that provide comprehensive lists of green building materials, including brief descriptions of the products and contact information for manufacturers. Manufacturers, in turn, can provide information on local distributors.

One of the most useful books is *GreenSpec: The Environmental Building News Product Directory and Guideline Specifications,* published by BuildingGreen. This superb resource contains a goldmine of information and is now available through BuildingGreen's on-line resource center. The on-line version offers more up-to-date listings than the print copy and can be searched more easily. It also offers links to in-depth reviews that have appeared in BuildingGreen's newsletter, *Environmental Building News.*

Another valuable resource is John Hermannsson's *Green Building Resource Guide.* In addition to listing and describing a wide assortment of green building products, Hermannsson has included information on pricing in the form of a price index. The price index lets you compare the price of a green building material to the conventional building material it replaces. Celbar is a type of insulation made from recycled newsprint by International Cellulose in Houston. Its price index is 0.5 to 0.8 compared to that of fiberglass batt. Translated, this means that Celbar costs 50 to 80 percent as much as the much less people- and environmental-friendly fiberglass batt.

Another superb resource is the Austin Green Building Program's web site, which provides an abundance of information on green building materials. The sidebar to the right lists on-line resources that provide similar information.

Several newsletters and magazines provide detailed analyses of green building materials. Each issue of *Environmental Building News (EBN),* for instance, includes an in-depth examination of one green building product or technique, such as green roofs or radiant-floor heating. The monthly newsletter also includes a wealth of information on new products and developments. *EBN* is an invaluable resource for architects, builders, and owner-builders. To

There isn't a single product that goes into a house—from the foundation to the roof—that doesn't have at least one green alternative.

On-line Assistance

Austin Green Building Program: www.greenbuilder.com/sourcebook/

The Center for Resourceful Building Technology's e-Guide (provides a searchable database of green building materials and their manufacturers): www.crbt.org

Oikos Green Building Product Information: www.oikos.com/products

catch up on issues as far back as 1992 (when the newsletter was first published), you can purchase *EBN Archives* on CD-ROM. The *Archives* are fully searchable and provide live links to manufacturers of more than 450 products that have been reviewed or described in the newsletter.

Yet another resource is *Green Building Advisor,* also published by BuildingGreen. Users enter data on the project they're about to undertake, and *Green Building Advisor* quickly assembles a comprehensive list of strategies to make the project as environmentally sound as possible. It offers advice on landscaping, energy efficiency, solar energy, green building materials, and much more. Any recommendation made by the program is backed up by layers of detailed information, available like so many things these days at the click of your mouse. It's like having several knowledgable green building experts at your side to provide guidance and answer all of your questions.

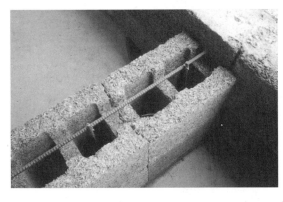

Figure 4-2.
Environmentally friendly Faswall blocks are made from sawdust and cement.

Source: K-X Faswall

Before you can use these tools, however, it is helpful to get an overview of what is available. To help you become better acquainted with the range of green building materials and techniques currently available to builders, the following section offers a checklist of green building options, organized by building component, for example, foundation or interior wall.

Don't be dismayed if the information leaves you wanting more. Many of these products and ideas are discussed more thoroughly in subsequent chapters, providing the greater depth of knowledge required to make sound decisions. If you are working with a builder, you can use the list to discuss options. If you are looking to buy a new house, you can use this list to compare your options. Readers interested in learning more may want to read the chapter on green building materials in my book *The Natural House,* then check out the green building directories mentioned above.

Figure 4-3.
Environmentally friendly Rastra block can be used to build beautiful homes like the one shown below.

Source: Dan Chouinard, Rastra

Green Product Checklist

Foundations

- **Fly-ash concrete and concrete blocks** substitute fly ash, a waste product from coal-fired power plants, for some of the Portland cement used to make concrete, producing a superior product that puts an abundant waste material to good use and reduces the amount of energy required to build a foundation.
- **Faswall blocks** for foundations and basements are 85 percent sawdust (waste) and 15 percent cement. Use of this product promotes recycling and the use of lower-embodied-energy materials (figure 4-2).

Figure 4-4.
Insulated concrete forms like these made by Greenblock in Woodland Park, Colorado, are filled with concrete to produce energy- and resource-efficient foundations and exterior walls of homes and other buildings. On the exterior surface of walls, the foam is typically stuccoed or finished with siding. Interior surfaces are typically stuccoed or covered with drywall.

Source: Greenblock

- **Rastra blocks** for foundations and basements are made from recycled plastic (polystyrene) and cement. Using them promotes recycling and the use of lower-embodied-energy material (figure 4-3).
- **Insulated concrete forms** (ICFs) are permanent foundation forms consisting of a foam board into which concrete is poured (figure 4-4). ICFs reduce labor and result in an extremely energy-efficient foundation. They also dramatically reduce the amount of high-embodied-energy concrete required to build foundations. A couple manufacturers use recycled plastic to make their foam.
- **Rubble trench foundations** usually consist of a concrete grade beam poured over a rock-filled trench. This technique, which is ideal for use in seismically stable areas, greatly reduces concrete use and can save significantly on foundation construction (figure 4-5).

Exterior Walls

- **Natural or recycled waste materials,** such as straw bales, adobe, rammed earth tires, cob, straw-clay, and waste paper (see box on page 76), use locally available, low-embodied-energy, nontoxic materials and offer user-friendly and relatively inexpensive ways to build walls, although some types of natural building can be quite labor intensive.
- **Structural insulated panels,** consisting of foam or straw insulation sandwiched between oriented strand board, reduce air infiltration and heat loss through walls and nearly eliminate framing required to build exterior walls. This product saves energy, reduces wood use, and increases comfort.

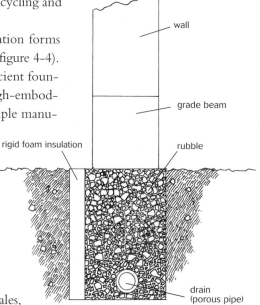

wall

grade beam

rigid foam insulation

rubble

drain (porous pipe)

Figure 4-5.
Rubble trench foundations (as shown here) are economical, perform well in nonseismic areas, and require far fewer resources than conventional foundations.

Source: Michael Middleton (from *The Natural House*)

- **Two-by-four wall construction with external foam insulation** performs as well or better than 2 x 6 walls with insulation located in wall cavities (that is, between studs). This technique reduces wood use.
- **Engineered lumber for studs, beams, posts, and headers** utilizes smaller-diameter trees and reduces the harvest of giant, stately trees from old-growth forests. Engineered lumber also makes more efficient use of the

---

BUILDING WITH BALES—PAPERBALES, THAT IS!

In the tiny town of Fraser, Colorado, near the Winter Park ski resort, Rich Messer and Ann Douden built the walls and foundation of their energy-efficient mountain home from two most unusual materials.

The couple created walls from huge bales of paperboard from laundry soap boxes. It's called poly-coated Kraft carrier board, but any type of paper would work. The paperboard was collected by Tri-R, a major recycling company in Denver, about 100 miles away. This material is quite clean, notes Robyn Griggs Lawrence in her article "Classy Trash," published in *Natural Home* magazine, but because it is coated with a thin layer of plastic, it is relatively difficult to recycle. Much of it ends up in landfills. Messer points out that for this reason paperboard is abundantly available in cities and towns, and it's inexpensive, too. Delighted to get rid of the stuff, Tri-R gave him the bales for free.

The huge bales of waste paperboard produce 36-inch-thick, heat-resistant walls that keep Rich and Ann warm and cozy in this cold, wintry climate. When covered with stucco, the walls appear a bit bumpy, but otherwise pretty normal. But what could support such heavy bales of waste?

Rather than build a conventional foundation, Messer decided to build the foundation of their new home out of another abundant waste: huge bales of waste plastic. This required twenty-eight bales containing post-consumer PVC plastic, mostly old toys, laundry baskets, and shampoo bottles, but any type of plastic would work. Rich paid $20 per bale, or less than $600.

The plastic bales were laid into a 5½-foot-wide foundation trench with compacted road base providing the underlying support. He then applied 3½ inches of foam insulation inside the plastic bale foundation to reduce heat loss.

Rich and a helper secured the roof to the paperboard bales by installing a concrete bond beam poured on top of them. (It is attached to the bales by rebar.) After the roof was installed, Rich blew recycled cellulose insulation into the ceiling cavity, yielding an R-50 layer of heat-resistant material to keep the house warm despite sometimes bitterly cold winter temperatures. He then sealed air cracks with leftover paper.

Besides having low fuel bills, the house is quiet and warm—unlike any house the couple has been in before. It creates "a serenity we didn't think was part of what we were building until we lived in it," says Messer. The only problem in getting more houses of this type built, he says, is financing. "Although many people have expressed interest in the building method, the biggest hurdle is bankers, who just don't like to hear about building houses with trash."

"To do so requires a willingness to take a risk," adds Messer. As forests decline and people look for other options, waste paper may become yet another source of building materials. Who knows, maybe someday you, too, will live in a house made of recycled trash. (For more on this house, see Robyn Griggs Lawrence's article in *Natural Home*, which is listed in the resource guide.)

trees we cut and is not likely to warp like standard dimensional lumber. Reduced warping, in turn, reduces discards. Because engineered lumber offers superior strength, studs can be spaced more widely, further reducing the amount of wood required to build a house.

- **Salvaged lumber** obtained from demolished buildings or recovered from lakes and rivers where it was lost in years past during transport to sawmills reduces the harvest of forests and puts a waste material to good use.
- **Sustainably harvested lumber** certified by a reliable, independent agency ensures that wood used to build a house comes from forests, even tropical rain forests, that are managed and harvested sustainably.
- **Insulated concrete forms, Rastra blocks, and Faswall blocks** for exterior walls have many benefits; see above.
- **Low- or no-formaldehyde oriented strand board** for exterior sheathing reduces the potential for contaminating indoor air.
- **Thermoply exterior sheathing** is made from recycled newspaper and a nontoxic binding agent. Use of this product reduces waste trashed in the landfills. Because it is lightweight and easy to install, Thermoply exterior sheathing can lower home construction costs. Because this product can be overlapped, Thermoply also reduces air infiltration, saving energy and making a home more comfortable.

Roofs

- **Engineered lumber** for framing roofs, for example, wooden I beams, replaces solid dimensional framing lumber. Engineered lumber provides superior strength yet requires 40 to 60 percent less wood than standard framing lumber, reduces the harvest of old-growth forests, and saves energy (figure 4-6). Most engineered lumber does, however, contain potentially toxic resins containing formaldehyde.
- **Low- or no-formaldehyde oriented strand board** offers the same advantages for roof decking as it does for exterior sheathing, listed above.
- **Sealed oriented strand board** reduces outgassing of formaldehyde.
- **One hundred percent recycled newsprint (e.g., Homasote) or a mixture of recycled newsprint and agricultural fibers** used for roof decking helps us eliminate waste, saves energy, and reduces our demand for natural resources.
- **Structural insulated panels** for building roofs reduce framing lumber and cut energy losses (see above).
- **Roofing felt** made from recycled paper, recycled paper and sawdust, or recycled PET (number 1) plastic promotes the use of recycled materials and reduces landfilling as well as demand for natural resources.
- **Metal roofing** made from recycled steel or aluminum utilizes wastes, saves energy in manufacturing, and, because it's durable, reduces roof maintenance and costly replacement, saving resources.

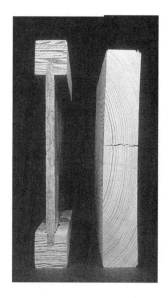

Figure 4-6.
The I beam made from engineered lumber (left) uses about half of the wood of a solid piece of lumber of the same dimension (right) and is made from smaller-diameter trees, thus eliminates the use of large-diameter trees from old-growth forests.

Source: Truss Joist Macmillan

Figure 4-7.
Engineered wooden I joists were first used for structural framework of floors (as shown here) but are now also used for framing roofs.

Source: Truss Joist Macmillan

- **Shingles made from recycled materials** such as sawdust (waste) and cement, or sawdust and recycled plastic, or recycled rubber, put waste to good use and provide long-lasting performance, reducing maintenance and costly replacement.

Ceiling and Wall Insulation

- **Wet-blown cellulose insulation** made from recycled newsprint and cardboard is nontoxic, uses an abundant waste product, reduces air infiltration, and saves energy.
- **Straw or straw-clay walls** use a low-embodied-energy, locally available, natural material.
- **Mineral wool insulation** made from a recycled waste product reduces waste heading to the world's overflowing landfills.
- **Rigid foam insulation** in structural insulated panels or applied externally to external sheathing (see above) reduces heat loss through framing members in a wall or ceiling (bridging loss) and cuts air infiltration and energy bills. Although most products are manufactured with ozone-depleting chemicals, there are some exceptions, such as Insulfoam.
- **Formaldehyde-free fiberglass insulation** is safer for workers and protects the occupants of a home from formaldehyde outgassing.
- **Encapsulated fiberglass batts** are safer for applicators.
- **Recycled-content fiberglass** uses recycled material, cutting waste and reducing the energy required to manufacture insulation.
- **Blow-in foam insulations** manufactured without ozone-depleting compounds and formaldehyde, such as Icynene spray insulation, reduce air infiltration and energy loss while protecting occupants of a home, applicators, and the environment from potentially harmful chemicals.

Windows and Doors

- **Well-made, energy-efficient windows**—usually argon-gas-filled, infiltration-resistant, double-paned, low-E (low-emissivity) glass with wood or other non-heat-conducting frames—dramatically reduce heat loss during the winter and heat gain during the summer, boosting comfort, saving energy, and reducing pollution. Metal cladding on exterior surfaces protects wood from the weather and reduces maintenance. (See chapter 6 for more information.)
- **Insulated window shades or thermal shutters** (internal or external) reduce heat gain during the cooling season and reduce heat loss during the winter.
- **High-quality, insulated, and airtight exterior doors,** either solid wood or steel or fiberglass doors with foam cores, reduce heating and cooling requirements, increase comfort, and cut energy bills. If doors are manufactured from wood, look for sustainably harvested materials.

- **Salvaged interior and exterior doors** can be made to be as good as new with a little sanding, painting, and staining, thereby reducing resource demand and cutting construction costs.
- **Interior doors manufactured from molded hardboard** made from recycled wood waste but without formaldehyde-based resins help us recycle more of our waste.

Exterior Siding and Trim (Interior and Exterior)

- **Natural plasters,** if applied correctly, protect walls and require little maintenance. Natural materials like earthen plasters often have a low embodied energy and cost less than manufactured materials such as cement and synthetic stuccos.
- **Metal siding** made from recycled steel or aluminum utilizes waste and provides many years of protection with minimal maintenance and repair. Durability, in turn, provides many environmental benefits.
- **Hardboard lap or shake siding** made from sawdust, resin, and cement provides low-maintenance protection, reducing costs, waste, and timber cutting (figure 4-8).
- **Molded fiber-cement boards or plastic lumber** for soffits, fascias, and other exterior applications provide the same benefits as hardboard lap or shake siding.
- **Finger-jointed trim** for internal trim uses scrap lumber and reduces waste.

Floors and Floor Coverings

- **Formaldehyde-free oriented strand board subflooring** protects indoor air quality.
- **Tile made from recycled materials** such as mine waste (feldspar) or automobile window glass has low embodied energy and supports the recycling industry.

Figure 3-5.
Siding made from cement and wood fibers on this house offers superior durability, lasting fifty years or more, and reduces maintenance and costly replacment.

Source: Dan Chiras

- **Locally produced tile,** even if not made from recycled materials, promotes the use of low-embodied-energy materials.
- **Hardwood flooring from salvaged wood** promotes the reuse of materials and diverts waste from our landfills.
- **Bamboo flooring** instead of hardwood flooring uses a fast-growing, renewable resource that is harvested with little damage to the environment. (Unfortunately, this product is currently shipped from China and Southeast Asia, a low-wage market with little or no environmental regulation.)
- **Natural linoleum** made from sawdust, cork, and natural resins in kitchens and bathrooms promotes recycling and helps protect indoor air quality.
- **Carpet made from recycled or natural materials** such as PET plastic, nylon, or wool puts an abundant waste to good use and reduces pollution, energy consumption, and resource extraction (figure 4-9).
- **Carpet pad made from recycled materials** such as clothing scraps, recycled newsprint, recycled nylon, or recycled rubber offers the same benefits as recycled carpeting.
- **Nontoxic adhesives** for carpets and tiles protect indoor air quality.
- **Stained or stamped concrete floors** reduce demand for resources (no need to install tile).

Interior Walls and Ceilings

- **Engineered or salvaged lumber** used for framing offers the same benefits listed above.
- **Drywall made from alternative materials** such as sludge from coal-fired

Figure 4-9.
Carpet made from recycled plastic pop bottles by Mohawk looks and performs like any other carpet and is cost-competitive with conventional carpeting.

Source: Dan Chiras

Figure 4-10.
Waste from coal-fired power plants (shown here) is chemically identical to the gypsum used to manufacture drywall, and is used to manufacture this product in England and Germany.

Source: Knauf Fiber Glass.

power plants, recycled and waste gypsum, and recycled paper helps reduce the steady flow of wastes to landfills and decreases the amount of energy required to manufacture drywall (figure 4-10).

- **Interior walls made from natural materials** such as cob or adobe are nontoxic and have low embodied energy.

Paints, Stains, and Finishes

- **Low- or no-VOC paints, stains, and finishes** help ensure good indoor air quality and protect applicators and workers at manufacturing facilities.
- **Mildewcide-, fungicide-, formaldehyde-, mercury-, and lead-free paints** offer the same benefits as VOC-free paints.
- **Milk or casein paints** promote renewable resource use and protect indoor air quality.
- **Water-based urethane** for floor finishes promotes good indoor air quality.
- **Salvaged latex paints** reduce waste.

Decks and Play Sets

- **Recycled plastic lumber** promotes the use of recycled materials and avoids exposure to toxic chemicals found in pressure-treated lumber.
- **Arsenic-free pressure-treated lumber** (treated instead with ammoniacal copper quaternary [ACQ] or copper boron azole [CBA]) for landscaping, raised-bed gardens, decks, and play sets reduces exposure to toxic chemicals.
- **Sustainably harvested redwood** promotes the use of sustainably grown, renewable resources.

Decks vs. Patios

Patios are usually more
environmentally friendly
than decks, according to the
Sustainable Buildings
Industry Council's *Green
Building Guidelines*. When
built out of bricks or traditional masonry, patios are
more durable and weather
resistant and typically
require less maintenance
than decks, they note. If you
must have a deck, SBIC recommends using recycled
plastic lumber or lumber
made from recycled plastic
and wood fiber. Reclaimed,
decay-resistant wood such
as redwood or sustainably
harvested decay-resistant
woods such as Pau Lope or
cedar are also good
choices. Be sure to avoid
pressure-treated lumber.

Accept the fact that all
products have an impact
and choose those that
meet the greatest number
of green building criteria.

Driveways

- **Recycled asphalt** promotes the use of waste from highway reconstruction.
- **Porous pavers** for walkways, driveways, and parking areas reduce surface runoff and flooding and promote groundwater recharge.

Green Building in an Imperfect World: Compromises, Trade-offs, and Caution

Green building materials, combined with an assortment of techniques and technologies discussed in this book, help us create better homes. In his book *Green Building in a Black and White World,* David Johnston points out that "every decision you make about the materials you use (in house building) either leads toward a more sustainable future or decreases the opportunities for future generations to live lifestyles that we have come to take for granted." Thus, the more green features included in a house, the greater the prospects of future generations and other species.

But a word of caution. No product is free from impact. Plastic lumber, for example, which is used for exterior trim, outlasts wood trim and is made from recycled milk bottles. Recycling plastic milk jugs, however, requires energy and produces some pollution.

In some instances, a product may meet only one or two sustainability criteria. Nonetheless, it is an improvement over conventional building materials. Some materials may meet many criteria but may be too costly to purchase or to ship to a building site. Other products may require special skills for installation and repair that are not available in your area.

My advice is not to get too bogged down in the imperfections of green building materials and the unavoidable trade-offs at this early stage in the evolution of the field. Accept the fact that all products have an impact and choose those that meet the greatest number of green building criteria, as outlined at the beginning of the chapter. By selecting the greenest products, we can make significant strides in reducing pollution, resource depletion, and other environmental impacts of home building, while stimulating the green building materials industry. In so doing, we help forge a sustainable path.

On the other hand, be aware of a problem called *greenwashing,* that is, promotional materials that make a product seem to be more green than it really is. In the past decade, many manufacturers have climbed onto the green building bandwagon. In their zeal to tap into this rapidly expanding market, some manufacturers have introduced products of rather dubious merit. Fiber-cement siding, for example, from James Hardie, a company based in Australia, is durable and virtually maintenance free. It uses less wood than standard wood siding. Unfortunately, the company acquires the wood from Australia and New Zealand and mixes it with cement to make its siding.

"Considering the energy-intensiveness of cement manufacturing and the overseas transport of wood fibers, fiber-cement siding does carry a high-embodied-energy burden," says Dan Imhoff, author of *Building with Vision*.

So, when shopping for green building materials, ask for supporting data. View products with a critical eye and beware of deceptive claims by manufacturers. But always remember that some trade-offs are inevitable.

Sometimes you may need to select a nongreen product to be green. Some products offer little advantage over conventional materials. For instance, I used tile made from recycled feldspar in my house. The product was shipped from the southeastern United States. In retrospect, using a locally produced tile from locally available clay might actually have been a better option from an energy standpoint.

Some products haven't worked out well, either, so view new products with healthy skepticism. Over the years, builders have been burned by innovative green building materials that did not perform as promised. The research center of the National Association of Home Builders posts information on field tests on its web site, listed in the resource guide. This site describes how products are supposed to work, how they should be installed, and their limitations. Field-testing of new products and materials will help you avoid potentially costly mistakes.

Note, too, that many green building products are cost-competitive with conventional materials; a few are even less expensive. Others cost substantially more. In such instances, remember that money saved in one area, for example, from making a house more energy efficient, allowing the installation of a smaller, less expensive heating and cooling system, can be used to offset the higher costs of other products, such as no-VOC paints, stains, and finishes. Keep track of materials and techniques that save you money, and apply savings to those items that cost more.

Green building is as much a way of thinking as it is a way of building. Fortunately, as you've seen in this chapter, the materials required to produce green homes are becoming more available by the day. All that is left is for us to embrace the idea of green building and use the products.

BUILDING NOTE

For years, pressure-treated wood used for decks, landscaping ties, and other outdoor applications was made with arsenic and chromium, which was banned in 2004 for residential applications. Today manufacturers are producing two types of arsenic- and chromium-free wood for docks, landscaping ties, fence posts, fence boards, hand rails, and other uses. One product is treated with water repellents and ACQ preservative that contains two biocides, copper and quat, to protect against decay and termites. Another, copper azole, contains copper and azole. These supposedly safer chemicals are also used to control fungi and bacteria in swimming pools and spas. Quats, for example, are commonly used in household disinfectants and cleaners. They are supposedly safe in environmentally sensitive areas, such as wetlands. For more on this product, log on to www.treatedwood.com.

CHAPTER 5

Wood-Wise Construction

Emeryville, California, is a fairly densely populated residential and industrial city located between Berkeley and Oakland. Here Larry Strain and his partner, Henry Siegel, built three homes on a vacant lot. But they were not ordinary homes. Siegel and Strain incorporated a variety of green building materials into the structures (figure 5-1). They also designed the homes to use 20 percent less wood by utilizing simple and effective framing techniques developed by the National Association of Home Builders Research Center. These techniques, such as spacing studs in walls 24 inches apart rather than the typical 16 inches, cut wood use without compromising the structural integrity of the buildings.

Known as *optimum value engineering,* these technique are among a variety of cost-effective methods that reduce the amount of wood used in new construction and remodeling. Before we examine optimum value engineering and other wood-saving methods, let's pause to answer the question: Do we really need to cut back on wood consumption?

Figure 5-1.
Built on a vacant lot in Emeryville, California, this home features many green building materials and techniques, including optimum value engineering.

Source: Siegel and Strain Architects

Protecting Forests: The Social and Ecological Imperative

According to the U.S. timber industry, more trees are growing in America today than before the Pilgrims landed on the continent. Although this assertion may be true, it is a bit deceptive. Many acres of forest have been cut down and converted to other uses since Europeans first established a foothold in North America. Approximately one-third of the nation's forested land is gone. Ninety-six percent of its old-growth forests have been eliminated. Although there may be more trees growing, they are smaller, younger trees, and many are growing in intensely managed plantations, not diverse forest ecosystems.

Forested land has declined dramatically in other countries as well. According to the Worldwatch Institute, since the advent of agriculture about 10,000 years ago, nearly half of the world's forests have been lost to development, including farming and human settlement. That's about 7.5 billion acres of timbered land that has been leveled to provide wood and make way for human development.

Every continent has experienced major deforestation. In East Africa, for example, 90 percent of the virgin moist forests have been destroyed. In the Philippines, deforestation has reached 97 percent. Europe has lost 70 percent of its forests. In Brazil, nearly 40 percent of the forests have vanished.

Deforestation continues today. According to the World Resources Institute, nearly 40 million acres of tropical rain forest, an area the size of the state of Washington, are leveled each year. Another 40 million acres are altered by road building or other human activities. Although most of the focus has been on deforestation in the rain forests of South America, Central America, Asia, and Africa, temperate deciduous forests—such as those in the southeastern United States—and northern coniferous forests—such as those in Canada and northern Russia—are also being heavily harvested.

Humankind's heavy reliance on forests might not be so alarming if we were making efforts to replant trees at a rate commensurate with cutting. But we are not. In less developed countries, for example, only one tree is planted for every ten trees cut down. In Africa, the ratio is one for every twenty-nine. Unsustainable harvesting is currently occurring everywhere throughout the world.

Making matters worse, timber cutting is often carried out with flagrant disregard for the environment. In Canada, for instance, where the logging industry wields enormous power, companies often violate national and provincial laws and regulations. They cut trees right to the banks of rivers. They clear-cut on steep, erosion-prone slopes. Moreover, clear-cuts are often larger than allowed by law, and companies reportedly fail to protect areas set aside for wildlife. Two-thirds of Canada's coastal temperate rain forest has been degraded as a result of heavy logging and development. In

According to the Worldwatch Institute, nearly half of the world's forests have been lost to development since the advent of agriculture about 10,000 years ago. That's about 7.5 billion acres of timbered land that has been leveled to provide wood and make way for human development.

Humankind's heavy reliance on forests might not be so alarming if we were making efforts to replant trees at a rate commensurate with cutting. But we are not.

Figure 5-2.
Large clear-cuts are the product of a wood-hungry society and destroy huge tracts of wildlife habitat, may increase soil erosion, and create a visual scar visible for many miles.

Source: Martin Miller, Visuals Unlimited

British Columbia, 140 stocks of salmon have been driven to extinction and 624 are at high risk, largely as a result of poor logging practices.

In less developed countries, the situation is even worse. Besides destroying habitat for many species, deforestation decreases vital fuel supplies for domestic use, mostly cooking. According to estimates of the United Nations Food and Agriculture Organization, 100 million people in twenty-six countries now face acute firewood shortages. In rural Kenya, shortages mean that some women spend up to 24 hours a week in a desperate search for wood.

Deforestation stems from a multitude of factors, ranging from shortsightedness and bad policy to ignorance and outright greed. Ultimately, however, deforestation results from the need to clear land to make room for farms and human settlements and to supply humans with wood and paper. In the United States, 60 percent of all the trees harvested each year are destined for construction, mostly home building. Paper production accounts for most of the rest of our nation's enormous timber appetite. With 1.2 million new homes built each year, and 85 percent of them made principally from wood, the devastation is enormous. Furthermore, timber harvesting is projected to increase dramatically throughout the world as the human population increases and as the global economy continues its burgeoning expansion. Given growth in population and economy, the prospects for the world's forests appear grim.

All of this is not to say that the situation is hopeless. Important efforts are underway to reduce our demand for wood, and many homebuilders and manufacturers have joined a global effort to reduce wood use to protect the world's forests. This chapter will outline a number of ways that these important goals can be achieved, beginning with conservation strategies.

"Each new home or remodel adds or detracts from the quality of life for its inhabitants, the local community, and the hundreds of places where the individual building components originated."

Randy Hayes
quoted in *Building with Vision*

Conservation Strategies to Reduce Wood Use

Conservation—using only what you need and using it efficiently—is vital to global efforts to reduce deforestation. This section examines seven components of the wood conservation strategy: (1) renovating existing homes, (2) building smaller homes, (3) building simpler structures, (4) creating more durable and adaptable buildings, (5) designing homes using optimum value engineering, (6) using engineered lumber and wood products, and (7) reusing and recycling wood waste.

Renovating Existing Buildings

Building a new home is a dream of many people. It offers an opportunity to custom design a home that meets one's individual needs and personality. But building anew can be costly, and it can place extraordinary strain on nerves and marriages. It also comes with a huge environmental price tag.

Staying put—that is, remodeling an existing home—or buying an older home in need of refurbishment is worth considering. To be sure, older homes have their problems. Some are damp and dusty. Others are drafty, uncomfortable, and inefficient. Many older homes are heated by noisy, inefficient furnaces that add directly or indirectly to local, regional, even global air pollution. Few older homes come with ventilation systems or vapor barriers to retard the flow of moisture into walls. Mold and mildew may be present in basements and wall cavities. These, in turn, can contaminate room air, causing an assortment of adverse health symptoms. And, of course, wiring and plumbing may be outdated and may require replacement. All of these problems, while fixable, can be costly, and sometimes the costs can be prohibitive.

But an existing house may have many beneficial attributes that speak in favor of remodeling. It may be located in a distinctive older neighborhood. It may be close to downtown, within walking distance of bus stops, shopping, museums, and other forms of recreation. An older home may be architecturally superior to a new home, and better built. It may have hardwood flooring, which is not only beautiful, but also easier on our health than the carpeting and vinyl flooring found in many newer homes. Furthermore, if a house is more than five years old and hasn't been recently remodeled, it likely has outgassed all potentially toxic pollutants. Remodeling an existing home also reduces land disturbance. And, of course, this strategy ensures that already harvested and processed building materials remain in use. By remodeling, you give the foundation, exterior walls, roof, and driveway a new lease on life.

Although a considerable amount of energy and a mountain of materials may be required to renovate a home, the demand for resources and the environmental impact you create are far less than when building a home from scratch. If you use green building materials and incorporate other earth-

> By remodeling a home, you give the foundation, exterior walls, roof, and driveway a new lease on life.

friendly ideas, such as passive solar heating and cooling, a remodel can have a positive impact on the future.

Building Small

Laurel Robertson and her husband and three children live in a 700-square-foot home in Austin, Texas. Far from feeling cramped and uncomfortable, the family thrives in this tiny space. How, you might ask, can five people flourish in such confined quarters when the average new home in the United States provides nearly 800 square feet per person?

For one thing, with its short winter, Austin permits a lot of outdoor living. A tent, sunshade, hammock, outdoor playset, treehouse, and a small grassy area allow activities to occur outdoors that might otherwise take place inside.

The Robertsons rely on other design features to ensure a comfortable life within this small, limited space. The house has an open floor plan and the kitchen/living room area has a tall ceiling. Both of these features create an illusion that the house is larger than it really is. Further conserving space, Laurel hangs baskets from pegs in the rafters of her kitchen for extra storage. "Out-of-season clothes," she adds, "get packed away to the attic so we can use smaller dressers." Each child has a small bedroom, measuring 8 by 12 feet, with shelves, a bed, a dresser, and a desk. "We get extra floor space by using a top bunk without the bottom," remarks Laurel, "and putting the dresser and closet beneath."

Smaller homes save resources. For example, they require less lumber and fewer windows, doors, toilets, carpets, and buckets of paint. Using less, they cost less to build. Many people who build small homes are able to pay for them outright and are thus emancipated from burdensome monthly mortgage payments.

Smaller homes offer many social benefits, too. As Robertson notes, living in smaller, compact homes brings families closer together. Architect Bob Theis of Berkeley, California, points out that carefully crafted smaller homes "guarantee a depth, richness, and intimacy uncommon in big houses." However, Theis notes, "This is not to say that a small house will automatically function well. Far from it. Where a poorly planned large house may cause inconvenience, a poorly designed small house can be sheer torture because there is so much less leeway." He adds, "A very small house requires that the spaces function like poetry, nesting and overlapping, each contributing to the whole in many ways."

If designed carefully, small homes can achieve great utility without creating a cramped, boxed-in feeling. In this era of suburban mansions, many families are discovering that they can live well in much smaller spaces. Many are finding that, with a little ingenuity, just half the square footage of the average new home, or a measly 1,200 square feet of well-planned living space, suffices. While some families may need a bit more, others require

Building a small home helps saves money and natural resources, but it is surely bucking the trend. Since 1950, the average new home in America has more than doubled in size, increasing from 1,100 ft^2 to 2,250 ft^2. In 1950 we got by comfortably with 290 ft^2 per person. Today, we have 800 ft^2 per person.

"A very small house requires that the spaces function like poetry, nesting and overlapping, each contributing to the whole in many ways."

Bob Theis

much less. What design features are required to create comfortable living space within the confines of a smaller footprint?

Designers of small houses find that recessed storage, such as bookcases, and built-in furnishings, such as tables, actually increase functional space. Nooks or bay windows equipped with cushioned seats also increase functional space.

Designers can also use tricks to make living spaces appear larger than they really are. Open floor plans, for example, enlarge our perceptions of space. Open floor plans also reduce or eliminate hallways that take up considerable space. Taller ceilings provide additional illusion, creating a more roomy feeling, as do windows.

Or designers can concentrate square footage in social centers of the house, for example, the kitchen or family room, where family members spend most of their time, according to architect Robert Gay in Tucson. This way, other rooms can be smaller.

Because people in our society also require separation, architects employ a number of design strategies to create real and imaginary boundaries. Changing floor levels, for example, creates a sense of separation in small homes. "A basic rule of thumb in design," says Bob Theis, "is that 1 foot of vertical separation between levels is perceptually equivalent to 10 feet of horizontal separation." As a result, level changes, even rather small ones, are extremely effective in "pulling spaces apart." In addition, level changes add to the visual diversity of a living space and contribute to a sense of spaciousness.

Separation can be achieved by room dividers such as partial built-in bookcases. They break up spaces, yet allow for an open feeling. Windows located in dividing walls also help open up living space. To make smaller spaces more functional, designers may reduce the number of doors or replace swinging doors, which take up space, with sliding "pocket" doors.

Outdoor areas for cooking, conversing, or just plain lounging in warm weather add functional space to a home in warmer climates. And they do it at a fraction of the cost of finished floor space, with far fewer resources. Be sure the spaces are well-defined and attractive. Also be certain to avoid those "dreary concrete terraces and windswept raised decks we see around us," reminds Theis. Easy access to outside living space is equally important.

Today, a plethora of books on building small are available. In particular, architect Sarah Susanka, author of the best-selling *The Not So Big House* and *Creating the Not So Big House,* offers numerous ideas and beautiful pictures that can help you.

Building Simply in Two-Foot Increments

Building small is an extremely effective means of reducing pressure on the world's beleaguered forests and saving money. Simple designs require far fewer resources, such as concrete for foundations and wood for framing, than more complex designs. Because of this, simpler designs also result in significant economic savings.

When building a conventional wood-frame home, wood use can be reduced by adjusting measurements to conform with the dimensions of standard building materials. "Since wood products are typically milled or produced in increments of two feet, designs based on a two-foot module result in the least waste from off-cuts," note Anne Edminster and Sami Yassa, authors of *Efficient Wood Use in Residential Construction*. Designing a room with even 2-foot dimensions, say 10 feet by 20 feet rather than 9 feet by 17 feet, will dramatically reduce cutoffs that frequently end up in dumpsters.

Building for Durability, Quality, and Adaptability

Durability is another vital element of green building. Producing homes that last for one hundred to two hundred years, rather than fifty, and require less maintenance greatly reduces resource demand, environmental destruction, and costs. But what is a high-quality home?

The definition of a well-built home depends in part on its location. In rainy climates, a well-built home is equipped with adequate overhangs to protect the walls from driving rains. A good foundation prevents moisture from seeping into the house. Foundation drainage, discussed in chapter 2, also helps. In earthquake zones, a well-made home must be designed to withstand periodic seismic activity. In other words, it must be firmly attached to its foundation, which is firmly attached to the ground. In hurricane- and tornado-prone regions, a well-built home requires special attention to wall materials. Concrete block and poured concrete hold up much better than wood framing. The roof must also be securely attached to the walls.

To know the special requirements of your area, talk to experienced and knowledgeable builders and structural engineers. They can tell you what to look for. As pointed out in chapter 2, be sure not to build in high-risk locations, for example, locations underlain by highly expansive soils, areas prone to mud slides, unstable slopes, or in regions plagued by avalanches or floods.

Quality and durability are also influenced by numerous other more subtle details, such as the type of materials used to build a house, the quality of the materials, the skill of the builders, and the use of flashing where roof lines abut walls and around skylights.

Unfortunately, most of us are pretty clueless about home construction, because building a home is a complex venture, far beyond most of our personal experience and knowledge. So what do you do?

If you are building a home, hire a builder who comes highly recommended. Talking to builders, rather than their salespeople, may help you assess their qualifications, but there are some smooth talkers who oversell their abilities. Better yet, you may want to chat with members of their work crews. They're often candid about what they're doing—whether they are putting up homes quickly and inexpensively or building for the long term.

If you are purchasing a home, you may want to hire a knowledgeable builder, remodeler, or structural engineer as a consultant. A few hours of

Complex home designs, while often visually appealing, require more dimensional lumber for framing than a simpler design offering the same square footage. Complex designs also tend to result in more mistakes during construction and more wasted wood. They also cost more to design, build, and finance.

his or her time can save you a lifetime of trouble. Studying home building certainly helps.

Individuals can also help protect forests by designing homes to accommodate and adapt to changes in our society. A single-family home designed for a young family, for instance, might also be fashioned so that it can be subdivided after the children leave home, creating a duplex for two older couples. A house could also be designed so that it is easy to create an apartment for an elderly grandparent or for a college student who attends a nearby school. I'll cover this topic in more detail in chapter 7.

Using Optimum Value Engineering

Wood use can also be reduced by avoiding structural overdesign, that is, designing homes much stronger than they need to be. Overdesigned buildings require significantly more resources to construct.

As environmental awareness increases and the cost of building homes continues to escalate, a growing number of architects and builders are beginning to use a technique referred to as *optimum-value engineering* (OVE). Sometimes referred to as *advanced framing techniques,* OVE is a design process that seeks ways to reduce the amount of lumber (and other materials) used to build a structure while maintaining the structural integrity of the building. Put another way, optimum value engineering seeks structural strength with minimal wood use. It cuts down on resource use and labor and can save a substantial amount of money. OVE also improves the energy performance of buildings, reducing heating and cooling costs.

Designers practicing optimum value engineering rely on rather simple techniques, for example, spacing studs 24 inches apart, rather than the industry-standard 16 inches. OVE designers may also space floor joists and roof rafters (framing materials in floors and roofs) farther apart to reduce lumber use without sacrificing strength. They also align framing members in the floors, walls, and roof of a building, a technique that transfers roof loads (the weight of the roof) directly onto framing members (figure 5-3a). *In-line framing,* as it is called, reduces the need for a double top plate, a horizontal piece of lumber that bridges the wall studs and attaches to the roof.

Using two-stud corners, rather than three-stud corners, is another way to cut down on wood use (figures 5-3b and c). Aligning window openings with stud spaces, another OVE technique, reduces the amount of framing lumber, as does proper sizing of the framing over smaller doors and windows, called *lintels* and *headers.* Many framers overbuild lintels and headers, installing doubled 2 x 10s when doubled 2 x 4s or doubled 2 x 6s are all that's needed.

Building homes designed by OVE uses no special equipment or fasteners, other than drywall fasteners, but it does require some retraining of crews. OVE can reduce wood use in homes by 10 to 20 percent. For a 2,400-square-foot home, OVE will save about $1,000 just in materials. Labor savings are estimated at 3 to 5 percent.

Saving Wood and Saving Money

American Value Homes of Richmond, Virginia, uses optimum value engineering when building its 2,100-square-foot single-family detached homes. Its builders rely on in-line framing, increased spacing for studs in non-load-bearing walls, efficient roof designs, and other measures. Since the company began using OVE methods, wood waste fell from 1.5 pounds per square foot to 0.5 pounds per square foot. Framing material costs fell from $3.75 per square foot to $2.00 per square foot.

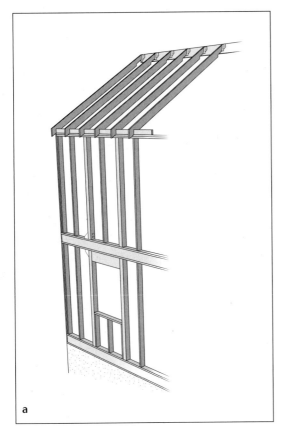

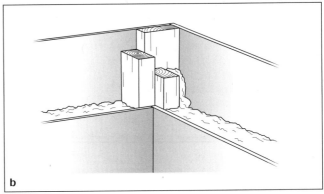

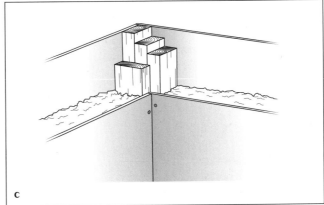

Additional savings accrue to the homeowner. For example, using less framing lumber in walls increases the amount of insulated wall space while reducing bridging loss—heat transfer through the studs to the exterior during the winter. Reducing bridging loss also reduces heat gain in the summer, lowering cooling bills. These improvements reduce fuel bills and make homes more comfortable.

Using Engineered Lumber and Prefabricated Components

Many builders reduce wood use and save money by building homes with engineered wood products. Engineered wood products are made from reconstituted wood—sawdust, shavings, or peelings—that are glued together to produce sheets of wood or dimensional lumber that can be used in place of 2 x 4s, 2 x 6s, and the like. Other engineered wood products, such as lam beams (short for laminated beams), are made by gluing smaller dimensional lumber such as 2 x 4s or 2 x 6s together. They're commonly used to replace large-dimension solid timber. Another engineered wood product is finger-jointed lumber. It's produced by gluing smaller pieces of wood together lengthwise. Snug finger joints between the individual pieces help create a secure bond. Finger-jointed lumber can be used to manufacture window and door frames, interior trim, and studs.

Engineered lumber offers many advantages over conventional wood

Figure 5-3 a, b, c.
By aligning studs with roof rafters and situating windows between studs, as shown in a), builders can save substantial amounts of wood without sacrificing the structural integrity of their homes. Two-stud rather than three-stud corners, as shown in b) and c), also reduce the amount of wood required to build a home and reduce heat transfer through thermal bridging.

Source: Lineworks

The Benefits of Engineered Lumber

Engineered lumber not only replaces old-growth timber but also generally makes better use of trees, capturing 50 to 80 percent of the wood fiber, compared to 40 percent in conventional lumber production. Consequently, fewer trees need to be cut down to provide wood for a house.

I joists were first introduced for use in floors. Because they don't creak or make noise, they were promoted as part of a silent floor system. Since then, however, builders have begun using wooden I beams for framing roofs.

Environmentally Friendly Cabinets

Neil Kelly produces a line of "total resource responsible and healthy cabinets" made from straw-based particle board and certified woods (lumber derived from sustainably managed forests). In addition, this company uses environmentally friendly finishes (low-VOC and water-based).

products. One of the most important is that it is made from smaller, faster-growing trees or wood waste. For years, plywood and dimensional lumber—especially 2 x 6s, 2 x 10s, posts, and beams—came from old-growth trees cut from the world's ancient forests. Engineered lumber not only protects old-growth forests, it better utilizes trees, capturing 50 to 80 percent of the wood fiber, compared to 40 percent in conventional lumber production. Consequently, fewer trees need to be cut down each year to provide wood for homebuilding.

Another advantage of engineered lumber is that it is stronger than standard dimensional lumber, so framing members can be more widely spaced. Engineered wood is also less prone to warping than larger solid pieces of dimensional lumber. This feature reduces waste and costs. Because it is manufactured, much longer pieces—even thirty-foot-long ones—can be made for specialty applications.

Engineered lumber can be used to build nearly every component of a home: floor joists, wall studs, posts and beams, roof trusses, subflooring, exterior sheathing, wood paneling, doors, cabinets, and countertops. One of the most popular products is the I joist. As shown in Figure 4-6 on page 77, the vertical member of an I joist is typically made of oriented strand board (or some similar product). The flanges are made from laminated veneer lumber. I joists require 50 to 60 percent less lumber than dimensional framing lumber of similar dimension.

Particle board is another engineered wood product that is widely used in new home construction. Particle board is made from sawdust and glue pressed together to form large sheets. These sheets, while not very useful on their own, are typically covered with a veneer of hardwood to manufacture cabinets, doors, and paneling. The advantage of this process is that it uses waste (sawdust) to make the bulk of the structure with a minimal amount of hardwood to produce an end product that many individuals find appealing.

Although engineered lumber does have its advantages, bear in mind that most products are made with glues or adhesives that outgas formaldehyde, which is toxic (see chapter 4). Several companies have begun to re-engineer engineered wood products to address outgassing and to reduce wood use even further. Agriboard Industries in Electra, Texas, for example, manufactures a substitute for particle board made from straw and a nontoxic, emission-free binding agent, according to the manufacturer. Agriboard is lighter than particle board and more resistant to moisture. It is easy to cut. Neil Kelly Company in Portland, Oregon, uses this formaldehyde-free product to manufacture some of its cabinets.

Phenix Biocomposites of Mankato, Minnesota, manufactures competitively priced alternatives to oriented strand board and plywood made from wheat and soybean straw. Their product is more water resistant than wood-based panels and requires less binder. Another company, Homosote, daily

Figure 5-4.
Prefabricated trusses like these reduce wood waste on a job site. Trusses made from certified lumber result in even greater environmental benefits.

Source: Dan Chiras

recycles 150 to 200 tons of newspapers, producing from them, among other products, a substitute for plywood and oriented strand board.

Green builders also use prefabricated building components to reduce resource use and costs. Some contractors, for instance, build homes using stud wall panels and prefabricated roof trusses that are assembled in factories that specialize in building these essential components (figure 5-4). The framed walls and trusses are custom-built according to the blueprints, then shipped to the site where they are put in place. The result is a high-quality product assembled at low cost. Quality controls, in-plant reuse and recycling, and experience in framing walls all translate into less wood waste per project.

Structural insulated panels (SIPs) are also preassembled and shipped to building sites. As you may recall from chapter 4, SIPs consist of foam sandwiched between oriented strand board. They're used to build exterior walls, roofs, and floors. Because they're sturdy, they require very little conventional framing. This, in turn, dramatically reduces the amount of wood required to build a house. In fact, SIPs reduce exterior framing lumber on a typical house by 25 to 50 percent. They also reduce the amount of time required to build the exterior walls of a home by more than a third. Reduced labor costs can increase a builder's profitability by an estimated 16 percent.

Combining prefabricated roof trusses and structural insulated panels for exterior walls may reduce total wood use on a building site by up to 25 percent and can reduce framing costs by several thousands of dollars per house. Even small builders can profit from the use of these materials. For production builders, that is, companies that produce hundreds of homes each year, the savings can be quite spectacular.

BUILDING NOTE

Structural insulated panels reduce exterior framing lumber on a typical house by 25 to 50 percent. Combining prefabricated roof trusses and structural insulated panels for exterior walls may reduce total wood use on a building site by up to 25 percent and can reduce framing costs by several thousands of dollars per house.

Reducing Wood Waste

As noted previously, home construction generates a huge amount of waste. Waste from a single home typically ranges from 3 to 7 tons. A large portion of the waste from a building site is wood. According to Tracy Mumma and Steve Locken of the Center for Resourceful Building Technology, builders typically waste enough studs on twenty new homes to frame an entire house.

Waste gobbles up landfill space and costs us dearly, both monetarily and environmentally. The authors of *Efficient Wood Use in Residential Construction* remind us that we pay for wasted materials two times, once when we purchase them and again when we dispose of them. To avoid wasting natural resources as well as their hard-earned money, green builders try to carefully estimate lumber required for a job, then attempt to use these materials to their full potential by relying on optimum value engineering and other techniques discussed in this chapter.

Many green builders also set aside scrap wood that is cut from larger pieces, sorting the pieces by size and type, and ask their workers to check the scrap pile when they need small pieces. It's far more economical in the long run to spend a couple of minutes to locate a 4-foot length of 2 x 4 from the wood pile than to cut one from an 8-foot piece. Although sorting through scrap piles may increase labor costs a bit, the savings in materials and disposal costs can easily offset it. The benefits to the forests, although not part of the economic calculus, make this strategy even more valuable.

Some builders reduce waste by sharing the savings on lumber returned to the lumberyard (for a refund) with their workers. Others ask subcontractors, such as framers, to purchase the lumber needed for the job, a measure that makes them more frugal.

Recycling Strategies to Reduce Wood Use

The conservation strategy outlined in the previous section can help us make significant inroads into the nation's wood use. Builders can also help out by recycling wood waste and using recycled or reclaimed lumber.

Recycling Waste Wood

Many progressive builders have begun to recycle waste wood from projects. Smaller builders may donate it to workers. On my job site, one worker hauled away unusable scraps to be burned in his woodstove. Bosgraaf Builders in Holland, Michigan, which builds about fifty-five single-family homes and duplex condominiums each year, hauls their scrap wood to the office parking lot, where it is offered free to the general public. They also advertise wood scrap in the local newspaper and occasionally deliver scraps to area residents. Other builders save good scrap lumber for new projects. Still others have the wood hauled away by a commercial wood recycler. In California, Centex Homes recycles much of its waste wood through a local company that converts the waste material into paper, garden mulch, particle board, and other useful products. Builders among us should note that commercial construction recycling outfitters often provide roll-off dumpsters to collect wood scraps during construction for recycling. Their charge is often competitive with that of trash haulers.

Much more can be recycled besides wood. Concrete, metal, glass, and cardboard are all recyclable and produced in significant quantity at building sites. Those interested in exploring construction waste recycling can obtain a copy of *Residential Construction Waste Management: A Builder's Field Guide* by the National Association of Home Builders Research Center.

Use Salvaged and Reclaimed Lumber

Builders can reduce wood use and protect forests by using reclaimed lumber acquired from warehouses, defunct factories, bridges, barns, homes, military bases, and other structures undergoing demolition. Reclaimed lumber can be used for siding, flooring, stairwells, posts, and beams. Surprisingly, billions of board feet of reusable lumber are available in North America from these sources. A large warehouse contains one million board-feet of useful lumber, offsetting the need to harvest one thousand acres of forested land, according to *Efficient Wood Use in Residential Construction*.

Reclaimed wood, often old-growth timber, can produce stunning results. The cabinets and vanities in my house were made from a 150-year-old barn slated for destruction. A local cabinetmaker remilled the lumber and built me exquisitely beautiful cabinets at a price competitive with that of any customcabinet maker in the area.

Numerous suppliers obtain wood from the mud in the bottom of rivers and lakes. These logs sank to the bottom of rivers and lakes during transport to mills in the late 1800s and early 1900s. Goodwin Heart Pine Company in Gainesville, Florida, for instance, specializes in salvaging heart pine and cypress logs from that region. After recovery by the company's divers, the wood is dried and cut into boards for a variety of purposes.

Some companies specialize in urban lumber. Don Seawater of San Luis Obispo, California, makes a living recovering urban blowdowns (trees that have been blown down in cities and suburbs), driftwood, and other "casualties." He cuts them into marketable lumber.

Marcus von Skepsgardh and Jennie Fairchild of the Protect All Life (P.A.L.) Foundation in Oakland (www.recycletrees.org) salvage redwood, pine, and eucalyptus that are cut down in the city. According to Fairchild, they commonly work with arborists or contractors who are removing trees for clients. P.A.L. appears on the scene shortly after the trees have been taken down, loads the logs onto the truck, and then transports them back to their yard for processing. Some arborists or contractors deliver cut trees to their yard. "We also work with public parks and city agencies," notes Fairchild. "They periodically undertake tree removal projects in public spaces and call us to take the trees. For instance, the City of San Francisco recently undertook a project in which they removed hundreds of 150-year-old cypress trees. They called us to recycle the trees."

Still others salvage wood from previously harvested forests. Brian Usilton has spent more than a decade hauling redwood logs from previously harvested forests in Mendocino County, California. Rejected by earlier loggers because they were not as high quality as other parts of the tree, these logs are in surprisingly good condition and still produce usable lumber. Green Mountain Wild Woods in Woodstock, Vermont, salvages disease-killed butternut from forests. Killed by an airborne fungus, these trees produce rot-resistant softwood useful for floors and other applications.

According to Dan Imhoff, author of *Building with Vision,* "Salvaged lumber is often tightly grained old-growth wood." It is so beautiful, in fact, that it is often featured in prominent locations to show it off.

Unfortunately, salvaged and reclaimed wood is not a cost-effective or "practical choice for the average householder who needs a few pieces for a shelving unit or small project," notes Erica Carpenter, cofounder of a California-based company, Jefferson Recycled Woodworks, that distributes reclaimed wood products throughout the country. It is generally too costly for such applications. However, salvaged lumber is practical for custom builders who use it for wood flooring, timber framing, and fine cabinetry.

Reclaimed and salvaged lumber can be purchased from numerous outlets in the United States and other countries. Currently, there are around 650 suppliers in the United States. In addition, you may be able to find salvaged lumber in the classified ads of your local newspaper. Whatever your source, be sure the wood is acquired in ways that are not destructive to the environment. Check out the Natural Resource Defense Council's publication, *Efficient Wood Use in Residential Construction,* for a list of reclaimed wood dealers in the United States. *GreenSpec* and other similar books also list suppliers.

Alternative Materials

Many builders are turning to alternative materials, including steel and a host of natural materials such as straw bales, to build homes with less wood. Let's take a quick look at steel and natural materials. More detailed discussions are presented in chapters 8 and 9.

Steel

A large-sized steel mill can produce two million tons of galvanized steel a year that can be used to make studs to replace 2 x 4s and 2 x 6s. Two million tons of steel studs, in turn, would satisfy nearly one-third of the United States' demand for wall framing materials and would permit us to "retire" 20 million acres of forested land from production.

Steel studs are stronger and more durable than wood studs. They don't warp and they don't burn. They can even be made from waste, thereby reducing the embodied energy and environmental impact of building materials.

Although the light-gauge steel studs required to frame a house can reportedly be made from the steel acquired by recycling four old cars and saves over forty trees, steel is still a rather high-embodied-energy material— its embodied energy is at least 20 percent more than that of wood. In addition, even though steel studs can be made from abundant recycled scrap and can in turn be recycled over and over, the mining, manufacturing, and transport of this material results in considerable energy use and pollution. Although steel studs are strong, they transmit heat in and out of a house, wasting energy year-round. A study performed by Oak Ridge National Laboratory showed that a wall with steel studs placed 16 inches apart has an R-value (R-value is a measure of resistance to heat transfer) 20 to 40 percent lower than a comparable wall framed with wood.

Light-gauge steel studs resist termite damage and are therefore widely used in areas where termites pose a threat to wood-framed structures, for example, in Oahu, Hawaii. However, in colder climates, moisture condenses on fasteners in the walls and may accumulate in and damage insulation. Steel studs also carry an electrostatic charge that tends to cause dust to collect on walls. Although this phenomenon also occurs in wood-framed walls, it is more pronounced in steel-stud walls. Because of this, steel-stud walls need to be cleaned and repainted more often.

Straw and Earthen Materials

In 1995, I built a home out of straw bales and rammed earth tires (figure 5-5). Although the south-facing exterior wall and some interior walls were framed using 2 x 6s and the roof was framed with wooden I beams, the building still used 50 to 60 percent less wood than a conventional home of identical proportions. You, too, can achieve reductions of this magnitude or even greater by building interior and exterior walls from natural or recycled materials such as straw bales, adobe mud, automobile tires, and recycled newspaper.

Not only do natural and recycled materials such as straw and tires reduce wood use, they are often locally obtained and thus reduce the embodied energy of a home. Some alternative building techniques, such as rammed earth tires, help put waste products to good use. In addition, natural and

Figure 5-5 a, b, c. Seeking to reduce wood use in homes, many individuals throughout the world—including the author—are building homes from straw bales (a) and other materials, even used automobile tires (b). The final product (c) can be quite attractive, not at all what you'd expect when someone tells you she's building a home from automobile tires or straw bales.

Source: Dan Chiras

Figure 5-6.
Although many individuals are turning to alternative materials to build their homes, wood still plays an important role in framing roofs and interior partition walls, as shown here on the author's rammed earth tire home.
Source: Dan Chiras

alternative homes are often ideally suited for passive solar heating and cooling. Thick straw bale walls, for instance, provide excellent insulation and rammed earth tire walls provide thermal mass to absorb heat generated by sunlight streaming into a home. See chapter 9 for more information on these techniques.

Using Certified Wood

The suggestions presented so far make dramatic inroads into wood use and could go a long way toward protecting the world's endangered forests. But they do not eliminate wood use. Even straw bale, cob, adobe, and other natural building techniques require wood for roof construction, door and window frames, and perhaps interior walls (figure 5-6).

To minimize the impact of the wood required to build a home, many builders use lumber and wood products obtained from sustainably managed and harvested forests. The closer to home, the better. But how do you know if wood is sustainably grown and harvested?

In 1994, the Forest Stewardship Council (FSC), headquartered in Oaxaca, Mexico, launched an international effort to promote sustainable wood production. This independent, nonprofit organization, free of business and government affiliations, created a set of standards for forest management that promote sustainable timber production. Their standards include region-specific guidelines to ensure proper management in a wide variety of forest types, tree species, and climates. The FSC does not inspect forests itself but, rather, relies on nine accredited international organizations. They inspect and certify timber management practices of companies harvesting wood

FSC certified wood offers
buyers a sense of satisfaction that their homes are
being built from wood that
has been produced in ways
that are gentle on the
Earth.

throughout the world on tree farms, on plantations, and in natural forests. As of May 2003, 391 forests have been certified by the Forest Stewardship Council, covering 69 million acres, nearly 8 million acres of which are in North America.

Although the FSC has been criticized by U.S. activists for a variety of reasons, such as permitting the use of herbicides to control weedy species when replanting land after harvest, its certification is generally considered to be the most rigorous of all certifying organizations. In fact, an FSC stamp of approval is a fairly high achievement. If all forests were managed under FSC guidelines, the world's forests would be in much better condition than they are now, for these standards ensure that forests are managed for values other than timber production, for example, wildlife habitat. Wood is tracked from the forest to the mill to the building supply outlet so that buyers are sure they are getting what they pay for.

The success of the Forest Stewardship Council certification program depends not just on independent examiners, but also on the availability of retail outlets—stores that stock certified lumber. Several companies have taken a lead in developing this market, including EcoTimber of Berkeley, California. EcoTimber began operations in 1992 and has grown to a multi-million-dollar business over the years, offering certified lumber as well as reclaimed and salvaged lumber.

Late in 1999, Home Depot, which supplies about 10 percent of America's lumber, announced that it would discontinue the sale of old-growth timber in its stores across the nation by 2002 while phasing in lumber and wood products from FSC-certified forests. In the summer of 2000, Lowe's, its major competitor, announced plans to aggressively phase out wood products from endangered forests while supporting FSC-certified lumber and wood products. Although both have fallen far short of their goals, their continuing commitment to certified lumber could dramatically increase the availability of sustainably produced lumber and wood products and spur more timber companies to join the burgeoning movement.

Certified wood offers buyers a sense of satisfaction that their homes are being built from wood that has been produced in ways that are gentle on the Earth. In addition, FSC certification guarantees that wood is produced in ways that respect the rights of indigenous peoples and maintain the economic and social well-being of local communities. For these reasons, certified wood typically costs more than conventionally produced lumber.

Wade Mosby, vice president of marketing for Collins Pine, one of the largest suppliers of FSC-certified lumber in the United States, notes that his company's forests are managed on 140-year rotations and its operations produce far fewer impacts. Although Collins Pine wood costs more, the trees it harvests tend to be of higher quality, with clearer, tighter grains. The slight additional cost, he argues, is "more than made up for in quality and performance." Longer harvest rotations, notes writer Dan Imhoff, "yield older, straighter wood. Straighter wood means fewer on-site rejections and a tighter building envelope overall. Using lumber that doesn't warp or twist lessens the chance of gaps developing in the framing structure, which can compromise energy efficiency and comfort as well as the durability of a building."

When building or buying a green-built home, look for FSC certified wood for your framing lumber, plywood, finish products (trim), and particle board. Some companies, such as Hayward Truss of Santa Maria, California, even use certified wood for preassembled trusses.

Combining Measures

As green architect Sim Van der Ryn remarks, "Wood, in the form of dimensional lumber, is simply a wonderful material: easy to work with, warm to the eyes and hand, natural, reasonably durable." It is, he says, "the material that defines the trade of carpentry and the standard material for houses and light buildings in North America." But, he goes on to say, we must reduce the "amount of newly harvested wood in homes by 50 to 80 percent" so "our children and grandchildren may get to experience forests that are more than the sylvan equivalent of cornfields." Van der Ryn concludes by saying that, fortunately, "we have a multitude of choices available to us in designing and building without destroying forests or polluting the planet. The building products industry . . . is beginning to change because of economic and environmental pressures."

Although the number of green building products that reduce wood use is growing, their share of the market is small compared to mainstream products. Still, even though the number of builders that integrate wood-saving

The Benefits of Combining Measures

According to builder and writer Steve Chappell, a traditional timber-frame home uses at least 30 percent less wood than a comparable stick-frame home. Timber-frame homes use large posts and beams to create the framework of a building. They are often made from locally harvested trees, which reduces the embodied energy of this exquisitely beautiful building material. Because the lumber is rough cut and minimally processed, further savings accrue to the builder. If obtained from sustainably managed forests, the benefits of timber-frame construction are even greater. Timber-frame construction also results in strong, durable buildings.

ideas is small, a green revolution has begun in the building industry.

Today, builders are finding that combining the many wood-saving tech-niques I've discussed in this chapter produces great savings. Building smaller homes, relying on optimum value engineering, using engineered lumber or, better yet, natural materials such as strawbales, and buying wood from cer-tified forests combine to provide unrivaled benefits to the builder, the homeowner, and the life-support systems of the planet.

Can wood-wise construction really help solve the world's environmental problems? "Absolutely," says Randy Hayes, head of the Rain Forest Action Network. "By carefully choosing construction materials and building approaches, you can influence the fate of the world's forests and help orchestrate a U-turn toward an ecologically sustainable society." Wise wood use will play a huge part in this fundamental redirection.

CHAPTER 6

Energy-Efficient Design and Construction

Habitat for Humanity is an international nonprofit organization that helps the working poor buy safe, decent, affordable homes. Relying principally on volunteer labor and many donated building materials, Habitat for Humanity has built more than 125,000 houses in more than 80 countries, including 45,000 in the United States, since it was founded in 1976 by Millard and Linda Fuller. House-building projects are sponsored by local chapters, called affiliates, and each house is sold to a preselected family that participates in construction alongside community volunteers. Families are given zero-interest loans to further ease the burden, and proceeds from loans finance future projects, creating a continuous supply of capital to build more houses.

In Tucson, Arizona, the local affiliate of Habitat for Humanity is building small, affordable homes for the needy (figure 6-1). These homes, averaging about 1,200 square feet, sell for a meager $50,000 . But the story doesn't end here. The Tucson affiliate of Habitat, like others throughout the country, has taken the idea of supplying habitat for humanity a major step forward. They're building homes that have less negative impact on the planet, too. One of their main focuses in Tucson is energy efficiency, which not only helps reduce environmental impact, but ensures much lower fuel bills, freeing up more money for food and other necessities. So confident are the

Figure 6-1.
This home in Tucson built by Habitat for Humanity comes with a fuel price guarantee thanks to rigorous attention to energy efficiency in all aspects of its design and construction and the use of energy-efficient appliances.

Source: Jim Moline

folks at Habitat that their homes come with a remarkable guarantee: that heating and cooling costs will run no more than $300 a year. That's $25 per month, or 81 cents per day!

Habitat provides the low-fuel-price guarantees through a local partner, Tucson Electric Power. The forward-looking utility works closely with the nonprofit organization to ensure owners "guaranteed comfort at a guaranteed price." Tucson Electric Power also partners with commercial homebuilders in Tucson, even those who build large houses over 5,000 square feet. These mammoth homes come with a guaranteed daily fuel bill of around $2.50 to $3.50 a day.

Guaranteeing heating and cooling bills is not a new idea. Chicago's Perry Bigelow has been doing it for years, as have builders in Montana and Colorado. Tucson Electric Power, however, is the first utility to offer this program. As you might expect, they're not doing it merely out of the kindness of their hearts. In fact, their impetus for this benevolent gesture is that, by working with area builders to produce super-energy-efficient homes with guaranteed low utility bills, the company will establish a loyal customer base, crucial in the newly deregulated energy market. In other words, the utility hopes that customers who buy homes with guaranteed low utility bills will stay with the company as new energy suppliers enter the marketplace.

So far, Tucson Electric Power has experienced great success in its efforts to promote energy-efficient home building. During the first year, for instance, the company had to pay only about $115 to the owners of two of the first eleven homes in the program, the amount by which household bills exceeded the guarantees. What are their secrets for success?

To begin with, homes built by Habitat for Humanity and area contractors are so efficient that they exceed one of the most rigorous energy standards in the United States, the Energy Star Performance Standards, guidelines established by the U.S. Environmental Protection Agency.

Another reason for the success of this program is that Tucson Electric employs building inspectors who ensure that homes are built according to the energy specifications. Inspections begin early in the construction process, too, as many seemingly insignificant details play a pivotal role in determining the overall energy performance and comfort of a home. Proper installation of windows and insulation, for instance, are as important as the products themselves in boosting the energy efficiency of a home.

For this program to work, the utility also sets some limits on heating and cooling. It asks homeowners to restrict thermostat settings to 72°F in the winter and 75°F in the summer. Consistently violating these guidelines disqualifies a family from the program.

Another key to the success of the program is that the utility monitors monthly bills. If a customer's bills are higher than expected—or a customer complains that the house is hotter or colder than expected—the home is inspected and the problem is remedied.

BUILDING NOTE

"Careless construction techniques often create situations in which energy-efficient materials perform little better than their inefficient counterparts."

Christina Farnsworth
"Southwest Utility Offers Energy Cost Guarantee," *Home Energy*

This chapter will explore home energy efficiency from a variety of angles. We'll begin with a look at ways you benefit from building an energy-efficient home.

Why Build an Energy-Efficient Home?

It's safe to assume that an energy-efficient home will probably cost somewhat more, perhaps another $1,000 to $2,000. So why spend the extra money?

The answers are many. First, money invested in energy-efficient design and construction can result in substantial initial cost savings. For example, by adding a little extra insulation, which might cost $1,000, you can install a much smaller heating and cooling system, easily saving double or triple what was invested in the added insulation.

Second, buying or building an energy-efficient home may qualify a homeowner for a lower-interest loan, which will reduce the monthly mortgage payment. Various lenders, such as the Federal Housing Administration, Countrywide Home Loans, and Chase, offer energy-efficiency mortgages.

Additional financial incentives are available from some local utilities. In fact, some of the nation's forward-looking utilities offer rather sizable rebates to those who install efficient appliances or insulation.

A third advantage of an energy-efficient home is the most obvious: It saves money on utility bills, a benefit that begins the first month you move in. How much you save depends on how well built your home is and how many energy-efficient appliances and light fixtures were installed. Special attention to energy conservation can result in savings amounting to tens of thousands of dollars over the thirty-year life of a mortgage. When combined with passive heating and cooling techniques, which are discussed in chapters 11 and 12, the monetary savings can be even greater.

Energy-efficient homes are also more comfortable year-round than your average house—much more comfortable. They'll keep you warm in the winter and cool in the summer. If tightly sealed, they will eliminate those annoying drafts that plague so many poorly built homes. Energy-efficient homes are often quieter than standard homes, too. The extra insulation added to improve energy performance acts as a sound barrier, drowning out the annoying racket of modern society. Moreover, energy-efficient design and construction may increase resale value, a factor that may become even more important in the future as energy becomes more expensive.

Last but not least, saving energy in homes helps protect the environment. By cutting down on energy

---

Why Buy or Build an Energy-Efficient Home?

- Save money on monthly fuel bills
- Experience greater comfort levels
- Reduce unwanted noise from outside
- Save money on your mortgage
- Reduce maintenance costs
- Increase resale value and ease of resale
- Protect the environment

use, we reduce the nation's demand for oil, coal, and natural gas. Reduced resource extraction protects fragile environments. Reduced processing, transportation, and combustion of fossil fuels mean less air pollution.

The question, then, isn't "Why build an energy-efficient home?" It's "Why not?"

Energy Use in Homes

As you can see, there are many good reasons for buying or building an energy-efficient home. Fortunately, there are also numerous ways to curtail energy use in your home without any sacrifice in comfort or convenience. In fact, you'll most likely increase comfort levels!

Before you can design and build such a home, however, you need to know where the energy is used. By identifying the "big-ticket items," you can design the most efficient, environmentally friendly home possible—for the best price.

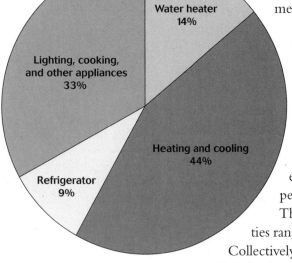

As illustrated in figure 6-2, heating and cooling consume most of the energy required by an average home—nearly 45 percent. Savings here, even modest ones, can have an enormous impact on your monthly energy bill.

The next major energy consumers are water heating and refrigeration. On average, water heating is responsible for 14 percent of a home's annual energy use. Refrigeration is responsible for another 9 percent.

These big-ticket items are followed by a mix of activities ranging from lighting to cooking to television watching. Collectively, they are responsible for one-third of a home's annual fuel bill.

Figure 6-2. Where do your energy dollars go?

Source: U.S. Department of Energy

With this information in mind, let's look at ways to save energy, starting with the biggest household consumers: heating and cooling.

Heating and Cooling a Home Efficiently

Heating and cooling a home begins with the site and the orientation of a house on the site. As discussed in chapter 2, a properly oriented home on a good site can dramatically reduce year-round energy demand.

Landscaping also influences the heating and cooling requirements of a home. Carefully placed trees can make a huge difference in heating and cooling costs, creating economy and comfort. (See chapter 15 for more on this topic.)

Planning and Design

Once you have selected a site and determined the optimal orientation, your goal should be to create a design that provides year-round comfort with minimal energy use. Such a design requires a holistic or systems perspective. This approach, visually referred to as *integrated design* or *whole-house design,* seeks to create a home that functions optimally on any given site in a given climate. It keeps in mind the sometimes-overlooked fact that building components, like the parts of our bodies, frequently interact. Changes in one component often have profound effects, both positive and negative, on others. Moreover, builders using integrated design constantly deal with issues of cost, comfort, and environmental benefits. The goal is to attain maximum levels of performance simultaneously in each and every category at minimal environmental and economic cost.

Integrated design is a far cry from conventional home design, in which components such as heating and cooling systems are considered independently at the "appropriate" time during the design and construction of a home. The results of this linear, sometimes fragmented approach can often be disastrous. At the very least, the home will function a little less efficiently. In the very worst cases, owners could be strapped with a lifetime of high energy bills—all because the builder and subcontractors failed to consider the components together.

To be effective, integrated design must begin early on—as early as the pre-design stage before blueprints are drawn. So that the parts of a house will function synergistically to achieve superb comfort and maximum economy, builders often convene meetings with the owner, architect, and subcontractors to discuss the goals of the project. During these meetings, members of the design team brainstorm ways to ensure maximum efficiency. As the design evolves, each member of the team is given an opportunity to pinpoint areas where energy savings could be increased.

Design meetings also afford an opportunity for team members to "critique" each other's designs and modify their piece of the whole to achieve greater energy efficiency. As Jeannie Leggert Sikora points out in her book, *Profit from Green Building,* "Simple design decisions can have a major impact on the energy . . . efficiency of a new home and its ultimate affordability." For example, placing ducts in unconditioned spaces, such as attics or unheated basements, greatly reduces the efficiency of a heating system.

Although early planning involving a team may require more time than the old-fashioned approach, such efforts are worthwhile. You'll find that some of the greatest savings come with decisions made early in the process. Simple design features that cost little, if anything, such as proper orientation of a building, can greatly boost the energy efficiency and comfort of a home.

Integrated design requires that builders work closely with subcontractors during the construction process as well, so that careless acts don't eliminate

Designing an energy-efficient home requires more time and more careful consideration of the site, local climate, sunlight availability, winds, and other factors than conventional construction requires. These factors help determine which energy-efficient features are practical and cost-effective.

Integrated or whole-house design seeks optimal performance and comfort in a home at minimal economic and environmental cost.

Some of the greatest savings come with decisions made early in the process. Simple design features that cost little, if anything, such as proper orientation of a building, can greatly boost the energy efficiency and comfort of a home.

Remember that one design does not fit all situations. Home designs that work well in a hot, arid climate, for example, could perform poorly in hot and humid or cold and rainy climates.

potentially important building elements. In cold climates, for instance, vapor barriers are often stapled to the framing to prevent moisture from indoor air from seeping into the insulation. A careless plumber or electrician can rip the vapor barrier or tear it out completely to fix a mistake. Failing to repair the damaged vapor barrier could have severe consequences down the road. Plumbers may also tear insulation out of section of a wall to run a pipe they should have installed earlier, forgetting to replace the insulation when they are done. The result of their inattention is an uninsulated wall cavity that loses significant amounts of heat.

If you are buying a green-built home, select one built by a contractor who practices integrated design. If you're hiring a builder, look for one who is open to this approach. Be sure builders work closely with subcontractors and frequently inspect their work to ensure a high-quality product. If you are building a home yourself, learn more about integrated design and put it into practice. To be effective, you will need to know as much as you can about the ways builders achieve energy efficiency, the subject of the rest of this chapter.

Creating an Energy-Efficient Building Envelope

Energy leaks in and out of homes all the time through ceilings and roofs, walls, windows, and foundations. Which way energy flows depends on the season. In the winter, when outside temperatures are lower than indoor temperatures, heat tends to flow outward. In the summer, heat tends to flow into a house.

How much energy escapes by different routes depends primarily on the design and construction of a building. Table 6-1 lists some average values for homes. As shown, windows and air filtration, that is, leakage through various cracks and openings in a house, are the largest sources of energy loss. It stands to reason, then, that attention to these areas can pay huge dividends.

Energy efficiency requires that the outer skin or building envelope—the walls, windows, foundation, and roof—is sufficiently insulated and airtight.

According the American Council for an Energy-Efficient Economy, each year approximately $13 billion worth of energy is lost in American homes through cracks and holes in the exterior of houses, around windows and doors, near foundations, or where pipes and electrical wiring penetrate the building envelope. This leakage increases annual fuel bills on average about $150 per household. Although this is not much on an individual level, the individual costs add up to a rather significant sum. And remember, energy waste means that more energy must be generated, which results in a string of environmental consequences ranging from habitat destruction to global warming.

Table 6-1	
Energy Loss from Homes	
Walls	15–20%
Roofs	10–20%
Foundations/floors	10–15%
Windows	20–40%
Air filtration	20–40%

As you learned in chapters 4 and 5, there are numerous construction techniques and materials that help promote healthy interiors and reduce wood use. Some of these techniques and materials also reduce the heating and cooling loads (requirements) of a house. Here's a recap of ideas that help create an energy-efficient envelope, and some new ideas as well.

- Use optimum value engineering, which results in a more efficient building envelope by reducing bridging losses and increasing the amount of insulated wall space.
- Build foundations and exterior walls from insulated concrete forms.
- Build exterior walls, floors, and roofs from structural insulated panels, which produce a well-insulated envelope and help reduce air movement into and out of walls.
- Frame exterior walls with 2 x 6s to create a deeper wall cavity that permits installation of more insulation than in 2 x 4 walls, or build 2 x 4 exterior walls with rigid foam insulation over the exterior sheathing.
- Although it is more costly, use double-wall construction (two 2 x 4 walls) for exterior walls; this permits greater insulation levels and reduces bridging losses.
- Build walls from straw bales and other natural materials, such as straw-clay and cordwood; they provide superior insulation.
- Install energy-efficient windows and doors.
- Install shades or thermal shutters to cover windows at night and on cloudy days to cut down on energy losses in the winter and heat gain in the summer.
- Install radiant barriers in attics or radiant barrier sheathing for roof decking to reduce summertime heat gain in hot climates (discussed below).

In order to be effective, most of these energy-efficient techniques require the installation of adequate amounts of insulation, our next topic.

Insulation. Insulation blocks the flow of heat across walls and other components of a building. How well insulation operates is measured by its R-value. R-value is a measure of heat resistance by a material. Put another way, R-value is a measure how well or how poorly a material like insulation or an assemblage of materials—for instance, a wall with wood, drywall, and insulation—resists the flow of heat through it by conduction. The higher the R-value, the greater the resistance. Thus, a ceiling with an R-value of 60 is better at conserving heat than one with an R-value of 40.

Heat-resistant materials such as insulation reduce the movement of heat out of a home in the winter and into a home during the summer and therefore help homeowners save energy year-round. Contrary to what many people think, a well-insulated home is just as important in hot climates as it is in cold climates. In fact, it may be even more important in hot climates because air-conditioning bills in such locations often exceed winter heating bills in colder climates.

A well-insulated building envelope reduces energy consumption, cuts fuel bills, and helps reduce the environmental impact of our homes during the heating and cooling seasons.

Figure 6-3.
Wet-blown cellulose insulation is a favorite of many green builders. Made from recycled newsprint and treated with a nontoxic flame retardant, this material provides excellent insulation and, unlike dry-blown cellulose does not settle in wall cavities, creating spaces.

Source: Central Fiber Corporation

Figure 6-4.
Encapsulated fiberglass batts contain fibers and are much safer and much more pleasant to install than unencapsulated batts.

Source: Johns Mansville

Insulation comes in many forms, but it is generally classified into four categories: loose-fill, blankets (batts and rolls), rigid foam, and liquid foam. Each has its own applications. For instance, loose-fill insulation, such as cellulose consisting of ground-up newspaper, blows nicely into attics, creating a thick blanket that resists heat movement (figure 6-3). Blanket insulation, such as fiberglass batts, fits nicely in floor and wall cavities between framing members (figure 6-4), while rigid foam insulation works well on foundations and basement walls and over exterior sheathing. Liquid foam, like fiberglass, also makes a nice wall insulation in wood-frame homes.

When deciding on the type of insulation to use, builders must also consider the impacts of the various products on health and the environment. For years fiberglass insulation batts were manufactured with a binding agent containing formaldehyde. In contrast, most rigid foam insulations were fairly safe for homeowners, but they were manufactured with ozone-depleting chemicals.

Sensitive to the environmental and health concerns of customers and installers, many manufacturers are now producing a new generation of insulation productions. Insulfoam, for example, produces a rigid foam insulation manufactured without ozone-depleting chemicals. Johns Manville, a major fiberglass manufacturer, has begun to produce fiberglass using a nontoxic acrylic binder that replaces the binding agents containing formaldehyde.

Owens-Corning has introduced a formaldehyde-free fiberglass product called Miraflex. It contains two different types of glass fiber that expand and contract at different rates as they heat and cool. Because of this, the fibers tend to curl and twist and thus bind together naturally. As a consequence, Miraflex requires no formaldehyde binders. Some fiberglass manufacturers are also producing encapsulated batts, that is, fiberglass batts encapsulated or covered by plastic. This product reduces worker exposure to fiberglass fibers, thought to cause lung cancer. According to an industry spokesman, all fiberglass insulation manufacturers are selling products made at least in part—usually around 30 percent—from recycled glass.

Despite these improvements, many green builders prefer recycled cellulose insulation. Made from recycled newspaper (and sometimes small amounts of cardboard), this material is blown wet or dry into wall cavities and attics. Treated with a flame retardant, cellulose costs a bit more to install than fiberglass, but it provides more insulation per inch. It also puts a prevalent waste material to good use.

Not to be left out, manufacturers of liquid foam products are also cleaning up their act by producing environmentally and people-friendly spray-in foam insulation. Icynene, for instance, is free of formaldehyde and ozone-depleting chemicals. It blows into open and closed wall cavities, then expands to fill the space. This product not only is resistant to water, it helps reduce air infiltration.

When insulating a home, be sure to pay attention to spots that are difficult to reach. And be sure that insulation is installed correctly by a competent professional. Improper installation can reduce its beneficial effects. To learn more about insulation, you may want to read the chapter on energy efficiency in my book *The Solar House*.

Windows. Good windows are also essential to energy efficiency. As noted earlier, windows can be a major source of heat loss in the winter and heat gain during the summer. Although window design has come a long way in the past two and a half decades, even the most energy-efficient windows can't match the R-value of a well-insulated wall. A wall may have an R-value of around 20, while energy-efficient windows on the market today frequently have R-values of 3 to 4. Even with an insulated curtain covering it, a window may attain an R-value of only 7 to 8.

With so much wall space covered by windows these days, it is important to select the most energy-efficient ones you can afford. When shopping for windows, you will find that there are many different types. The most basic distinction between window types is that some open and others don't. As a general rule, those that open leak more air and lose more energy than those that don't. While all houses require openable windows, it's important not to go overboard. I've found that you can usually achieve good ventilation in a home by strategic placement of openable windows, usually one or two on the north side and one or two on the east, west, or south side.

Windows also come in single-, double-, and triple-pane varieties. Nowadays, single-pane windows are used only in warmer climates, but even here this practice is foolhardy, as single-pane windows can gain a lot of heat during the summer, greatly increasing cooling bills.

Double-pane windows are standard in new construction in most cold-weather markets. But it is not the two panes of glass that account for their efficiency. Rather, it is the air space between the glass. Why? Because air is a poor conductor of heat.

Even better than air is argon gas. It is a poorer conductor of heat and, when used to fill the space between panes of glass in a window, creates an even higher R-value window. Window manufacturers also increase the efficiency of their products by applying a clear, thin coating of tin or silver oxide on the inside surface of the glass or by inserting special films between the panes. These retard heat movement through the glass. The result is a product known as a *low-E window* (*low-E* stands for low emissivity of heat, aka heat transfer).

Efficiency can also be achieved by selecting the right type of window sash and frame (the sash is the material that holds the glass in place; the frame houses the sash and attaches to the wall). Window sashes and frames are made

"When it comes to insulation, more is almost always better. General insulation guidelines for efficient passive solar homes are R-30 walls and R-60 roofs in temperate climates, and R-40 walls and R-80 roofs in extremely hot or cold climates."

Ken Olson and Joe Schwartz
"Home Sweet Solar Home: A Passive Solar Design Primer,"
Home Power magazine

BUILDING NOTE

Aspen Glass in Boulder, Colorado is now manufacturing windows with R-values of around 10.

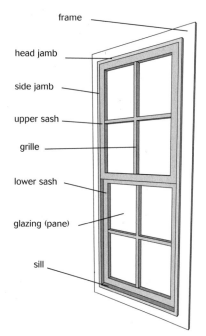

frame
head jamb
side jamb
upper sash
grille
lower sash
glazing (pane)
sill

**Figure 6-5.
Components of
a standard window**

Source: Lineworks

from a variety of materials, each with its own advantages and disadvantages. The most energy-efficient ones are generally made from wood. To protect wood in sashes and frames from moisture and sunlight, manufacturers typically install a metal or vinyl cladding over the outside surface of the wood.

Windows sashes and frames are also made from fiberglass, vinyl, or a composite material consisting of vinyl and sawdust. These materials offer many of the same advantages as wood, especially low heat conduction and durability. However, synthetic materials may pose risks to workers in the factories where they're made. For example, vinyl windows are manufactured from polyvinyl chloride (PVC). Vinyl chloride, the raw material from which PVC is made, is a carcinogen. PVC windows may also outgas toxic chemicals into homes.

Another serious drawback of vinyl is that it expands and contracts significantly in response to changes in temperature. Over time, notes green building expert Alex Wilson, "the expansion and contraction . . . can loosen seals, cause cracks at corners and on flanges." This may lead to premature failure and can cause leakage, which decreases the energy efficiency of the window. Fortunately, vinyl window sashes and frames are frequently heat welded to prevent damage, and failures are rare. If you choose vinyl frames, or buy a home with them, be sure that the corners are heat welded.

Over the years, a number of window manufacturers have used metal sashes and frames. While inexpensive, these products can create an energy nightmare, leaking enormous amounts of heat through the highly conductive frames. Be wary of this material. If you are considering metal windows, be sure that they've been insulated internally to reduce energy losses.

When shopping for windows, you should also be on the lookout for ones with warm edges. Warm edges are created by placing nonconductive material, known as a *spacer,* between the panes of glass around the periphery of the double- and triple-glazed windows. Spacers reduce heat conduction at the edge of the glass where heat loss is most significant and can improve the efficiency of a window by 10 percent.

Glazing spacers also reduce condensation around the cold edges of windows. Moisture condensing on windows can drip onto sashes and sills, causing the wood to deteriorate. In cold climates, moisture accumulating on windows can freeze. When the ice melts, it also drips onto sashes and sills, causing considerable damage over time. Because they reduce condensation, warm edges reduce maintenance and greatly extend the life span of windows.

What to Look for in Windows

• Double- or triple-paned glazing
• Nonconductive sashes and frames
• Metal cladding on exterior surfaces
• Low-e glass
• Argon-filled air spaces
• Glazing spacers (warm edges)
• Low air infiltration

When buying windows, there are many other factors to take into consideration. One of the most important is their efficiency. Fortunately, most manufacturers provide information on the energy efficiency of their windows, often via stickers that contain all or some of the following efficiency measurements: (1) U-value, (2) air infiltration, (3) solar heat gain coefficient, and (4) visual transmittance. Of these, only U-value is required.

U-value is a measure of heat transmission through a material. It is the opposite of R-value, which, as just noted, is a measure of the resistance of a material to heat flow. U-value is calculated by dividing R-value into one (U-value = 1/R). Thus, an R-value of 3 would yield a U-value of 0.33 (1/3 = 0.33). The lower the U-value, the better, because a lower U-value means less heat loss.

U-values for windows are often reported for the middle of the glass as well as for the entire assembly, which takes into account heat loss around the edges. As a rule, whole-unit U-values of greater than 0.3 (which is an R-value of 3.3 or better) are required for most energy-efficient homes, although some manufacturers are developing windows with U-values as low as 0.1 and 0.05.

The second important measurement is air infiltration. Air infiltration is determined by the type of window and the quality of construction. As noted earlier, openable windows permit more air infiltration than nonopenable ones. Generally, the higher the quality, the lower the air infiltration. Air leakage is measured in cubic feet of air per minute per square foot of window surface (cfm/ft^2). Look for windows with certified air leakage rates of less than 0.30 cfm/ft^2, recommends energy expert and author Paul Fisette.

Next is the solar heat gain, that is, the amount of solar heat transmitted through a window when the sun is shining on it. Window manufacturers report this factor as the solar heat gain coefficient or SHGC. It varies between 0, meaning no solar gain, to 1, indicating 100 percent solar gain. (For recommendations, see sidebar on this page.)

The final measure to be concerned with is visual transmittance (VT), the amount of light a window transmits. Visual transmittance is reported as the percentage of light passing through a window compared to the amount passing through an open hole in the wall the same size as the window. A VT of 80 percent means that the window permits 80 percent of the visible light to enter. For reference: Tinted windows permit 15 percent VT, while clear glass permits up to 90 percent VT. For most of us, a window with 60 percent VT appears clear; below 50 percent appears dark.

If you're building your own home, you may want to study windows in more detail or work with a reputable glass supplier who is familiar with your house plans. Owner-builders and professional home builders will find RESFEN, an inexpensive computer software developed by the Lawrence

Solar Heat Gain Coefficient Recommendations

For passive solar homes, south-facing glazing should have a high solar heat gain coefficient with sufficient U-value to conserve heat at night and during cloudy periods and low infiltration for maximum efficiency. The colder the climate, the more solar gain is needed, and the higher the SHGC should be. Paul Fisette recommends SHGCs for hot climates under 0.4. For intermediate climates, SHGCs between 0.4 and 0.55 work well. In cold climates, the SHGC should be greater than 0.55.

Figure 6-6.
Radiant barriers reduce heat
gain during the summer
and reduce heat loss
during the winter.

Source: Florida Solar Energy Center

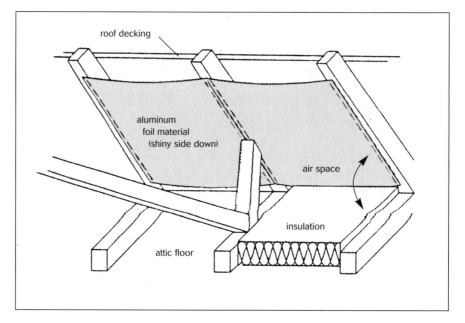

Berkeley Laboratory, to be useful. It is used to assess window performance in homes. This program helps a designer minimize energy loss, maximize comfort, control glare, and maximize daylighting (natural lighting). It also assists in selecting windows for specific applications, for example, north-facing and west-facing walls, and is available from the National Fenestration Rating Council (see the resource guide).

Radiant Barriers. Radiant barriers also help improve energy efficiency, especially in extremely hot climates, such as Arizona, southern California, and Florida.

Two types of radiant barriers are used. One type consists of a thin film of aluminum with a plastic or paper backing. This material is tacked to the framing members in attics, as shown in figure 6-6. To reduce labor costs, some builders are now using a radiant barrier on the roof decking, that is, the sheathing nailed to the roof frame. Known as *radiant barrier roof sheathing*, this product is made from plywood or oriented strand board with a thin radiant barrier glued in place. The aluminumized surface is usually placed facedown.

Radiant barriers block heat absorbed by roofs, preventing it from entering a home through attics and ceiling cavities, causing discomfort and raising utility bills. They also help retain heat in the winter, although the benefit is minor compared to their reduction in summer heat gain.

By reducing heating and cooling loads, radiant barriers reduce energy consumption and utility bills. In Florida, for instance, a homeowner can expect an 8 to 12 percent reduction in cooling bills after installing a radiant barrier, according to the Florida Solar Energy Center.

Creating an Airtight Building Envelope

In addition to installing appropriate levels of insulation, energy-efficient windows, and radiant barriers, green builders usually weatherize their homes to make them airtight.

As noted previously, significant amounts of heat pass through the building envelope through cracks around windows, electrical outlets, and places where electrical wires and pipes penetrate exterior walls and foundations. Hot air moving out during the winter wastes considerable amounts of energy, hiking energy bills unnecessarily. Similarly, hot air movement into a house during the summer makes it less comfortable and jacks up cooling bills.

Movement of air in and out of our homes, known as *infiltration* and *exfiltration*, is responsible for losses ranging from 20 to 40 percent of our total annual heating and cooling bills. For instance, if you're paying $200 a month to heat and cool your home, you're wasting approximately $500 to $1,000 a year because of air leaks.

The key to an airtight house is to make sure that cracks in the building envelope are caulked, sealed, and weatherstripped. Sealing a home is inexpensive and relatively easy. However, most energy-efficient builders I know prefer to hire a weatherization subcontractor after the house has been framed and insulated and windows and doors have been installed. In addition to caulking and weatherstripping, these specialists may also apply a vapor barrier on walls, as shown in figure 6-8, to further reduce infiltration and exfiltration.

Once a home is weatherized, most subcontractors run a test to check their work. Known as the *blower door test,* this simple and relatively inexpensive operation is designed to determine the air exchange rate, that is, how much air will move into and out of the house. To perform the test, the windows and doors are closed and the front door is replaced with a fan mounted on nylon fabric which is fitted tightly in the door opening. When in place, the fan is switched on and outside air is blown into the house. An instrument measures the amount of air that enters the home under pressure. This data is then used to estimate natural air movement under normal operating conditions.

Ideally, a well-sealed home should permit 0.35 to 0.5 air changes per hour. That is, it should permit one-third to one-half of the air in the house to be replaced every hour. If the air exchange rate is higher than this, cracks are identified and sealed.

Green builders use the phrase "build it tight, ventilate it right." So, once a house is sealed, they install mechanical ventilation systems. Ventilation systems replace stale indoor air with fresh outdoor air and help eliminate

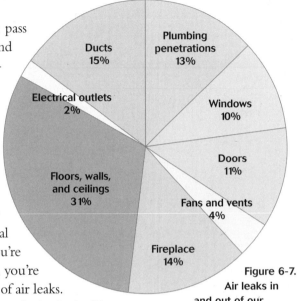

Figure 6-7. Air leaks in and out of our homes through every tiny crack and opening in the building envelope. This pie chart shows where most air escapes and where we should pay the most attention when sealing a house.

Source: U.S. Department of Energy

BUILDING NOTE

Airtight drywalling ensures that all points of air leakage are sealed, and much of this is done before the drywall is even put in place. This technique does not employ vapor barriers. Builder Perry Bigelow of Chicago uses the airtight drywalling technique in all of his homes. Because they are so well insulated and airtight, Bigelow guarantees annual heating bills of under $400.

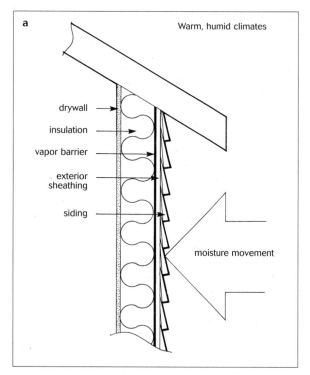

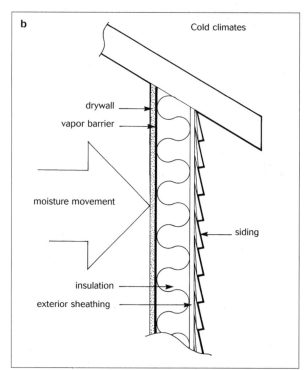

Figure 6-8.
Vapor barriers are helpful in nearly any home. They prevent moisture from penetrating the wall and thus help keep insulation dry. Even a tiny amount of moisture in insulation dramatically reduces its R-value. In warm climates, moisture barriers are typically installed just beneath the exterior sheathing (a). In cold climates, moisture barriers are usually installed beneath the drywall (b).

Source: David Smith

indoor air pollutants. As noted in chapter 3, mechanical ventilation permits much more controlled exchange of air than is possible in a leaky home. To reduce energy losses in such instances, many builders install heat recovery ventilators (HRVs) as described in chapter 3.

Passive Solar Heating and Cooling

Passive solar energy is a system designed to heat a home primarily through sunlight and without mechanical heating systems. Passive solar heating relies on south-facing glass to let winter sun into a home, insulation to hold heat in, overhangs to block the sun during the cooling season, and several other design features.

Making a home airtight and energy efficient can dramatically reduce energy use and eliminate costly fuel bills. But even greater strides can be made in reducing energy use by buying or building a passive solar home. Passive solar heating—defined as a system designed to heat a home with sunlight and without expensive mechanical heating systems—relies on south-facing glass to let winter sun into a home, insulation to hold heat in, overhangs to block the sun during the summer, and several other design features (figure 6-9).

Houses can also be cooled without expensive, energy-hungry, and noisy air conditioners or evaporative coolers. This technique, known as *passive cooling,* is achieved by an assortment of simple yet highly effective measures. The tools of the trade include proper orientation, adequate insulation, and shade.

Interestingly, many of the measures required to create an energy-efficient

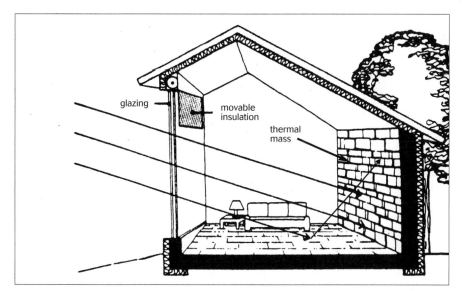

glazing

movable insulation

thermal mass

Figure 6-9.
Passive solar homes allow the low-angled winter sun to enter through south-facing windows. Visible light is absorbed by solid surfaces and converted to heat.

Source: U.S. Department of Energy

house also contribute to the twin goals of passive heating and cooling. If you are going to considerable effort to create an energy-efficient home, you should definitely consider taking the next step and using renewable resources for heating and cooling. We'll explore passive solar heating and cooling in more detail in chapters 11 and 12.

Energy-Efficient Heating Systems

If you've built a superefficient home and designed it to be heated and cooled passively, you'll very likely have little use for a mechanical heating and cooling system (although building departments will probably require that you install one). Even if you haven't taken advantage of passive heating and cooling, you'll need a much smaller system, saving substantial amounts of money. Whatever the case, be sure to install an energy-efficient heating and cooling system. This will reduce your utility bills and dramatically decrease the environmental impact of your home.

Heating systems vary considerably, from simple wood-burning stoves and masonry heaters to complex mechanical systems, such as radiant-floor heating.

Wood-burning technologies are simple and straightforward. They rely on a renewable resource, and many new woodstoves and masonry heaters are quite efficient and clean burning. I'm particularly fond of masonry heaters, massive stoves that burn wood efficiently and produce hours and hours of steady, comfortable heat from relatively little fuel (figure 6-10).

Wood-burning stoves and the like, however, do produce pollution and require considerable operator involvement. Except for pellet stoves, they also lack automatic controls that permit a house to be heated when its occupants are away for extended periods.

If you need automatic heat, you may want to look into mechanical systems such as radiant-floor heat and hot-water baseboard. These systems rely primarily on natural gas or fuel oil. Water is heated in a boiler and transported

Figure 6-10 a.
Masonry heaters are
high-mass woodstoves made
from masonry materials
such as bricks and adobe.
They burn wood efficiently,
produce very little pollution,
rarely need cleaning, and
provide long hours of
comfortable heat with
very little wood.

Source: Biofire, Inc.

b. In a masonry heater, hot
gases flow through an
elaborate internal flu system,
transferring heat into the
masonry. Heat then slowly
radiates into the room.

Source: Nicholas Lyle and Kristin Musnug

Many homes use electricity
to provide heat. Electric
heat is costly and inefficient. Most of the nation's
electricity is generated by
coal. So, unless the electricity comes from clean
sources, such as wind or
solar energy, it usually
comes with a pretty high
environmental price tag.

via pipes to the rooms of the house. In a radiant-floor system, hot water typically travels through pipes in the floor, releasing heat into adjacent rooms. In
a hot-water baseboard system, heat is released into rooms from special radi-
ators, usually installed along the base of the walls.

In the past few decades, many manufacturers have introduced new boilers for radiant-floor and hot-water baseboard systems that can achieve efficiencies in the 90 percent range, making these systems economical and relatively clean burning. High-efficiency furnaces are also available, but they
tend to operate in the 80 percent efficiency range.

Many modern boilers and furnaces also feature sealed combustion chambers to reduce indoor air pollution (figure 6-11). As noted in chapter 3, a
sealed combustion chamber ensures that pollutants from fossil fuel combustion do not escape into the room. Many modern boilers and furnaces also
feature power venting—a fan that draws fresh air into the combustion
chamber and forces exhaust gases (containing pollutants) out through
another pipe. These are often called induced-draft models. Although heaters
and boilers with sealed combustion chambers and power venting cost a bit
more, they save on energy bills. In addition, they typically outlast less effi-

cient models because they produce fewer damaging by-products that tend to clog burners and reduce the life of the unit, according to Jeannie Leggett Sikora, author of *Profit from Green Building*.

Forced-air heating systems are the most popular and least expensive of the mechanical heating systems on the market today. Like radiant-floor and hot-water baseboard systems, they typically rely on natural gas or fuel oil (although some use electricity) to produce heat within a combustion chamber in the furnace. Hot air produced in the furnace is then transported throughout the house by a series of ducts.

Forced-air systems warm rooms quickly, unlike baseboard hot-water and radiant-floor systems, which require a longer period to bring a room up to temperature. Blowing air, however, tends to increase exfiltration, which makes these systems less efficient than radiant-floor and hot-water baseboard systems. In addition, blowing air can be uncomfortable and noisy fans can be annoying.

Despite these drawbacks, some forced-air systems are fairly efficient. To ensure maximum efficiency, be sure that the ducts are well sealed. Quality builders seal ducts with mastic, a durable material, not duct tape, which tends to give out in a few years and cause leakage. Also be sure that ducts are located in conditioned space—that is, heated areas—and not cold attics or under floors. Running ducts in conditioned space can significantly reduce energy losses, by as much as 30 percent.

Central placement of furnaces or boilers is also desirable from an energy standpoint. Central locations minimize the length of runs, reducing ducts and pipes and saving money. It also results in less energy loss as heat travels from the furnace or boiler to the room where heat is required.

Two other options to consider are heat pumps and solar hot-water heating systems. These provide a way of supplying heat to a home with minimal impact on the environment. Using refrigeration technology, heat pumps extract heat from the ground or the air, even when outside air temperatures are low. Heat extracted from the outside is then transferred into the house, where it is distributed via ducts as in a forced-air system or via pipes containing hot water in a radiant-floor or hot-water baseboard system (figure 6-12). Besides using renewable fuel, heat pumps are superefficient.

Solar hot-water systems consist of panels, usually mounted on the roof, that gather sunlight energy and transfer its heat to a fluid that is pumped into the house, where the heat is transferred to water used for domestic purposes,

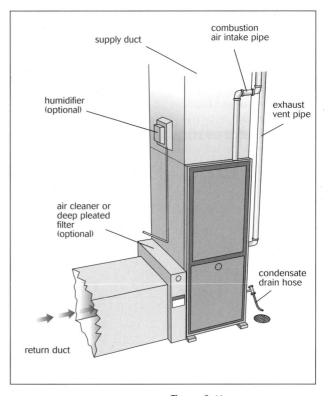

Figure 6-11.
In an induced-draft furnace or boiler, outside air is drawn into the combustion chamber. Exhaust gases are forced outside. Because the combustion process is so efficient and most of the heat is transferred to the house, exhaust gases can be vented through plastic pipe.

Source: Lineworks

Efficiency Ratings of Furnace and Boilers

Furnaces and boilers come with a sticker listing the annual fuel utilization efficiency, known as AFUE, which rates the percentage of useful energy obtained from the fuel. Look for models with annual fuel utilization efficiencies in the high 80s to mid-90s.

Figure 6-12.
A ground-source heat pump draws heat from the ground during the winter and transfers the heat into the interior of a home, so there's no combustion and very little outside energy use. In the summer, the unit draws heat out of a house and transfers it to the earth, cooling the interior.

Source: David Smith

such as washing clothes or bathing, or to provide space heat. New models like the Thermomax shown in figure 6-13 are fairly inexpensive and extremely efficient, producing heat even on cloudy days. They make economic sense in many climates.

Heating systems are quite complex and choosing the right one can be a challenge. If you want to learn more about these systems and the criteria used for selecting one, consult my book *The Solar House* or study some of the books and articles listed in the resource guide.

Energy-Efficient Cooling Systems

Energy-efficient design also greatly reduces summer cooling loads in a house. As you will see in chapter 12, the color of a house, its orientation, and the vegetation that grows around it, especially shade trees, help keep a home cool. Combined, these techniques greatly reduce, and may even elim-

Figure 6-13.
The Thermomax solar water heat is one of the most efficient models around and even captures a significant amount of solar energy on cloudy days.

Source: Dan Chiras

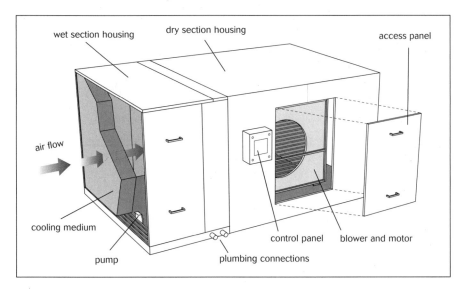

Figure 6-14.
Evaporative coolers blow outside air across a moistened pad. The cool, moist air is then distributed throughout the house.

Source: Premier Industries; drawing by Lineworks

inate, the need for a mechanical cooling system. These systems fall into three groups: (1) evaporative coolers, (2) air conditioners, and (3) heat pumps.

Evaporative coolers mount in windows or on rooftops. As shown in figure 6-14, outside air is drawn into the unit by a fan. As it passes through the unit into the house, the air flows over a constantly wetted mesh (labeled cooling medium). The cool, moist air then enters the house, usually at a central location, providing comfort.

Evaporative coolers are fairly inexpensive and generally do not require extensive ducts to distribute cool air throughout a house. They're so effective that they are also used in large office buildings. However, evaporative coolers are only effective in dry climates.

Air conditioners are mounted on rooftops, alongside homes, or in walls and windows. Outside air is drawn into the unit, dehumidified and cooled, then blown into the house. Large centralized air-conditioning units distribute the cool air throughout the home using an extensive duct system, often shared with a central heating system.

Central air conditioners work well, but they can be quite costly to install and operate. They also utilize ozone-depleting chemicals that can leak into the atmosphere. They're effective in a wide range of climates, especially hot, humid regions.

Heat pumps, mentioned earlier in the chapter, also double as cooling units. But they work in an entirely different fashion from evaporative coolers and air conditioners. They cool a home by drawing heat out of the interior and dumping it outside. Heat pumps work well in a variety of climates, from hot and humid to hot and dry. They also operate fairly efficiently. However, some heat pump systems (the ground-source heat pumps) can be costly to install.

Which mechanical cooling system is greenest? Heat pumps are one of the most environmentally benign means of cooling a home and are popular in the southeastern United States. Next on the scale of environmental acceptability is

BUILDING NOTE

Be sure that air conditioners are not situated in sunny areas, as this greatly reduces their efficiency. Shady spots work best.

Table 6-2

Heating and Cooling System Shopping Guide

System	Rating			Special Considerations
Natural Gas and Oil Systems	Look for the FTC (Federal Trade Commission) EnergyGuide label with an AFUE (Annual Fuel Utilization Efficiency) rating for gas- and oil-fired furnaces and boilers. The AFUE measures the annual efficiency. Energy Star furnaces have a 90 AFUE or greater.			Bigger is not always better! Too large a system costs more and operates inefficiently. Have a professional assess your needs and recommend the type and and size of system you should purchase.
Air-Source Heat Pump	Look for the EnergyGuide label that contains the SEER (Seasonal Energy Efficiency Ratio) and HSPF (Heating Seasonal Performance Factor) for heat pumps. The SEER measures the energy efficiency during the cooling season and HSPF measures the efficiency during the heating season. The Energy Star minimum efficiency level is 12 SEER or higher.			If you live in a cool climate, look for a heat pump with a high HSPF. If you purchase an Energy Star heat pump, you are getting a product that is in the top 25% for efficiency. Contact a professional for advice on purchasing a heat pump.
Central Air Conditioners	Look for the EnergyGuide label with a SEER for central air conditioners. The Energy Star minimum efficiency level is 12 SEER. Energy Star central air conditioners exceed federal standards by at least 20%.			Air conditioners that bear the Energy Star label may be twice as efficient as some existing systems. Contact a professional for advice on sizing a central air system.
Room Air Conditioners	Look for the EnergyGuide label with an EER (Energy Efficiency Ratio) for room air conditioners. The higher the EER, the more efficient the unit is. Energy Star units are among the most energy-efficient products.	What Size to Buy? Area in ft² 100–150 150–250 250–350 400–450 450–550 550–700 700–1,000	Btu/hour 5,000 6,000 7,000 9,000 12,000 14,000 18,000	Two major decisions should guide your purchase: buying a correctly sized unit and buying an energy-efficient unit. If the room is very sunny, increase capacity by 10%. If the unit is for a kitchen, increase the capacity by 4,000 Btu per hour.

the evaporative cooler. Remember, though, that it works best in hot, arid climates and consumes a fair amount of water. Last on the list is the air conditioner.

Air conditioners require more energy to operate and, like heat pumps, use ozone-depleting chemicals. Fortunately, many manufacturers have improved the performance of their units by making them considerably more energy efficient and by installing better seals to prevent leakage.

When shopping for an air conditioner, be sure to select the most efficient model. To help you in your search, room air conditioners come with stickers that list the energy efficiency rating (EER). Models with EERs over 10 are recommended. Central air conditioners come with stickers that list the seasonal energy efficiency rating (SEER). It is similar to the EER. Look for units with ratings of 12 or above. If you live in a humid area, choose a model that is effective at removing humidity from indoor air; people feel more comfortable at lower humidity levels.

Higher-efficiency air conditioners cost more, but the additional investment is often quickly reimbursed through lower utility bills and, possibly, by rebates from local utilities. When shopping for an air conditioner, look for a model with a "fan-only" switch. This feature allows the fan to run at night without the air conditioner portion operating to provide nighttime ventilation. Another useful feature is an automatic delay fan switch. This feature keeps the fan running for a short while after the cooling mechanism shuts off. This, in turn, lengthens the cooling cycle, making optimum use of the appliance, saving money, and reducing overall energy use. Variable-speed compressors and fans also help reduce energy use. They do this by matching fan and compressor operation to cooling demands. In contrast, single-speed models operate at one speed, regardless of demand. Variable-speed units can achieve SEER ratings of up to 17. They're also quieter and provide better dehumidification.

Table 6-2 lists guidelines for buying heating and cooling systems. You may

Size Matters!

Buying an efficient heating and cooling system is important, but you can also save money by sizing these systems correctly—or making sure your builder sizes them properly. Undersized heating and cooling systems cannot keep up with daily demands and thus tend to run constantly as they struggle to heat or cool a home. Oversized heating and cooling systems can satisfy demand, but they waste a lot of energy doing so.

SIZING HEATING AND COOLING SYSTEMS

Buying an efficient heating and cooling system is important, but you can also save money by sizing these systems correctly—or making sure your builder sizes them properly. Undersized heating and cooling systems cannot keep up with daily demands and thus tend to run constantly as they struggle to heat or cool a home. Oversized heating and cooling systems can satisfy demand, but they waste a lot of energy doing so. One reason they are wasteful is that they cycle on and off frequently. An oversized air conditioner, for instance, will rapidly cool down a house, then shut off. As soon as the house warms up, the system kicks in. Although this may not sound serious, start-ups require additional energy. In addition, oversized heaters and coolers may not operate long enough to reach optimal efficiency. As a result, these systems typically provide comfort at a much higher cost. How big should a heating or cooling system be?

As a general rule, heating and cooling systems should be no more than 25 percent larger than the calculated heating and cooling requirements for a home. Don't forget that an airtight, energy-efficient home needs less heating and cooling than a conventional home— sometimes much less. Don't let a heating contractor or overzealous salesperson talk you into a larger unit "just to be safe."

Another way of saving energy with mechanical heating and cooling systems is to turn thermostats down at night. This simple measure can cut heating and cooling bills by around 10 percent. Setting temperature lower at night can be done manually or, as many prefer, automatically with a programmable or automatic setback thermostat. Not only do programmable thermostats allow us to turn down the heat at night, they can be programmed to turn the heating systems down when we are away from our homes.

Programmable thermostats allow similar adjustments to the cooling system, saving considerable amounts of energy and money without sacrificing comfort. Choose a model with a manual override—a way to alter the temperature if your use patterns temporarily change—and sure to select a model that comes with the Energy Star label.

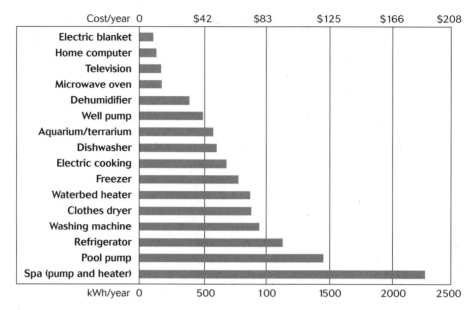

Cost/year	0	$42	$83	$125	$166	$208
Electric blanket						
Home computer						
Television						
Microwave oven						
Dehumidifier						
Well pump						
Aquarium/terrarium						
Dishwasher						
Electric cooking						
Freezer						
Waterbed heater						
Clothes dryer						
Washing machine						
Refrigerator						
Pool pump						
Spa (pump and heater)						
kWh/year	0	500	100	1500	2000	2500

want to ask your local utility for advice on the most energy-efficient models on the market. Also be sure to inquire about rebates or other financial incentives the utility may offer. If you want to learn more, pick up a copy of *Consumer Guide to Home Energy Savings* by Alex Wilson, Jennifer Thorne, and John Morrill. It provides detailed, but highly understandable, information on the operation of mechanical cooling systems and lists the most efficient models on the market today.

Energy-Efficient Appliances

After heating and cooling systems, water heaters, refrigerators, and freezers are the next largest energy consumers in a home, followed by washing machines, dryers, dishwashers, televisions, and computers (table 6-3).

Many green builders routinely install energy-efficient appliances, taking care of this detail for the buyer. When building or remodeling a home yourself, you can find information on energy-efficient appliances by consulting *Consumer Guide to Home Energy Savings*. You can also look for the EPA Energy Star logo on appliances when shopping (figure 6-15), which certifies an appliance to be among the most energy efficient in its class. As you may know, all major appliances sold in the United States come with a brightly colored energy sticker. As shown in figure 6-16, it notes how much energy a particular appliance uses in a year and how it compares to similar models. Be aware, however, that the superefficient models may not be included in the range. That's why *Consumer Guide to Home Energy Savings* is so handy.

Table 6-4
Major Appliance Shopping Guide

Appliance	Rating	Special Considerations
Programmable Thermostats	For minimum Energy Star efficiency, thermostats should have at least two programs, with four temperature settings each; a hold feature that allows users to temporarily override settings; and the ability to maintain room temperature within 2°F of desired temperature.	Look for a thermostat that allows you to easily use two separate programs; an "advanced recovery" feature that can be programmed to reach the desired temperature at a specific time; a hold feature that temporarily overrides the setting without deleting preset programs; and the Energy Star label.
Water Heaters	Look for the EnergyGuide label that tells how much energy the water heater uses in one year. Also look for the FHR (First Hour Rating) of the water heater, which measures the maximum amount of hot water the heater will deliver in the first hour of use.	If you typically need a lot of hot water at once, the FHR will be important to you. Sizing is important—call your local utility for advice.
Windows	Look for the NFRC (National Fenestration Rating Council) label that provides U-values and SHGC (solar heat gain coefficient) values. The lower the U-value, the better the window's insulative properties.	Look at the Climate Region Map on the label to be sure that the window, door, or skylight you have selected is appropriate for where you live.
Refrigerators and Freezers	Look for the EnergyGuide label that tells how much electricity, in kilowatt-hours (kWh), the refrigerator or freezer will use in one year. The smaller the number, the less energy it uses. Energy Star-labeled units exceed federal standards by at least 20%.	Look for energy-efficient refrigerators and freezers. Refrigerators with freezers on top are more efficient than those with freezers on the side. Also look for heavy door hinges that create a good door seal.
Dishwashers	Look for the EnergyGuide label that tells how much electricity, in kilowatt-hours (kWh), the dishwasher will use in one year. The smaller the number, the less energy it uses. Energy Star dishwashers exceed federal standards by at least 13%.	Look for features that will reduce water use, such as booster heaters and smart controls. Ask how many gallons of water the dishwasher uses during different cycles. Dishwashers that use the least amount of water will cost the least to operate.
Clothes Washers	Look for the EnergyGuide label that tells how much electricity, in kilowatt-hours (kWh), the clothes washer will use in one year. The smaller the number, the less energy it uses. Energy Star clothes washers use less than 50% of the energy used by standard washers.	Look for the following design features that help clothes washers cut water usage: water level controls, "suds-saver" features, spin cycle adjustments, and large capacity. For double the efficiency, buy an Energy Star unit.

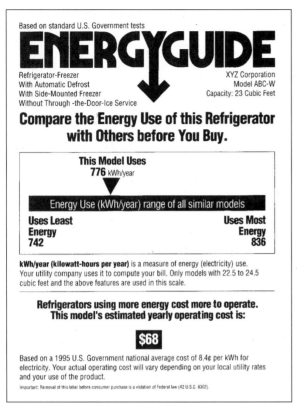

Figure 6-16.
The Energy Guide label found on all appliances lists important data that helps consumers identify the most energy-efficient and cost-effective models on the market. Note, however, that Energy Star appliances are not typically included in the comparison—so buying one of the most efficient units in its class may not be the best option unless it carries the Energy Star logo as well.

Source: *Consumer Guide to Home Energy Savings*

When comparing appliances, be sure to note other features such as water use and capacity. You should also consider noise levels and the type of fuel an appliance requires. Water heaters, for instance, operate on natural gas or propane, electricity, or sunlight. A solar hot-water heater is one of the most sustainable options, although it requires tanks and an extensive system of pipes.

You may also want to consider a tankless or on-demand water heater, as shown in figure 6-17. On-demand water heaters generate hot water for domestic use as it is needed. In a gas-fired on-demand unit, for example, water flows through the heater as soon as a hot water faucet is turned on. A gas burner ignites, heating the water instantaneously to the desired temperature. So long as water flows through the unit, the flame stays on. When the faucet is shut off, water flow stops and the flame shuts off.

Unlike conventional water heaters, which store large quantities of hot water in tanks, awaiting periodic use and losing heat, tankless water heaters heat only as much water as is needed at any one time. Larger units can accommodate several uses simultaneously—for example, two showers and a washing machine. Because they heat water on demand, tankless heaters reduce annual fuel bills for water heating by about 20 percent.

While a great deal of information on energy-efficient appliances is available these days, don't expect to find the most efficient appliances at local discount stores. These outlets tend to carry the cheapest and hence the most popular units. To purchase energy- and water-efficient appliances, you will very likely have to shop at an appliance store. To locate state-of-the-art appliances, such as the energy-miserly SunFrost refrigerator, you may have to shop at specialty suppliers like Real Goods, a mail-order supplier, and renewable energy suppliers. You can find them on the Internet and in the resource guide in this book.

Be wary of convenience features, such as ice and cold-water dispensers on the doors of refrigerators. They not only jack up the price of the refrigerator, they use more energy. Also, look for appliances that allow you to adjust for different uses. For example, some dishwashers allow you to adjust for the size of the load or to turn off the drying cycle (the heat coil that dries dishes after they've been cleansed).

Two more things to keep in mind are operation and replacement costs. Remember that inexpensive bargain-basement appliances may cost less initially, but they require much more energy and water to use, adding to your fuel and water bill. Inexpensively built units may need more maintenance and will give out earlier than well-made appliances. In the long run, the bargain may end up costing you a lot more. It certainly costs the planet more!

And finally, do your research. It is easy to be swayed by a salesman pointing out the benefits of his or her product. Ask for documentation. Don't take a salesperson's word for it. Consult *Consumer Guide to Home Energy Savings* and compare models you are looking at to those listed in the tables. *Consumer Reports* and *Consumer Reports Annual Buying Guide* are also helpful references. Produced by the Consumers Union, they rate appliances on the basis of reliability, convenience, and efficiency. Talk to as many people as you can, especially people who repair and install appliances, before you make up your mind. Salespeople are often happy to point out deficiencies in a competing product.

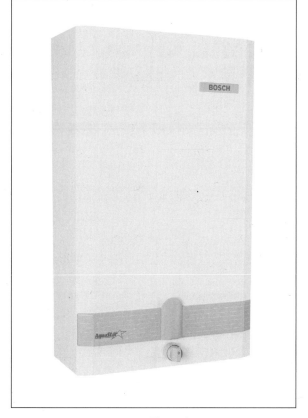

Figure 6-17.
Tankless water heaters save energy and water and are a great investment.

Source: Controlled Energy Corporation

Energy-Efficient Lighting

Energy-efficient lighting also helps reduce monthly energy bills. One of the easiest ways to save on lighting is to resist the temptation to over-illuminate, that is, to provide more light in a room than is necessary. Overlighting is a problem in many new homes. In many homes I've visited, electricians installed a dozen or more recessed lights in living rooms and family rooms. Each light fixture is equipped with a 125-watt incandescent light bulb and all of the lights are on one switch. In such instances, two or three hundred watts of strategically placed lighting would usually provide sufficient illumination.

Reducing the number of light fixtures and putting lights or banks of lights on separate switches so occupants can adjust lighting to meet their needs both help reduce electrical demand.

Rooms can also be designed for task lighting, that is, with light fixtures located in parts of a room where they are needed most, for example, over a desk or reading area. Task lighting saves money initially by reducing costs for lighting fixtures and installation and saves money over the long term by reducing energy bills.

Wall switches with built-in timers or motion sensors are also useful for families that have problems turning off the lights after leaving a room. Even greater savings can be made by replacing standard incandescent light bulbs with energy-efficient fluorescent compact light bulbs (CFLs). Compact fluorescent

Inexpensive bargain-basement appliances may cost less initially, but they require much more energy and water than efficient models. Inexpensively built units may also require more maintenance and will very likely give out much sooner than well-made appliances. In the long run, the bargain may end up costing you a lot more. It certainly costs the planet more!

BUILDING NOTE

One of the easiest ways to save on lighting is to resist the temptation to overillu-minate, that is, to provide more light in a room than is necessary.

light bulbs are ideal for areas in which lamps are on for long periods each day, such as living rooms, kitchens, and bedrooms. According to *Energy Savers,* a booklet produced by the U.S. Department of Energy, replacing four standard light bulbs in high-use areas with CFLs can reduce lighting costs by half. Compact fluorescent light bulbs can be used in recessed light fixtures, table lamps, floor lamps, torchiere lamps, floodlights, and even spotlights.

Compact fluorescent light bulbs are color adjusted so they produce a friendlier, more pleasing light than standard fluorescent bulbs. Although prices have fallen considerably since their introduction in the 1980s, CFLs do cost more than standard incandescent light bulbs. In retail stores, they sell for around $10 or $12. From bulk sellers such as Costco, you can buy them in packs of four to eight costing about $3 to $4 per bulb, although my expe-rience is that they're not as good as the more expensive CFLs. Don't be alarmed at the price, however. Even though CFLs cost much more than ordinary light bulbs, they use only one-fourth as much energy to produce the same amount of light. Moreover, CFLs last six to ten times longer, up to 10,000 to 12,000 hours, than ordinary light bulbs (most are rated for around 1,000 hours). In my house I have CFLs that I installed in the late 1980s, and they're still working fine.

Because they are so efficient and last so long, compact fluorescent light bulbs actually pay back the initial higher cost several times over. Lights of America, a leading manufacturer of CFLs in the United States, advertises a lifetime savings of $44 for its 75-watt-equivalent bulb if you are paying 8 cents per kilowatt-hour for electricity. This bulb uses only 20 watts to produce lighting equivalent to that of a 75-watt incandescent bulb and costs about $12. You save $32! The 115-watt-equivalent bulb, which uses only 22 watts to pro-duce the light of a 115-watt incandescent bulb, is advertised to last up to 12,000 hours and will save nearly $90. It costs about $14. Your savings: $76.

When shopping for CFLs, look for models that come with replaceable bulbs, that is, bulbs that can be separated from the ballast, the electronic device found at the base of the bulb. Ballasts often last up to 65,000 hours—six to seven times longer than the bulb—so replacing a burned-out bulb can save you money and reduce the demand for resources.

You may also want to consider full-spectrum CFLs for reading lamps. These light bulbs produce light that is similar to sunlight at noon; it increases visual acuity and can, according to the manufacturer, help you concentrate longer with greater comfort. The bulbs are supposed to reduce eyestrain while creating a balanced and relaxing environment. I use one by my computer and have been extremely impressed. It's much easier to read under than the standard compact fluorescent it replaced.

Halogen lamps are also more efficient than incandescent lights. They use about 50 to 70 percent as much energy as a standard incandescent bulb. Nevertheless, they still are fairly inefficient and produce a lot of waste heat. They're pretty costly, too. When you have a choice between a compact fluorescent light bulb and a halogen light bulb, choose the CFL. A torchiere using CFLs, for instance, uses 60 to 80 percent less energy than its halogen counterpart.

Using compact fluorescent light bulbs not only reduces electrical consumption without sacrificing light, but they produce much less waste heat than standard light bulbs. Reduced heat production, in turn, helps reduce summertime cooling loads.

Another way of reducing electrical consumption for lighting, known as *daylighting,* is using natural light from windows and skylights to illuminate. My passive solar house has numerous south-facing windows that let in sunlight during the heating season. These windows also permit light to enter year-round, greatly reducing the daytime use of electricity. In fact, I rarely have to turn on a light during the day, even on cloudy days.

Savings can also be obtained by installing efficient outdoor lighting. One way to improve on efficiency is to install CFLs in porch lights, floodlights, or other outside lights that are left on for long periods. If you live in a cold climate, however, be sure to buy lamps with cold-weather ballasts.

Solar electricity can be used to provide electricity to outside lights, too. It's especially useful in supplying remote lights on your property, for example, lights on outbuildings or at the end of your driveway. I installed a solar path light along the front walkway to my house to illuminate it at night (figure 6-18). This device, purchased from Real Goods, contains a small solar cell that produces electricity during the day. The electricity is stored in a tiny battery. At night, the light switches on when it senses motion—for example, when we open the front door or a car pulls up on the driveway.

Figure 6-18.
Solar path lights or accent lights, as they're sometimes called, like this one at the author's home, contain a small solar cell that generates electricity that is stored in a battery. At night, the lights illuminate driveways and walkways.

Source: Dan Chiras

Energy Efficiency and Green Building

Energy efficiency is an essential element of green building. As you have seen in this chapter, energy efficiency can be achieved in many ways, from installing insulation and high-quality windows to weatherizing a home to installing energy-miserly appliances. Building with low-embodied-energy materials, discussed in chapter 1, is also part of the energy-efficiency equation.

Energy efficiency reduces energy use in our homes, saves money, and helps protect the planet. It is a gift we give to ourselves, our children, their children, and the millions of species that share this lovely planet with us.

THE ENERGY STAR PROGRAM

Several times in this chapter I've referred to the Energy Star program, run by the U.S. Environmental Protection Agency and the Department of Energy. In the 1980s, these agencies initiated an effort to cut down on energy use in computers and other electronic equipment. Recognizing that energy conservation was the best way to control pollution, the agencies later focused their attention on appliances, such as refrigerators and washing machines.

Working *with* manufacturers, rather than at odds with them, the EPA and DOE have helped stimulate the production of much more efficient electronic devices and appliances. To help consumers identify stellar products, they invented the Energy Star logo, a sticker that is affixed to those products deemed to be the most efficient in their class.

After enjoying considerable success in this arena, the EPA and DOE then took the ideas that worked so well in appliance and electronics manufacturing into home building. Working with home builders, they developed a list of recommendations to improve the energy efficiency of homes.

Builders who participate in the program can obtain an Energy Star seal of approval for new homes that use at least 30 percent less energy than similar homes built using the Model Energy Code, once thought to be a fairly advanced residential energy code but one that few states and municipalities have adopted.

The Energy Star program is a voluntary collaboration between builders and the EPA and DOE. Builders are free to choose the materials and techniques needed to achieve savings. No one dictates what they must do. In other words, energy performance is the goal, and how it is achieved is more or less up to the builder.

To qualify for the Energy Star rating, a home has to be certified by an independent third party, a home energy auditor who checks details, such as insulation and types of windows, to see if the home's energy performance meets the rigorous criteria. The auditor also performs a blower door test, described in the chapter.

When buying an Energy Star house, you can be assured you have purchased a home that is efficient to heat and cool, economical, and comfortable. When building your own home, you may want to hire a contractor who participates in the Energy Star program. If you are building your home yourself, follow the guidelines in this chapter and consult other books on energy-efficient building listed in the resource guide.

Energy-efficient construction can also qualify you for special lower-interest loans. PHH Mortgage Services and other companies, for instance, often offer slightly lower interest rates for Energy Star homes. The loans are, on average, about one-eighth of a point lower than mortgages offered to those who purchase standard homes. Therefore, even if an Energy Star home costs a little more, and they often do, the homeowner still comes out ahead. Consider the example shown in table 6-5.

As illustrated, the Energy Star home costs $2,500 more. However, because of the lower mortgage rate, the mortgage payments are nearly identical. But energy efficiency measures cut down on monthly fuel bills. As a result, the Energy Star homeowner saves $22 per month, or $264 per year.

Besides offering slightly lower-interest loans, participating mortgage companies also allow homeowners to qualify for larger loans, sometimes called "stretch loans." Stretch loans provide 10 to 24 percent more money than might otherwise be offered, thanks to energy conservation measures in a home. In addition, some lenders offer cash-back bonuses at closing that may exceed or even eliminate any additional up-front costs required to purchase an Energy Star home.

Table 6-5

Cost Comparison: Standard Home vs. Energy Star Home

	Sticker Price	Up-Front Cost	Mortgage Amount	Interest Rate	Monthly Mortgage Payment	Monthly Energy Cost	Total Cost per Month
Standard Home	$160,000	$16,000	$144,000	8.0%	$1,057	$100	$1,157
Energy Star Home	$162,500	$16,281	$146,250	7.875%	$1,060	$75	$1,135

By now it should be pretty clear that energy efficiency makes good sense. It reduces energy use, saves money, and helps protect the planet. It is a gift we give to ourselves, our children, their children, and the millions of species that share this lovely planet with us. Energy efficiency is a no-brainer. Who wouldn't want to buy or build a home that has a higher appraisal value and is cheaper to operate, healthier to live in, more durable, and easier on the environment?

CHAPTER 7

Accessibility, Ergonomics, and Adaptability

In September of 1998, I was working on the clerestory windows of my house, fifteen feet above a tile floor. It was late on a Friday afternoon. I was done writing for the week, and looking forward to finishing this project, which I'd been pursuing in my spare time for the past couple of weeks, then slacking off for a few days. I set up the ladder and walked down to the living room to crank up the volume on my stereo. The next thing I remember is waking up in the multiple trauma unit of Swedish Hospital.

As best as I can tell, after turning up the stereo, I climbed the ladder to sand the sash and frame of a window that had been damaged by moisture. Then I began to apply a varnish. Sometime later, I don't know when, the ladder slipped out from under me and I crashed to the tile floor, spilling varnish all over my kitchen and breaking my hip and two ribs as well as a few bones in my forearm and wrist. The doctors pinned the fractured bones of my forearm back in place a few days later and then put me in a cast with metal rods, screws, and clamps to stabilize the whole mess. I spent the next two weeks in the hospital recovering from these injuries and the concussion I suffered in the fall when my head hit the tile.

When I came home, I was in a wheelchair. That was when I learned an important lesson about designing for accessibility: Homes need to be made accessible for people with a wide variety of abilities and disabilities. I hadn't given the idea a moment's thought during the design and construction of my home.

From the vantage point of a wheelchair, however, it became clear that the house, which seemed fine to me as a healthy, ambulatory adult, was a nightmare for a person with impaired abilities. Why?

My house consists of three levels. The main living area, kitchen, and dining room are all on one level. The hallway leading to the bedrooms is raised eight inches, and the bedrooms are raised another sixteen inches. Getting a wheelchair up those steps was impossible.

The silver lining behind this unfortunate experience was, of course, that I learned a great deal about making a home accessible and hence safe and usable by all who might visit or live here. I hope that the information in this chapter will help others think about the necessity of making homes user-friendly to a wider range of ages and abilities. Over the years, I have also learned about making houses ergonomic—that is, more convenient and efficient.

Designing for accessibility and ergonomic efficiency supports an important goal of green building: making our homes more people-friendly. As Sam Clark points out in his book, *The Real Goods Independent Builder,* the benefits of these approaches are houses that are more convenient, more efficient, and safer and that work well for more people. These approaches are also more environmentally friendly, because the flexibility they provide means that you may not need to remodel your home as you age, if your health takes a turn for the worse, or if you have an accident. They also make renovation easier if it becomes necessary to address a family's changing profile, for example, when children leave home. Your home, in essence, is more adaptable, and that saves resources in the long haul.

Designing for Accessibility

The accessibility of a home, as the term is used here, refers to the ease of entrance, of maneuverability within the building, and of exit, not just for the able-bodied but also for those whose age, health, or physical condition makes these functions more challenging.

When most people design homes they design for healthy adults and children, as if our lives never change. An architect, for example, might design a house for a family of four: two young children and their parents. Little thought is given to the occasional mishap that relegates a family member to a wheelchair or crutches, or to our inevitable aging.

When designing a house for accessibility, much of what is needed is pretty obvious if you take the time to imagine what life would be like if you were confined to a wheelchair or required to use crutches or a walker. If you are building or buying a house, take a few moments to look over the design and ask yourself the following questions: How easy would it be to get into and out of the house? How easy would it be to move about within the house? Where would a sick or aged family member sleep? Could they get into a shower or bathtub easily? Could a person in a wheelchair prepare meals? Could he or she do laundry easily? Could he or she access sinks? The answers to these questions will open your eyes—and those of your designer and builder—to the possibilities of accessible design.

Unfortunately, building codes generally do not require new homes to be constructed or existing homes to be remodeled with accessibility in mind, although there are some jurisdictions (Atlanta, Georgia, and Austin, Texas) that now require certain new homes to include some basic accessibility features. When building a home, study the building codes of these cities. You can also hire an accessibility expert to review your plans. Someone who understands accessibility codes and has experience designing and/or building homes for accessibility will help you identify places in your design where maneuverability, access, and safety could be improved.

> The accessibility of a home refers to the ease of entrance, of maneuverability within the building, and of exit, not just for the able-bodied but also for those whose age, health, or physical condition makes these functions more challenging.

Figure 7-1.
Steps are difficult to for individuals on crutches and impossible for individuals bound to wheelchairs. Expensive ramps must be installed to accommodate family members when accidents strike. By including a ground-level entrance, builders can make a home instantly accessible at no extra cost. It pays to think ahead!

Source: Lineworks

The front door is where accessibility all begins. An inconspicuous ramp or, even better, ground-level access to a home allows people in wheelchairs or on crutches to enter and exit a home easily and safely (figure 7-1). When I was injured, my chiropractor, Doug Petty, who is also a good friend, graciously built a ramp over my front steps so that I could get in and out of my house. Even then, it was too steep to use on my own, given the slope of the existing entryway. Designing for ease of entrance and exit saves work and resources later on.

Wheelchair access can also be facilitated by designing a home with slightly wider doorways. While exterior doors are usually wide enough to accommodate a wheelchair, standard interior doorways are usually quite narrow, often around 30 inches. For ease of movement, indoor openings of 34 to 36 inches are recommended. Slightly wider hallways also make it easier for people in wheelchairs to navigate, as do turn-around zones in hallways and other high-use areas.

Bathrooms should be made large enough that a wheelchair-bound person can comfortably transfer from the chair to the bathtub or toilet. Grab bars can be installed when a house is built to facilitate transfer. (If they are not installed initially, be sure to leave room for them.) Tubs should be raised 3 to 4 inches so they're better aligned with wheelchair seats. A transfer seat makes the trip from chair to bath easier, too. Showers should be built larger—at least 36 inches square inside—for easy access via a wheelchair. The floor of the shower should be flush, or nearly flush, with the bathroom floor.

Laundry rooms should be large enough to allow access in a wheelchair. Storage should be easy to reach. Sinks in bathrooms and countertops in kitchens should be built at wheelchair height and with ample leg room under the counter (figure 7-2). Be sure to insulate hot water pipes to avoid any potential burns resulting from unintended contact by wheelchair users.

"Perhaps the biggest obstacles to mobility are steps and steep slopes. A few stairs are an obstacle for many elderly people, and even one step is an almost absolute obstacle for wheelchair users."

Sam Clark
*The Real Goods
Independent Builder*

Figure 7-2.
Installing special sinks and
toilets like the ones shown
here during construction
saves remodeling costs
should a family member
become wheelchair-bound.

Source: Lineworks

In two- and three-story houses, be sure to include a ground-level bedroom or a room that could be converted to one. A guest bedroom, den, or home office, for example, could be converted to a bedroom for a temporarily or permanently disabled family member.

Making a home more accessible also means paying attention to the types of handles and knobs installed on doors and cabinets. Those whose manual dexterity has declined—for example, as a result of arthritis or injury—will find that D-pulls work well on cabinets and are easier to manipulate than knobs. Lever handles are more convenient and less frustrating for everyone. You may want to omit thresholds on interior and exterior doors and minimize them on exterior doors.

Vertical-axis (front-loading) washers are also a good idea. They're readily accessible by someone in a wheelchair, as well as being much more energy and water efficient than conventional top-loading models.

Some features that promote safety and convenience can be added later. For example, you can install an intercom system in your home if the need arises. However, installing the wiring when the house is built will make the job of installing the system easier and less costly later. Prewiring a home for an intercom system requires very little additional cost. Grab bars can be added to bathrooms as required, too. You can also add new thermostats with large, raised numbers for visually impaired members of your household.

Unlike other areas of home building, accessibility lacks an extensive reference list. If you want to learn more, I recommend you obtain copies of *Adaptable Housing* by the U.S. Department of Housing and Urban Development, *Elder House* by Adelaide Altman, *Building for a Lifetime* by Sam Clark, and *Residential Remodeling and Universal Design: Making Homes More*

Comfortable and Accessible by the National Association of Home Builders Research Center. I've listed these and other references in the resource guide, along with a few booklets on universal bathroom and kitchen planning. The National Association of Home Builders also publishes a catalog of accessible products called the *Directory of Accessible Building Products.*

Accessible design increases quality of life with little, if any, additional resource demand. Sam Clark notes that almost any project, including most new homes, can be designed to be more accessible and that doing so "need not be expensive or obtrusive in any way." Small considerations mean that anyone can come to visit without fear of being rendered useless, including aging parents with failing knees or bad hearts that make stair climbing difficult, if not impossible. Ultimately, accessibility "means you can stay in your house as you age or as your health changes." Clark adds, "But it also makes a better house for all now. The kitchen will work much better for all users. Storage will be more efficient and capacious. It will be easier to move things around in the house and move about it yourself." Accessibility also influences the quality of the environment, making resource- and energy-intensive remodeling projects and other adjustments unnecessary.

> Accessible design increases quality of life with little, if any, additional resource demand. Moreover, doing so need not be expensive.

Ergonomic Design

Many years ago, I taught a course on whitewater kayaking through the Colorado White Water Association. We held our annual spring training at the Colorado Outdoor Center along the banks of the Arkansas River. The gentleman who built the facility was about five feet tall. For the most part, the facilities worked well. He used standard dimensions for almost everything. However, when he installed the showerheads, he placed them in a position that was appropriate for him—which turned out to be about chest-high for me. I had to practically get on my knees to get my hair wet!

Similar problems occur in many homes. If you have ever had to mount a stool to reach the upper shelves of your kitchen cabinets, crawl on your hands and knees to get to an electrical outlet, or endure back pain when working at a kitchen counter that was too low, you know what I'm talking about.

The reason that our homes don't always function as efficiently as they could is that people range in size and most homes are designed for the average adult—that is, for one who is of medium height, active, young, and healthy. If you don't fit that description, your home may be less convenient than it could be, and you may even suffer pain as a result.

Enter the study of ergonomics. *Merriam Webster's Collegiate Dictionary* defines ergonomics as an "applied science concerned with designing and arranging things people use so that the people and things interact most efficiently and safely." Ergonomics is concerned not just with immediate dangers but also

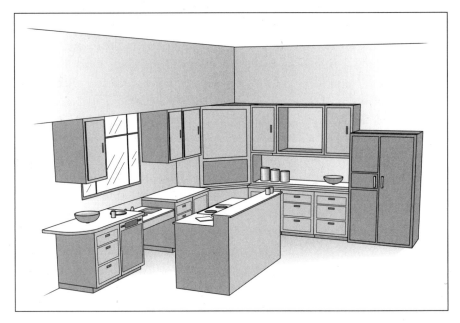

with long-term dangers posed by using "things" in our environment. The ergonomic keyboard I use, for instance, has reversed severe upper back and neck pain resulting from nearly twenty-five years of writing at a computer.

Most efforts to make homes more ergonomically sound focus on kitchens, electrical outlets, and light switches. Let's start in the kitchen.

Locating countertops at convenient heights reduces back strain and makes our actions more efficient. But remember, one size does not fit all. I'm six feet tall and my partner, Linda, is five feet two inches. Counters are often too low for me and too high for her.

Wise kitchen design might suggest that countertops be situated at two levels, one for mom and one for dad, with the chief cook and bottle washer receiving the greatest priority. And if you want your children to join in meal preparation, you can build in a countertop for them as well. Because they inevitably grow up, an adjustable one is a good idea.

At little or no extra cost, most light switches, outlets, and cabinets can be placed in an "optimal reach zone"—an area 22 to 44 inches from the floor—that works well for a broad range of people. Studies show what is obvious but overlooked: It takes more energy to reach way down or kneel down to reach something. It is also more difficult or more dangerous to reach for things over your head, as I found out recently while trying to get a cooler off a high shelf in my garage. When I pulled on the cooler, a cooking pan on top of it, which couldn't be seen from my vantage point, came down and hit me in the face, breaking my reading glasses in two. Table 7-1 lists standard and ergonomic locations of various house components.

Making a home ergonomically sound also means arranging it so traffic flows smoothly and so that things we need can be obtained efficiently. Pay special attention to how kitchens are laid out so that meal preparation,

Table 7-1
Standard Height vs. Ergonomic Height

Item	Standard Height (in inches)	Ergonomic Height (in inches)
Table	30	29
Kitchen counter	36	variable
First upper shelf	55	48–50
Top shelf	75+	70
Outlet	12–18	24
Switch	48	44
Door (width)	30–32	34–36

Source: Sam Clark, *The Real Goods Independent Builder*, White River Junction, Vermont: Chelsea Green, 1996.

which can be laborious, is made easier. Storage spaces in a home should be convenient and easy to access as well.

Making a home ergonomic is relatively easy and involves awareness and common sense. For those who want to learn more, I highly recommend *Humanscale 1-2-3* by Niels Diffrient, Alvin Tilley, and Joan Bardagjy (see the resource guide).

Designing for Adaptability

As we grow older, not only our health but also our relationships and our need for space change. Singles living in tiny apartments marry and move to their first home to make room for their combined possessions. Married couples bring children into the world, increasing their need for space. But in time, our children graduate from high school, then move on, first to college or trade school, then to full-time employment and families of their own. Families of four become families of two again, and the additional space required to house the gang is no longer necessary. Rather than move out of the house and a neighborhood you love (with the flower and vegetable gardens you've toiled over for two or three decades) into a smaller home that matches your reduced need for space, wouldn't it be terrific if you could simply convert part of the house to an apartment—say, for a college student, a struggling artist, or perhaps an elderly relative—and stay put?

Well, you can. People all over the country are doing this. Some are even converting garages to delightful living quarters. In Golden, Colorado, many homeowners have built upstairs apartments over garages to accommodate students who attend the nearby Colorado School of Mines.

As the human population expands, converting basements, bedrooms, and even extra garage space to apartments could help society meet its need for housing—and it helps us achieve this goal economically. Furthermore, conversion achieves these goals without the extensive resource demands of building new houses. Adapting buildings, rather than demolishing them, also

As the human population expands, converting basements, bedrooms, and even extra garage space to apartments could help society meet its need for housing—and it helps us achieve this goal economically. Conversion achieves these goals without the extensive resource demands of building new houses.

makes sense from an energy standpoint. According to William Bordass of William Bordass Associations, the embodied energy in a building "is equivalent to five to ten years of operational energy."

Advance planning can make conversion easier and more resource efficient. For example, if the plumbing for a kitchenette is "stubbed in" in advance, it is much easier to convert two bedrooms and a bath to a small apartment with a kitchenette, a living and dining area, a bedroom, and a bath, which is ideal for a single adult or a married couple with modest personal holdings. The basement could be converted to yet another apartment. You may find as you grow older that you need assistance; the apartment could become a residence for your caretaker.

Pay attention to egress issues, that is, how future apartment dwellers are going to enter and exit, when designing an adaptable home. With advance planning, you should be able to provide a private entrance for your future renters. While we're on the subject of privacy, you may also want to consider installing sound insulation in interior walls between your living space and that of the future renters. Everyone will be happier if you do.

Adaptable design requires many other considerations. For example, when choosing a site or a new home, be sure there's room for building expansion in case you want to add an apartment to the house. Jessica Boehland recommends constructing walls so windows or doors can be added easily. She also recommends sizing and proportioning spaces to accommodate a variety of uses. When building, she asserts, parts should, wherever possible, be connected mechanically, not chemically. In other words, components of a house should be screwed or bolted together, rather than glued or welded, for ease of dismantling and reconfiguration. Generally, the simpler the design, the more easy it is to alter a building. Recording the location of services such as gas and electrical lines and structural elements such as load-bearing posts during construction also makes adaptation easier. For more of her "adaptability" ideas, you may want to read her article, "Future-Proofing Your Building," in the February 2003 issue of *Environmental Building News.*

Designing a home to be adaptable adds expense and complicates the process, to be sure, but as Boehland notes, "The practice can reap impressive benefits." It "effectively installs an insurance policy into a building. By equipping a building for change, we set the stage for its easy reorganization, accommodation of new technologies and services, and eventual adaptation to new, yet unimaginable uses."

Convenience, Efficiency, Safety, and More

Accessible and ergonomic design render a home more convenient, more efficient, and safer for a wide range of people, no matter what their age or health status. Adaptable design makes a house more useful to you as your family grows and changes. All three design approaches are important elements of green building, making our homes more user-friendly over the long haul. Small changes in the design that add very little to the total cost of the home can increase the lifetime utility, comfort, and safety of a home while lessening resource demand and expenses.

CHAPTER 8

Using Concrete and Steel to Build Green

In Pueblo, Colorado, contractor Judy Fosdick is building homes made from concrete. Although homes made of concrete are not new—even, for instance, Thomas Edison built homes for his employees using this material—Judy is helping pave the way for a new generation of environmentally friendly homes. Her buildings meet rigorous energy-efficiency standards and utilize solar energy for wintertime heat. She has also developed designs for small, affordable, passive solar concrete homes that could be used in any subdivision.

Fosdick's company, Tierra Concrete Homes, produces houses made from precast concrete panels made on-site (figure 8-1). It's an idea developed and patented by her husband. These panels are set on an insulated concrete foundation.

With the interior walls and floors also made from concrete, you'd think that her homes would be cold, impersonal, and inefficient. They are not. Rather, Fosdick's homes are warm, comfortable, and extremely energy efficient. Her company has won several prestigious awards for energy-efficient home building, including the EPA's Most Energy-Efficient Homes of the New Millennium Award.

In Fosdick's houses, concrete walls and floors provide thermal mass that absorbs solar heat on sunny winter days and slowly radiates the heat back into the living space at night or

Figure 8-1 a & b.
Judy Fosdick and her husband build homes from precast concrete panels (a), shown here being lifted into place. Although concrete may seem like an unlikely green building material, it is durable and could outlast ordinary wood-frame homes by centuries, making it a great option. Below is a completed home (b).

Source: Tierra Concrete Homes

In 1908, Thomas Edison filed a patent application for a home built entirely out of concrete. To prove the viability of this idea, he had eleven cast-in-place concrete homes built for his employees in Union, New Jersey.

With winds often exceeding 100 miles per hour and their occurrence increasing as we unwittingly alter the planet's climate with a continuous outpouring of greenhouse gases and the ongoing destruction of carbon-dioxide-absorbing forests, hurricanes could become a major factor in the evolution of home building over the next few decades.

during cold, cloudy spells. In the summer, this thermal mass provides passive cooling. The exterior walls are protected on the outside by a thick (4-inch) layer of rigid foam insulation, which is typically covered with stucco, although brick and stone can also be used.

Fosdick is part of a growing legion of contractors building green homes from fairly conventional materials. Although concrete may seem like an unlikely candidate for an environmentally friendly home, due to the fairly high embodied energy of the material, Fosdick is quick to point out that first impressions can be wrong . . . very wrong. In her words, "When you consider transportation costs of lumber and the one-hundred-times-greater life span of concrete, concrete is much more environmentally friendly." Fosdick goes on to say, "Concrete gets harder with age, its life span is infinite, and it can be recycled." Concrete has other advantages, as well, that make it an option in green building, as does another conventional building material, steel.

Green Building with Gray Concrete

If you are like me, concrete seems more ideally suited for bridges, highways, and factories than for houses. It seems impersonal and unsightly. Visiting a concrete home, like those Judy Fosdick is building, however, quickly dispels the image. In a finished home, there is little evidence of concrete. As in a stick-frame home, carpeting or tile covers the floor. Wallboard or plaster covers the walls. If built and oriented correctly to take advantage of the sun's free heat, and if insulated well, concrete homes stay warm in the winter and

Figure 8-2. Concrete houses like this home in Florida are an excellent choice in coastal areas subject to hurricanes.

Source: Dan Chiras

cool in the summer on their own. Little fossil-fuel energy is required to provide comfort. Huge economic savings accrue to the homeowner.

Concrete homes are being built all over the country, but they are especially popular in coastal states such as Florida (figure 8-2). One reason for their popularity in such locations is their supreme resistance to hurricanes. With winds often exceeding 100 miles per hour and their occurrence increasing as we unwittingly alter the planet's climate with a continuous outpouring of greenhouse gases and the ongoing destruction of carbon-dioxide-absorbing forests, hurricanes could become a major factor in the evolution of home building over the next few decades.

Hurricanes are not the only weather phenomenon that could affect the direction of home building in the years to come. Meteorological studies show that tornados are also on the rise, having doubled in frequency in the past twenty years. Tornados are causing tens of millions of dollars of damage to homes each year. Home builders and home buyers in the path of twisters would be wise to rethink the materials from which they build—or rebuild—their homes. While the average stick-frame home has little chance against a category 5 hurricane or a tornado, concrete homes can withstand these assaults, incurring little, if any, damage. In the wake of devastating storms, while neighbors struggle to pick through the debris of their shattered homes, the owner of a concrete home is able to resume life with little, if any, change in routine. The only repair required might be windows and the only picking up involved in getting back to normality is removing trees that may have been toppled by the storm.

Durable as it is, concrete does have its drawbacks. For one, it has a rather high embodied energy. Embodied energy, as explained in chapter 1, is the amount of energy required to produce a material. In this case, it is the energy required to mine and process the raw materials and manufacture concrete, a mixture of Portland cement, aggregate (small rocks), sand, and various chemical additives. Energy required to transport raw materials and the finished product is also part of the total embodied-energy equation.

For concrete, one of the biggest contributors to its high embodied energy is Portland cement, which is the binder, that is, the material that holds everything together. Portland cement consists primarily of limestone, silica, and alumina, which are mined, crushed, and heated to over 2,000°F. The product is then mixed with gypsum, which controls the set time of cement. When concrete cures, it creates a hard, durable material.

Mining and manufacturing of Portland cement not only requires a fair amount of energy, it produces considerable amounts of environmental damage and pollution. In fact, the cement industry is one of the largest producers of carbon dioxide in the United States, and the second major producer in Great Britain. Worldwide, manufacture of cement is responsible for 8 percent of all carbon dioxide emissions.

To address these problems, many cement manufacturers have begun

While the average stick-frame home has little chance against a category 5 hurricane or a tornado, concrete homes can withstand these assaults with little damage.

CONCRETE PRODUCTION

Concrete is a mixture of Portland cement (a binding material) and aggregate (sand or small rocks or both). The components are mixed with water, then set up, producing a solid material with great compressive strength. Once mixed, concrete can be poured into forms to create a wide variety of buildings as well as roadways, bridges, and dams, many of massive proportions.

Concrete is produced in massive quantities—about one ton per capita worldwide! Although concrete is extremely useful, its production causes many environmental problems. Each component, for example, is mined from the Earth in environmentally damaging open pit mines. The mining and manufacturing of concrete requires significant amounts of energy. In fact, worldwide concrete production is responsible for about 8 percent of the global carbon dioxide emissions. Although concrete contributes to many environmental problems, it does offer a feature that helps offset its impacts: durability. Durability means that concrete can outlast other building materials and may actually turn out to be a better material, for reasons explained in the text.

Unfortunately, concrete cannot be recycled in the truest sense of the word. That is, it cannot be smashed and wetted and repoured. However, it can be crushed and reused in this state, for example, to create artificial reefs or retaining walls.

---

substituting fly ash, a waste product from coal-fired power plants, for a portion of the Portland cement. Fly ash is typically dumped in landfills in the United States. However, according to environmental building expert Alex Wilson, fly ash rarely comprises more than 25 percent of the cementitious materials of "green" concrete. Although this is a definite improvement over conventionally made concrete, it still leaves much to be desired.

Cement in concrete is also hazardous to workers. It can burn the skin. Dry dust containing silica can cause silicosis, a chronic, incurable lung disease that develops twenty to thirty years after exposure to silica dust. Coughing, shortness of breath upon exertion, and tightness in the chest are the most noticeable symptoms, but the disease can progress to respiratory failure. Chromate found in trace amounts in concrete has been shown to increase the rate of stomach cancer in cement workers in England. Certain additives in concrete can cause skin ulcerations and burns or other skin conditions such as eczema.

Concrete also requires internal reinforcement. On its own, concrete has what engineers call *compressive strength*. That is, it can withstand considerable compression pressure. However, concrete is really quite brittle, so it can possibly snap under lateral pressure, meaning that it has little *tensile strength*. A bridge made out of concrete alone would crumble under the pressure. To improve its tensile strength, builders reinforce internally by installing carefully spaced steel rods, called *reinforcing bars* or simply *rebar*. Rebar is used in foundations and concrete walls. In earthquake zones, additional rebar must be used. Reinforcement adds to the labor and materials costs of building a home, as well as its environmental impact, but it greatly increases the strength and durability of a concrete building.

So with all this going against it, why would anyone recommend concrete as a green building material?

With safety precautions to protect workers in factories and on the job site, some experts think that concrete has a place in green home building. Because concrete houses outlast conventional stick-frame houses by centuries, the high embodied energy and pollution generated by concrete production must be amortized over the lifetime of the structure. When you do so, the benefit becomes clear. If a concrete house lasts five times longer than a stick-frame house, its pollution and environmental destruction must be compared to that generated by the production of five typical homes.

Dome Homes

One extraordinary resource-efficient and durable concrete home is the concrete dome home. Because most of us have been brought up in homes with straight walls that intersect at 90-degree angles, the thought of living in a dome may seem odd. As figure 8-3 shows, dome structures can be quite attractive. Moreover, the floor plan shown in figure 8-4 demonstrates that rooms can be laid out nicely. They provide adequate space with ample room for furniture.

Domes offer a number of significant advantages. For one, they provide the most square footage for the least amount of material. Concrete dome homes require 200 to 300 percent less concrete and use 300 to 400 percent less rebar than a conventional concrete home made from the same material (table 8-1). In addition, concrete domes require half as much labor, and the domes are stronger and better able to withstand natural assaults, such as hurricanes and tornados. Domes are also inherently more energy efficient than the standard rectilinear home because they present less exposed surface area. The less surface area, the less heat loss in the winter and heat gain in the summer.

Figure 8-3.
Concrete dome homes like this one provide more living space with less material and are extremely resistant to winds and tornados. They can also be extremely beautiful.

Source: Monolithic Dome Institute

Figure 8-4.
The floor plan of a concrete dome home shows that it can be spacious, beautiful inside.

Source: Monolithic Dome Institute

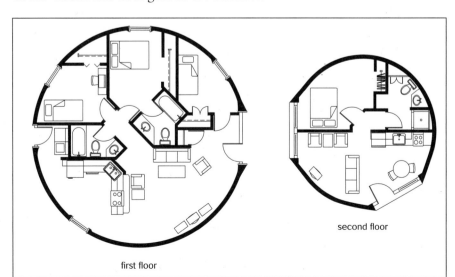

second floor

first floor

Table 8-1

Comparison of Standard Concrete Homes to Dome Concrete Homes

Type	Size (feet)	Square footage	Interior volume (cubic feet)	Surface area (square feet)	Concrete required (cubic yards)	Rebar required (pounds)
Square	25 x 52	1,248	9,984	2,464	69	13,300
Dome	diameter, 40 height, 16	1,257	12,197	2,060	38	4,200

Advantages of Concrete Dome Buildings

Domes offer a number of significant advantages. They provide the most square footage for the least amount of material. They require half as much labor. And domes are stronger and better able to withstand hurricanes and tornados than rectilinear structures. Domes are also inherently more energy efficient that the standard rectilinear home.

David, Barry, and Randy Smith, three brothers from Idaho, are leaders in dome building. The brothers previously made a living by insulating metal storage buildings for potatoes. Today, they focus most of their attention on building concrete domes. When they first began making concrete domes in 1976, they followed in the footsteps of others, spraying "shotcrete"—a form of concrete that has been thinned to make it sprayable—on the outside of 12-foot-diameter "balloons" (inflatable membranes) attached to foundations. According to David, their first approach, the EcoShell, "had many problems." The application of the concrete to the exterior of a balloon (they call it an Airform) "was just plain miserable." Shotcrete could be applied only in good weather. And it was very difficult to apply the material. Curing of the concrete out in the weather was poor, and using a concrete gun outside often resulted in a mess. Nothing in the vicinity was safe from the spray—not the neighbors, the car, the bushes, or family pets.

Over the years, the brothers altered their system. They now apply the shotcrete from inside, and they can work year-round without coating the neighbor's yards with it.

To begin building the new form of dome, which they call EcoShell II, the brothers first attach the Airform to the foundation. They then attach an entryway (airlock) that allows them to gain access to the interior. A fan fills the Airform with air. Next comes a layer of material called Stikum. This sticky material is applied to the interior of the Airform. Then comes the first coat of shotcrete, followed in rapid succession by a second coat. When the shotcrete reaches ½-inch thickness, they apply steel reinforcing bar, using special hangers, to increase the tensile strength of the concrete. After the rebar is in place, the remainder of the shotcrete is sprayed on. The total thickness of concrete in the wall is 1¼ to 1½ inches.

After the concrete has been allowed to set for a couple days, the Airform is peeled off. Foam insulation is sprayed on the outside of the dome, and a waterproof coating is applied over the insulation to provide long-term protection from rain and snowmelt. Interiors are plastered or stuccoed.

The Smith brothers now have over four hundred monolithic domes to their credit, scattered over forty states and overseas, and are the undisputed kings of the dome. However, domes are not a new invention. Buckminster Fuller was a proponent of geodesic domes. The brilliant and talented Michelangelo extolled them, saying "The dome is a major work of art, the perfect blending of sculpture and architecture in displacing space . . . the most natural of all architectural forms."

Other contemporary architects and structural engineers are also impassioned dome builders. Dr. Arnold Wilson spent forty years teaching civil engineering at Brigham Young University in Utah. He has designed many large dome structures with spans as great as 240 feet. He envisions domes as large as 1,000 feet. In his view, the inherent strength of a dome is one of its greatest advantages. "Domes," he says, "are just too good of a thing not to gain in popularity. They can withstand just about any force, and they are economical to build and maintain."

Wilson predicts that domes will gain in popularity in the future. He argues that they could even provide a practical way of creating low-income housing, especially in less developed countries. Although acceptance of the round structure poses a significant challenge, domes offer so many benefits that their future could be bright indeed. There's even a nonprofit organization, the Monolithic Dome Institute (founded by one of the Smith brothers), dedicated to promoting dome construction. The institute offers conferences and workshops on dome construction and has published a book of plans for monolithic dome homes, *Dome Dwellings 97,* and a journal, *The Roundup: Journal of the Monolithic Dome Institute,* both available on their web site.

> "Domes are just too good of a thing not to gain in popularity. They can withstand just about any force, and they are economical to build and maintain."
>
> **Dr. Arnold Wilson**
> Brigham Young University

Houses Built of Steel

Like concrete, steel is an industrial material generally associated with automobiles, bridges, and skyscrapers, certainly not homes. Actually, steel is used in many new homes. As briefly mentioned in chapter 6, light-gauge steel studs are used to frame interior and exterior walls of homes. The practice is well entrenched in Switzerland, Austria, New Zealand, and Australia. Steel is fairly widely used in Ontario, Hawaii, California, Texas, and the Gulf Coast states, and it is often touted as a green building material. How can an energy-intensive material such as steel be considered good for the environment?

Proponents of steel, the steel industry, and many builders ascribe a long list of environmental benefits to using steel. One of the most notable is that light-gauge steel framing can replace wood studs in a home. As proponents are fond of noting, you can frame a house with the steel made from four old Fords or with wood studs made from forty-four trees. This poignant comparison illustrates one of steel's most redeeming values: It reduces wood use. With an average of four hundred studs in a new home, and 1.2 million new homes built each year, 85 percent of which are framed in wood, the advantage of steel is obvious. A large steel plant can manufacture more than two million tons of galvanized steel in a single year—enough to meet 30 percent of new home construction needs, protecting at least twenty million acres from the chain saw annually.

MAKING STEEL

Steel is a metal alloy, a combination of iron and other elements such as carbon, tungsten, and titanium that occur alone or in various combinations in the finished product. Iron, the principal raw material, comes from iron ore, extracted from the earth's surface in huge mines.

To make steel, iron ore is crushed and mixed with coke (carbon made from coal) and limestone. The mixture is then heated in a blast furnace. In the process, hot carbon combines with oxygen and limestone combines with the foreign materials, producing waste slag that floats and thus easily separates from the molten iron. Further treatment drives off unwanted impurities, and the addition of other elements, such as tungsten, to the molten iron, produce steel.

The production of steel requires an enormous amount of raw materials that must be extracted from the earth, causing considerable damage to the environment. It also requires a significant amount of energy, which produces air pollution. Recycling steel can help reduce the inputs and dramatically reduces its environmental impact.

Steel offers other significant advantages. It resists moisture and insect damage, for example, from termites. For that reason, steel studs are now used in 60 percent of the new homes built on the Hawaiian island of Oahu. Steel studs will not burn, either, although they would likely warp in a fire and become unusable.

Steel is recyclable and can be used over and over again in a process known by proponents as "endless recyclability." Because recycling uses about 35 percent less energy than making new steel, the total embodied energy of steel over the long haul is quite low.

Steel is also made from recycled waste, which comprises about 35 percent of the material. (Most of it is industrial scrap, with only 22 percent of the recycled material being post-consumer content.) Scrap metal for making new steel, such as that taken from old ships and cars, is quite abundant, as are global iron supplies.

Pliny Fisk and his colleagues at the Center for Maximum Building Potential in Austin, Texas, point out that unless a building made from virgin wood can survive 60 to 70 years, and many don't, it is likely to have a greater overall environmental impact than a building made from steel framing materials.

Steel is structurally stable, too. It won't twist or warp over time like wood studs. This, in turn, results in a more structurally sound and more energy-efficient building. (Energy efficiency, in this instance, results from the fact that the building envelope does not crack as much as one with an internal wood frame. In contrast, wood framing can warp, twist, and compress, creating cracks in the building envelope that leak air, increasing heating and cooling bills and reducing comfort.)

But the picture is not so clear as proponents would have you believe. According to Nadav Malin of *Environmental Building News,* the steel in a

A large steel plant can manufacture more than two million tons of galvanized steel in a single year—enough to meet 30 percent of new home construction needs, protecting at least twenty million acres from the chain saw each year.

2,000-square-foot home represents 20 percent more embodied energy than that of wooden studs. Steel production also produces more pollution and toxic emissions than equivalent wood production.

As a building material, Malin points out, steel's biggest drawback is its thermal conductivity—that is, its ability to conduct heat into or out of a house. Steel studs create significant thermal bridges that reduce the energy efficiency of a wall. According to a study performed by researchers at the Oak Ridge National Laboratory, the R-value of a steel-stud wall is 20 to 40 percent lower than that of a comparable wall framed with wood studs. To prevent this problem, builders routinely apply a rigid foam insulation to the exterior wall surface of steel-studded homes. While this offsets thermal bridging, it increases the cost of building as well as the material requirement and the environmental impact of building.

Steel studs also cause a problem commonly referred to as *ghosting,* the accumulation of dust on wall surfaces due to electrostatic cling. To be fair, the same problem occurs in stick-frame houses, though it is nowhere near as pronounced. In a steel-framed home, walls must be cleaned and repainted more often than in a wood-framed home.

Environmental impacts of producing steel also deserve serious consideration. Brazil is a major producer of the pig iron used to make steel, producing about one-sixth of the world's supply. Pig iron is made from iron ore smelted with charcoal. The charcoal, in turn, is made from trees clear-cut from tropical rain forests. Huge areas of virgin forest are cut down each year in Brazil to produce the charcoal needed to make iron.

When building or buying a new home or remodeling an existing one, should you look favorably on light-gauge steel studs?

The answer is "maybe."

To answer the question, you'll need to do some research. First, you will need to determine the recycled content of the specific product available to you. The higher the recycled portion—and the greater the post-consumer content—the better. You also need to ascertain where the iron ore used to make the steel comes from. If it comes from Brazil or another region where the tropical rain forests are leveled to make charcoal to smelt the ore, it would be difficult to justify its use. Yet another consideration is the distance of the building site from the steel mill. The longer the distance, the greater the embodied energy.

If the recycled content is high, the origin of the raw materials is acceptable, and the distance from the mill is minimal, steel could be a suitable material. Its durability, "endless recyclability," and structural stability will offset many other negatives. Moreover, if the light-gauge steel studs are engineered and designed to reduce thermal conductivity, as some newer ones are, steel studs could move a notch higher on the scale of desirability.

Many Avenues and Many Trade-Offs

If you are reading this book cover to cover, by now you have come to realize that there are many ways of building green. Trade-offs are common, and no product is free from some impact. Even materials that seem highly unlikely candidates for green building, such as concrete and steel, may turn out to be suitable for creating environmentally friendly homes. As you will see in the next chapter, there are ways to reduce the trade-offs and create homes that come as close to sustainability as humanly possible with currently available materials and techniques.

CHAPTER 9

Natural Building

For virtually all of human history, people have fashioned shelter from natural building materials such as grasses, wood, and mud. Even today, many people throughout the world still build with natural materials. Readers may be surprised to learn that approximately half of the world's people live in buildings fashioned from dirt. In China alone, there are an estimated 90 million earthen buildings.

Although the vast majority of the homes made from natural materials provide shelter for the poor, many more people in wealthy industrial nations are now turning to these materials to build their homes. Using straw bales, logs, and earth, this daring group of homeowners is defying convention. Although they may be raising an eyebrow or two at the local building department and eliciting a snigger or two from conventional builders, these pioneers could very well be charting a course to the future.

The reasons for this trend toward the use of natural materials are many. Natural homes can be clean and comfortable. Discard any notions of the primitive sod homes in which the pioneers of North America lived. Natural homes are durable and long lasting. With thick walls of earth or straw, they can foster a sense of deep and abiding security. They also offer unrivaled beauty. Many straw and natural homes have sensuous curved walls and rounded corners that provide, in the words of natural builder Ianto Evans, much needed "refuge from the right angle."

Homes made from natural materials provide comfort and aesthetics far in excess of what is offered by many modern homes.

Figure 9-1.
Author Carolyn Roberts built this beautiful straw bale home with the help of her sons and friends in the desert outside Tucson, Arizona. This home is not only beautiful, it's energy efficient, providing comfort with very little outside energy.

Source: Rick Peterson

Natural building also offers a chance to build with locally available materials. Because they are typically harvested near the building site, these materials have a low embodied energy. These materials are also harvested from renewable resources. Moreover, natural homes can be especially energy efficient. Thick straw bale walls, for instance, provide superior protection against outdoor temperature extremes. Natural homes are also ideally suited for passive solar heating and cooling (see chapters 11 and 12), which reduce fossil fuel consumption and lessen the environmental impact of our homes. As Ianto Evans, Michael Smith, and Linda Smiley write in their book, *The Hand-Sculpted House: A Practical and Philosophical Guide to Building a Cob Cottage,* "In areas where wooden buildings need air conditioners, earthen buildings right next door are cool and fresh all summer long. While neighbors struggle to pay utility bills, cob houses [one type of natural home] stay snug and warm in winter."

Not only are natural houses comfortable to live in, they are healthier, too. The air inside is cleaner and healthier because natural materials do not contain toxic substances that are released into indoor air.

Moreover, in this day and age, when homes are typically built by professionals and most often in an assembly-line approach, natural building provides a welcome alternative. Because most natural building techniques are relatively easy to master, natural home building provides us an opportunity to build our own shelter, customizing it to meet our needs and to express our personalities. As Evans explains, "Natural building is democratic. Most natural building techniques . . . are accessible to old women and little boys, the impractical, the handicapped, the impoverished. . . . Natural building empowers those who have been all their lives persuaded that they should leave building to 'professionals.'"

Natural building is gaining respect among building professionals, too. Each year, more and more architects and professional builders are exploring natural materials and techniques, which combine traditional approaches with state-of-the-art innovation. Hundreds of new straw bale and other natural buildings appear on the landscape each year.

Natural building is becoming more widely accepted, sometimes even embraced, by building departments, too, thanks in part to highly effective advocates such as David Eisenberg of the Development Center for Appropriate Technology and Bruce King, a structural engineer who founded the Ecological Building Network. While Eisenberg has been working for years with building code officials to help incorporate sensible guidelines and regulations for natural building, King and others are studying the structural properties of natural building materials and making that information available to a wide constituency. Their efforts are helping ease the entry of natural materials and techniques into mainstream building.

In this chapter, I'll explore most of the dozen or so natural building

Because they are suited for passive heating and cooling and are especially energy efficient, natural homes require far less outside energy to provide comfort. Not only do natural homes use less fossil fuel energy to operate, but because they're made from locally available materials, they also require substantially less energy to build than conventional stick-frame homes. Lowering energy consumption creates a cornucopia of environmental benefits, among them substantial reductions in greenhouse gas emissions.

options for owner-builders and contractors, examining how they are made as well as the pros and cons of each.

Building with Earth

Many natural buildings are made from minimally processed earth, excavated on-site or nearby.

Adobe

Adobe has been used to build homes, churches, and other buildings for centuries throughout the world. Many thousands of adobe structures remain standing today in China, the Middle East, North Africa, South America, Central America, and the United States. Many are still occupied.

Traditional adobe structures were fashioned from bricks made from local subsoils containing clay and sand, with straw added to increase their strength. This material is wetted, mixed, and then poured into block forms. After the bricks have dried a bit, they are removed and left in the sun to complete the process.

Adobe bricks are laid in a running bond (overlapping pattern for strength) on a foundation and mortared in place (using the same mud that's used to make the bricks). Walls are then typically finished with an earthen plaster to protect them from the erosive forces of rain and wind. Although cement stuccos have been used in place of earthen plasters in recent years, the results have often proved disastrous (figure 9-2).

BUILDING NOTE

Earthen plasters are ideally suited to adobe construction because they expand and contract at the same rate as the adobe bricks. This, in turn, reduces cracking. Earthen plasters are also well suited to adobe because they allow water vapor to escape through a wall while providing an effective barrier against liquid water. (Earthen plasters perform a little like Gortex rainwear.)

Figure 9-2.
Contrary to popular belief, cement stucco turned out to be a bad choice—an extremely bad choice—as a finish for adobe bricks. The materials expand and contract at different rates, causing cracking. Cement stucco also traps moisture inside, causing the adobe to "melt."

Source: Cedar Rose Guelberth

Contrary to what many think, adobe buildings are suitable for many climates, even colder ones, provided exterior walls are insulated to reduce heat loss. The same goes for many other earthen building materials.

BUILDING NOTE

Adobe, like other earthen building materials, is ideally suited to passive heating and cooling. It provides, thermal mass required for these natural conditioning techniques to work, although it has a slightly lower heat capacity than concrete and standard masonry products.

While U.S. builders construct million-dollar adobe homes, in less developed nations, such as Mexico, adobe is the material of choice among the poor. That's because the material is locally available and inexpensive labor is abundant.

Pros and Cons of Adobe. Adobe building is relatively easy to master. Once you get the mix right, work proceeds very quickly. Beginners find it quite forgiving. If you make a mistake, you can tear a section of a wall down and start again. If you don't want to make your own bricks, you may be able to purchase them from a local supplier (especially if you live in the desert Southwest), or you can rent a machine that makes bricks on-site at a pace that would make a traditional adobe brickmaker's head spin.

Most adobe building in this hemisphere has occurred in the warm climates of Central and South America and the southwestern United States. Although it performs optimally in desert climates with hot, dry summer days and mild winters, adobe can function well in many colder regions as well, provided walls are insulated to reduce heat loss. In New York State, for example, there are an estimated forty adobe homes. Paul Revere's home in Boston was built from adobe bricks.

Besides being suitable for a range of climates, adobe is amenable to many architectural styles. In the United States, however, most new adobe homes are built in the classic southwestern style.

Adobe, like other earthen building materials, is ideally suited to passive heating and cooling (see sidebar). Further adding to its long list of advantages, adobe is fireproof.

Despite its many benefits, adobe does have a few disadvantages. For one, adobe construction is fairly labor intensive. Making adobe bricks and building houses from them requires a great deal of time.

Because building with adobe requires an inordinate amount of labor, these homes can also be costly. In fact, in the United States, where labor costs tend to be fairly high, adobe homes are generally expensive buildings constructed by custom builders for a wealthy clientele. Bear in mind, however, that the exorbitant cost of many of these structures is not just the result of high labor costs. Many of these homes are large, complex buildings with fancy appointments—expensive tile, cabinetry, fixtures, and appliances. All of these factors add up to produce buildings with expensive price tags. Adobe itself is pretty cheap, and if you do the work yourself you can build an adobe home fairly economically.

Cob

Another earthen building material is English cob, also known as *cob* or *monolithic adobe. Cob* is the English word for a lump or rounded mass.

Cob homes are built from lumps of mud with the same constituents as adobe: sand, clay, and straw. Rather than using the mix to produce blocks, however, the mud is applied to the foundation directly, often by hand or by the shovelful (figure 9-3). The walls are then massaged into shape by hand. Cob construction therefore lends itself to sensuous curved walls, arches, and niches.

"A cob cottage is the ultimate expression of ecological design," write cob builders Ianto Evans, Michael Smith, and Linda Smiley in *The Hand-Sculpted*

Figure 9-3 a & b.
Cob is composed of the same materials as adobe, but it is placed on the walls in loaves (cobs) that are massaged in place. The result is a monolithic adobe structure often with thick, sensuous, curved walls.

Source: Dan Chiras

House. "Made of the oldest, most available materials imaginable—earth, clay, sand, straw, and water—cob houses are not only compatible with their surroundings, they are their surroundings, literally rising up from the Earth. They are light, energy efficient, and cozy, with curved walls and built-in whimsical touches. They are delightful. They are ecstatic."

Cob homes can be as much an expression of artistry as a place to live. Becky Bee, author of *The Cob Builder's Handbook,* says that cob construction is "like hand-sculpting a giant pot to live in." But don't think the walls are fragile. They're not. In fact, cob walls are usually at least 4 inches and as much as 24 inches thick and are literally as solid as a rock. When cob dries, it becomes as hard as sandstone. Cob walls are whitewashed, lime plastered, or coated with an earthen plaster to protect them from the weather.

Pros and Cons of Cob. Cob is ideal for owner-builders, as most of the work is done by hand or with simple hand tools. Cob building is great fun and, as noted above, permits extraordinary freedom of expression. Cob is durable, too. Many thousands of cob buildings in southern England have been continuously occupied for the past five hundred years! Cob is also ideally suited for passive solar heating and cooling because of the thermal (heat-absorbing) mass of interior and exterior walls. Furthermore, cob is suitable for many climates. In rainy areas, however, special care must be taken to protect the exterior walls from driving rains. Good overhangs and lime-plaster finishes work admirably in such instances. In cold climates, excessive heat loss through uninsulated walls makes cob a less practical alternative, although there are some ways to safeguard against this problem—for example, by making the walls extra thick or installing insulation inside the walls.

Like other natural building techniques, cob construction is gaining in popularity. More and more professional builders are available to build your home or help you out. Some, like Ianto Evans, Linda Smiley, and Michael Smith, offer hands-on workshops that bring in volunteers to help build your walls. As is the case with many other natural building techniques, there are numerous books and videos on cob building, which I've listed in the resource guide at the end of the book.

History of Rammed Earth Building

Rammed earth is an ancient technique. In fact, rammed earth buildings dating back to the seventh century B.C. have been discovered in China. Parts of the Great Wall of China, begun over five thousand years ago, were made from rammed earth. Ancient rammed earth buildings are also found in North Africa and the Middle East, where the practice continues today. Many rammed earth structures can also be found in France—two thousand years ago, it was the dominant form of building in the Rhone River Valley. Historians believe the Romans introduced this building technology to the picturesque valley.

Rammed Earth

Another traditional natural building material is rammed earth. It is typically a mixture of clay and sand and a little Portland cement. Some of the newer generation of rammed earth builders use sand stabilized with a small amount of cement, rather than clay. As its name implies, rammed earth is soil that has been rammed (or compacted) into wall forms. Workers build a wall by erecting wooden or steel forms on specially reinforced foundations. Once the forms are securely in place, moistened dirt is shoveled into them, about 6 to 8 inches at a time. It is then compacted, often with a pneumatic tamping device. Additional soil is added, then tamped, and so on and so on until the form is filled to the top. Soon after the tamping ends, the forms are peeled off, revealing gigantic blocks of earth, typically 12 to 18 inches thick and 6 to 8 feet long. A new form is then placed next to it and filled with dirt (figure 9-4). Once the walls are formed, the roof, windows, and door frames are installed.

Exterior surfaces of rammed earth walls may be left "raw" or may be coated with a protective layer of plaster. Interior surfaces are often left unplastered to reveal the rich, inviting beauty of this natural material.

Rammed earth building is growing in popularity in the United States thanks to the pioneering work of David Easton, builder and author of the book *The Rammed Earth House.* In recent years, rammed earth builders have emerged in California, Arizona, and New Mexico. In western Australia, one-fourth of all new homes are built from rammed earth.

Pros and Cons of Rammed Earth. Like other earthen structures, rammed earth homes are solid and fireproof and can withstand hurricanes and tornados far better than conventional wood-framed homes. Rammed earth is also ideal for passive solar heating and cooling because, like other earthen structures, it has tremendous thermal mass. Moreover, rammed earth homes are ideally suited for hot, dry climates. In colder climates, they require exterior insulation to prevent excessive heat loss during the winter. If proper attention has been paid to detail, these homes will stay unbelievably cool in the summer and warm in the winter—and they look great, too (figure 9-5).

On the downside, rammed earth construction requires extensive formwork and heavy equipment and is therefore more suitable for contractors who build a number of homes each year than for owner-builders. While a contractor can recoup the cost of expensive forms by using them over and over, an owner-builder cannot. In addition, finding a contractor outside of the desert Southwest and California may be difficult. Nonetheless, there are ways for individuals to produce rammed earth homes economically—for example, by hand-tamping and reusing lumber required to build the forms.

Rammed Earth Tires

Americans discard nearly 250 million automobile tires each year. Although some are recycled or shredded and burned, many of them end up in tire dumps, where they may spontaneously ignite, creating fires that smolder for years. What if those tires were used to make homes?

Preposterous, you say?

Figure 9-4.
Workers ram earth into a form using a pneumatic tamping device. Once the dirt is compacted, the forms are removed, producing a solid, massive, beautiful exterior wall.

Source: Dan Chiras

Figure 9-5.
An exquisite rammed earth home built by Pat and Mario Bellestri of Soledad Canyon Earth Builders.

Source: Soledad Canyon Earth Builders

Figure 9-6.
Rammed earth tire homes use tires as forms. Tires are laid flat on solid ground or a foundation. Dirt is shoveled into them and then compacted. Once the tire walls are completed, they are covered with earthen plaster. There's no odor and no fire danger.

Source: Dan Chiras

Americans discard nearly 250 million automobile tires each year—enough to produce 30,000 to 40,000 2,000-square-foot homes a year!

Not really.

I live in a house made from eight hundred used automobile tires gathered from local tire shops.

Building homes with tires, called *rammed earth tire construction,* is not entirely natural, but it does have many redeeming qualities worth serious consideration.

In rammed earth tire construction, used automobile tires—free of those potentially troublesome chemicals you smell in a tire shop, which have long since been released—are laid on compacted subsoil or a foundation. A small piece of cardboard is placed over the hole at the bottom of each tire. The tires are then filled with dirt from the site, which is then compacted with a sledge hammer or a pneumatic tamping device (figure 9-6). After considerable pounding or tamping, the tires are fully compacted. (A tightly packed 15-inch automobile tire will hold 300 to 350 pounds of dirt.)

After the first row is completed, a second row is laid on top of the first course in a running bond pattern (as with adobe bricks, overlapping for strength). A piece of cardboard is placed over the opening to prevent the dirt from escaping. One by one, the tires are filled with dirt and then compacted. Six to eight rows of tires make up a wall, as shown in figure 9-7.

The completed wall is finished with mud plaster or cement stucco and wary neighbors who've been anxiously watching a tire home being built will breathe a sigh of relief. They now know that the house isn't going to look that bad after all!

Tire homes are the innovation of maverick architect and builder Michael Reynolds of Taos, New Mexico. He's been building his brand of homes, called *Earthships,* since the mid-1970s.

Pros and Cons of Tire Homes. Rammed earth tire homes put an abundant waste material to good use, diverting it from dumps and providing years of useful shelter. They also reduce wood use. Filling tires with dirt from the site uses an abundant, locally available, natural building material.

Another advantage of rammed earth tire homes is comfort. Most such buildings are earth sheltered and passively heated and cooled, techniques described in chapters 10–12. The combination of earth sheltering, south-facing glass for solar gain, and highly insulated ceilings produces a home that stays warm in the winter and cool in the summer. This feels great, reduces energy bills, and saves homeowners thousands of dollars over the lifetime of the house.

Figure 9-7.
A completed tire wall on the author's home. Although neighbors were fearful that this might be the final product, we covered this wall with cement stucco to create the lovely curved interior walls.

Source: Dan Chiras

Figure 9-8.
One of the author's students tours a lovely Earthship in Taos, New Mexico.

Source: Dan Chiras

Properly built rammed earth tire homes, like the conventional earth-sheltered homes described in chapter 10, are bright and cheery. Earthships have an open design and can be quite elegant (figure 9-8).

Despite their many benefits, some disadvantages should be acknowledged. Perhaps most significant of these is that packing tires can be very laborious, although you can speed up this process by using a pneumatic tamper. Special precautions must be taken to protect earth-sheltered walls from ground moisture. And not all building officials will look kindly on this unusual form of architecture. Some people have trouble with the appearance of Earthships, asserting that they appear too "funky" for their tastes. With ingenuity, however, this unusual building technique can be used to produce more conventional-looking homes (figure 9-9).

Because Earthships incorporate so many principles of green building, I recommend that you study them carefully to appreciate Reynolds's integration of systems, which is aimed at creating totally independent and sustainable shelter. What you learn can be applied to virtually any type of home—and should be.

Earthbags

Earthbag construction is a newcomer to the natural building movement. Earthbags are a versatile and durable building material whose use has been pioneered principally by Iranian-born architect Nadir Khalili, who lives in California, and two innovative builders, Kaki Hunter and Doni Kiffmeyer, who make their home in my favorite mountain-biking town, Moab, Utah.

Earthbags are made from rejected polypropylene bags (the kind bulk rice comes in) or burlap bags. Bags are placed on a foundation, filled with

Figure 9-9.
The author's passive
solar/solar electric rammed
earth tire and straw bale
home. Not at all what some
neighbors anticipated when
they learned that the home
was going to be made from
these unusual materials.

Source: Dan Chiras

moistened clay-dirt or cement-stabilized dirt, and then tamped (figure 9-13). Soon, the bag flattens, forming a blocklike structure that dries as hard as rock.

Like adobe blocks and rammed earth tires, earthbags are set in a running bond, then covered with mud or lime-sand plaster, both of which adhere well to the surface.

Pros and Cons of Earthbag Construction. Earthbag construction, like rammed earth tire construction, is ideally suited to owner-builders. Earthbags can be used to make foundations in drier climates. They can also be used to build walls, sheds, and entire homes. Earthbags are ideal for mak-

Figure 9-10 a & b.
Earthbags are placed on the
foundation, filled with dirt (a),
and tamped (b). When com-
pacted, the dirt forms a
nearly bricklike block that is
ideal for exterior walls of
homes. Shown here is an
earthbag grade beam (an
above-grade component of a
foundation) for a small shed
that houses the author's
backup generator.

Source: Dan Chiras

ing round structures with dome roofs or creating vaults (figure 9-11). Their relatively high mass makes them ideal for passive solar heating and cooling.

Despite these and other benefits, earthbag construction is slow and laborious. A lot of pounding is required! This new technology is also largely untested, so obtaining approval from a local building department may be difficult.

Figure 9-11.
Earthbags are ideal for building domes and vaulted structures like this Honey House in Moab, Utah, built by Doni Kiffmeyer and Kaki Hunter.

Source: OK OK OK Productions

Cast Earth

Cast earth is the newest earthen building technology. Cast earth homes are made from a slurry of soil mixed with 10 to 15 percent heated (calcined) gypsum. The slurry is then poured into forms set on a foundation. After the material sets up, usually that same day, the forms are removed and the house is framed in. Cast earth walls are typically coated with plaster or stucco for protection and aesthetics (figure 9-16).

Pros and Cons of Cast Earth. Cast earth homes are beautiful and functional. Like other earthen building techniques, this method is ideally suited to passive solar heating and cooling. Also like other earthen building materials, cast earth is ideal for hot, dry climates. With external insulation, cast earth homes are also suitable for colder climates. Because the house is poured and the slurry hardens quickly, exterior walls go up rapidly. This process therefore lends itself to mass production.

The main problem with cast earth is availability. Unlike all other natural building materials, cast earth building is a proprietary process. You or I can't build a cast earth home. It must be built by a builder who has been certified by the originator of the idea, Harris Lowenhaupt, a Phoenix-based

Figure 9-12.
This beautiful cast earth home is yet another option for those wishing to build with less lumber.

Source: Cast Earth Affiliates

metallurgist who invented the process in 1993. (Lowenhaupt carefully guards the identity of one component in the mixture, a chemical that's added to the mix to prevent it from setting up too quickly.)

Straw Homes

In the mid-1980s, I got my first introduction to straw bale building. I was invited to some friends' house to view a video on straw bale construction. I have to admit, the idea of building a house from straw bales seemed odd, even silly, at first. However, the more I studied the technique the more feasible, even sensible, it seemed, and the more interested I became, so much so that when I built my new house in 1995, I added a straw bale living room.

Straw Bale

Straw bale building is a relative newcomer to the natural building scene. It began in the late 1800s in the Sand Hills of western Nebraska. Because wood was—to put it politely—scarce on the wind-swept plains and the soil was too sandy to make sod homes, early settlers used the only other resource they had to make homes: hay. They used not loose hay but, rather, baled hay, thanks to the recent invention of the mechanical baler. Later, they started using baled straw to build shelter.

Most settlers saw their bale homes as temporary structures and fully intended to build a "real home"—like those of their eastern cousins—when they could afford the wood. (Wood had to be shipped in at considerable expense.)

Over time, however, many of these straw bale pioneers found that their bale homes actually performed well, protecting them from the cold winter

Living Lightly, Living Independently

Much of the interest in natural building techniques such as straw bale that is currently growing in more developed countries like the United States, Canada, Australia, and Europe stems from a desire to live more lightly on the land, and to live independently.

BUILDING NOTE

Studies show that a load-bearing straw bale wall will support nearly seven times *more* weight than a 2 x 4 stud wall!

Figure 9-13. Straw bale buildings like this home in the Sand Hills of Nebraska put a local resource, straw, to use and could save millions of trees and huge amounts of fossil fuel because they're so energy efficient. They're also extremely comfortable to live in.

Source: Catherine Wanek

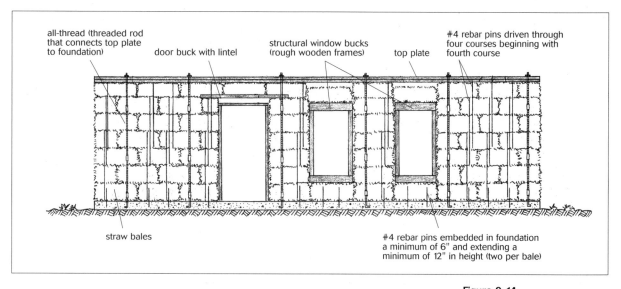

all-thread (threaded rod that connects top plate to foundation)

door buck with lintel

structural window bucks (rough wooden frames)

top plate

#4 rebar pins driven through four courses beginning with fourth course

straw bales

#4 rebar pins embedded in foundation a minimum of 6" and extending a minimum of 12" in height (two per bale)

Figure 9-14.
In a load-bearing straw bale wall, such as the one shown here, straw bales form the support for the roof and provide insulation as well.

Source: Michael Middleton

winds of western Nebraska. Abandoning thoughts of building wood-frame homes, many of them plastered their straw bale walls, converting their temporary shelters into permanent domiciles. Remarkably, several of these structures are still standing today (figure 9-13).

The straw bale is the basic building block of the homes that bear its name. Straw bale homes come in two basic varieties: load-bearing and non-load-bearing.

As shown in figure 9-14, a load-bearing wall is made principally of straw bales. They support the weight of the roof (roof load) and provide wall insulation. The bales are laid on a waterproofed foundation and stacked in a running bond pattern, like adobe blocks, to "lock" the bales together and increase the rigidity of the wall.

During wall construction, additional rigidity is provided by pinning the wall. Pins are made of steel rebar or natural materials such as bamboo, dowels, and sticks. They may be driven into the bales, a process called *internal pinning,* or attached to the inner and outer surface of the bale walls, a process known as *external pinning.*

In load-bearing straw bale walls, a wooden top plate is secured to the top row of bales. It provides a means of attaching the roof to the walls and also distributes the weight of the roof evenly on the walls. This, in turn, prevents buckling of the walls due to uneven pressure. To anchor the roof to the walls, however, the top plate in a load-bearing wall needs to be securely attached to the foundation.

The non-load-bearing straw bale wall consists of a supportive structure, typically a post-and-beam frame. Made of posts, logs, or concrete blocks, the frame bears the weight of the roof. Straw bales are piled on the foundation in a running bond pattern between the posts and the beams, where they provide insulation. Consequently, this technique also goes by the name the *in-fill method.*

BUILDING NOTE

Current thinking is that external pinning of straw bales is the best option. It gives the wall the most rigidity and resistance to lateral loads created by winds. Unfortunately, unless the pins are tightly cinched to the wall, they're often difficult to plaster over.

Figure 9-15.
This straw bale home in Kanata, Ontario, fits beautifully with its surroundings and stays warm in the winter and cool in the summer thanks to the thick walls of straw.

Source: Catherine Wanek

Pros and Cons of Straw Bale Building. Straw bale building is the most popular natural building technique, except for log home construction, and offers numerous benefits. One that many straw bale advocates promote is energy efficiency. Thick straw bale walls greatly curtail heat loss, especially when combined with well-insulated ceilings, energy-efficient windows and doors, and airtight construction. When designed for passive solar heating and passive cooling, a straw bale home is truly one of the most comfortable and efficient homes on the planet.

Recognizing the imperiled status of the world's forests, many people see straw bale construction as a means of reducing wood consumption. Except for roof framing and construction of interior walls, straw bale homes require very little wood. Using cob or other natural materials for interior walls can cut down on wood use even further, as can other materials and techniques outlined in chapter 5.

Further adding to its appeal, straw bale building provides a means of putting a waste material to good use. In many farm fields throughout the world, straw from crops such as wheat and rice is viewed as a waste product of grain production. After the harvest, farmers often burn their fields. This gets rid of the straw, which can gum up plows, and returns mineral nutrients to the soil. Building houses out of straw bales, therefore, seems like a good way to utilize this "waste" and to reduce clouds of air pollution billowing from the nation's farm fields.

Straw bale construction appeals to many because straw is a locally available resource in most parts of the world. Unlike wood, which must be shipped hundreds, sometimes thousands of miles, straw often can be supplied by a farmer within a short distance from the building site. Buying locally lowers the embodied energy of a straw bale home.

In recent tests, two-string bales (bales with two strings rather than three strings of bailing twine holding them together) laid flat on a wall were shown to provide about R-32, an impressive level of wall insulation. A 2 x 6 wall filled with fiberglass or cellulose insulation would weigh in at roughly R-20 to R-22.

Straw bale building is also a relatively simple procedure that can be mastered by many, unlike the complicated framing techniques required for standard stick-built homes, which may take years to master. It is conducive to owner-build projects, enabling many people to pursue the time-honored tradition of building one's own shelter.

Straw bale building, like other forms of natural building, is also seen as a way to provide inexpensive shelter for many, a positive benefit in a world where average new home prices have skyrocketed beyond the reach of people of ordinary means.

Another benefit of straw bale construction has become quite evident: its elegant beauty. Some folks like straw bale for a simpler reason: because it's fun. Often carried out communally through wall-raising parties or workshops, straw bale building builds personal and community relationships as well. The same can be said about many other natural building techniques.

Straw bale walls coated with plaster rarely have problems with mold and are quite durable—so long as common-sense building practices are followed. After plastering, straw bale walls are extremely fire resistant. Thick walls of straw also drown out external noise.

Straw bale building is an approved form of construction in many jurisdictions. Obtaining a building permit for a straw bale home is generally no more difficult or costly than for any other type of home. Mortgages and insurance are also available in many places, and straw bale resale values appear to be respectable.

Despite its many benefits, straw bale does have some shortcomings. Straw is the shaft of cereal crops, such as wheat and rice. Many farmers use enormous amounts of irrigation water and tons of pesticides to grow their crops. Using straw, rather than plowing it under, also robs the soil of nutrients. Straw bale building is not always as inexpensive as some would have you believe (see sidebar). And if not protected well from water, straw bale walls can mold and deteriorate.

Although straw bale does have some inherent disadvantages, its popularity continues to skyrocket.

Straw–Clay

Straw-clay has a long and successful history of use in Europe, primarily Germany, dating back over five hundred years, and has recently been introduced in the United States, thanks primarily to the efforts of New Mexico–based architect and builder Robert LaPorte.

Straw-clay is made from straw and a water-clay mixture known as *clay slip*. The two are mixed together by hand or machine until clay slip lightly coats the straw. The straw-clay is placed between forms attached to the exterior wall frames. It is then tamped by hand (figure 9-16). Once the forms are filled, they are removed, and the wall dries slowly. After the walls are thoroughly dry, they are coated with plaster, usually an earthen plaster.

BUILDING NOTE

While finished straw bale walls are extremely resistant to fire, they are very vulnerable to fire during construction. Loose straw on the site and the exposed walls can catch fire easily, so take care to avoid open flames and sparks. Clean up loose straw after each workday.

Economics of Straw Bale Building

Straw bale construction can range in price depending on how much work an owner does and how much he or she must contract out. Cost also depends on the complexity of the design. The more complex, the more costly. In addition, cost depends on details, for example, the type of tile, the type of cabinetry, the amount of finish work, and so on. Custom-made straw bale homes often cost well over $100 per square foot.

Wall panels made from straw-clay are typically nonstructural—that is, they act as a thick layer of natural insulation between structural members of the wall. The more insulation you need, the thicker the walls should be.

Pros and Cons of Straw-Clay. Straw-clay walls are ideal for owner-builders. They are fireproof and, when covered with plaster, look fantastic. Moreover, they provide excellent insulation and, therefore, have a valuable role to play in building energy-efficient, passively heated and cooled homes. When interior surfaces of exterior walls are coated with a 1- to 2-inch-thick layer of earthen, lime, or cement plaster, they provide a good source of thermal mass.

On the downside, straw-clay is fairly labor intensive. Also, although this seemingly unusual material has been used in Germany for centuries, it is not a widely known technique in North America, so obtaining approval from a local building department may be problematic.

Figure 9-16.
A workshop participant packs straw-clay into forms at a natural building workshop taught by the author in northern New Mexico.

Source: Dan Chiras

Other Forms of Natural Building

The list of natural building materials does not end with earth and straw. There are at least four others worthy of consideration: stone, papercrete, cordwood, and log.

Stone

Stone is one of the oldest of all natural building materials. It was used in Europe to build massive castles, aqueducts, walls, churches, barns, garden walls, sheds, roads, and walkways. Except for the addition of mortar, stone building has not changed since its beginnings.

Stone walls are laid on sturdy foundations, which are usually made of gravel or stone. Rocks are laid either dry or with cement or earthen mortar in a running bond. Mortar is not a glue to hold poorly fitted stones together but, rather, a bonding agent that holds well-fitted stones in place. It reduces air infiltration and prevents pests from entering a home.

Pros and Cons of Stone Building. Homes built from stone are beautiful and durable, potentially able to last for many centuries. Stone is a locally available material, too, often harvested at little cost—just labor! In addition, stone walls can support a great deal of weight.

MAKING STONE BUILDING EASIER: THE SLIPFORM METHOD

Piling stones precisely on top of one another in a running bond pattern is difficult work that requires considerable skill and money. If you are interested in building with stone but lack the experience, time, and financial resources required to build the traditional way, don't despair. There are options. One of them is known as the *slipform technique.*

A slipform is a wooden form mounted on a foundation. It is used to build stone foundations and stone walls.

In the slipform method, wooden forms are first placed upright above the foundation, then secured with braces and wires. After the forms are in place, squared and plumbed, stones are selected.

Stones are typically laid along the outer face of the form, one course at a time. Concrete is poured into the form after each course to fill the rest of the cavity. As a general rule, the largest stones are used for the lower portions of the wall, both for strength and appearance.

In this technique, walls are laid up in one-and-a-half-foot sections. After a section is completed, the form is raised—or "slipped" up—and work continues.

The slipform method is discussed in detail in Karl and Sue Schwenke's book, *Build Your Own Stone House Using the Easy Slipform Method.* Although this technique makes wall building easier, it does require construction of wooden forms—at the very minimum eight, but usually more. This, in turn, means more wood, more labor, and higher costs, all of which may be justifiable if they help you build better walls. Slipform stone walls also look surprisingly different from stones laid up the old-fashioned way. Furthermore, the inside surface of the wall is concrete and must be coated with plaster or some other material.

Stone construction is ideal for passive solar heating and cooling. Not only can stone be used to build exterior walls, this abundant and versatile material can be used to build interior walls, floors, fireplaces, and masonry heaters. These elements serve the functions for which they're designed but also provide internal thermal mass.

Uninsulated exterior stone walls function best in hot, arid climates. Like the earthen walls described earlier, unless they're insulated or extremely thick, external stone walls lose too much heat in cold climates to be practical.

Despite its many benefits, stone building does have drawbacks. For one, building with stone is slow and extremely difficult work, although there are ways to do it that reduce the labor and time (see the box above).

Stone is heavy and can cause hernias and significant back problems if one is not careful. (I'll spare you the motherly advice on picking up heavy objects.) Like concrete, stone exhibits great compressive strength but offers little tensile strength. In seismically active areas, steel reinforcement is required to compensate for this weakness. Steel rods are laid in the mortar joints between adjacent courses.

Another disadvantage of stone building noted above is that stone homes can be quite cold. Stone, like earthen materials, is a good conductor of heat. As a result, on cold days heat passes right through stone walls, creating a cold interior. In addition, moisture in the indoor air often condenses on the cold stone walls, creating a potential health problem for occupants, as moisture serves as a breeding ground for mold and mildew.

To avoid these problems, some builders create double stone walls. That is, they build two walls on the foundation separated by an airspace. Air is a poor conductor of heat and therefore retards the movement of heat from inside a home. Insulation can also be inserted into the cavity between the two stone walls to further reduce heat loss.

Insulation can also be provided by building exterior stone walls alongside conventionally framed walls. After the stone walls are finished and the roof is on, the interior walls are insulated and drywalled. The result is a wall with much greater thermal resistance (R-value) than an ordinary stone wall.

Papercrete

Papercrete is probably among the newest of the natural building materials. It is made from recycled newspaper, sand, cement, and water. To make papercrete, you begin by cutting newspaper into small pieces, which are then soaked in water for a day or so. Once the paper starts to break down, sand and cement are added, and the mixture is stirred to create a thick slurry.

The slurry can be poured into block forms—similar to those used to manufacture adobe bricks—or poured into wall forms mounted on the foundation, similar to those used in rammed earth or slipform stone construction.

Once poured, papercrete begins to set up, slowly transforming from a slurry into a solid block that offers both mass and insulation—not as much mass as a rammed earth or adobe wall, but much more than a straw-clay or straw bale wall. If blocks are made, they are mortared in a running bond pattern, then plastered.

Papercrete is easy to make—my eight-year-old son made some for a science project a couple years ago—and utilizes a ubiquitous waste. Blocks are lightweight and easy to lay up in a wall (figure 9-17). However, because papercrete is so new there is much to be learned about its performance in different climates. My guess is that it would function well in mild to moderate climates, but in severe climates it might require additional (external) insulation for optimal performance.

Cordwood Masonry

Cordwood masonry is a method for building low-cost, natural homes from logs laid perpendicular to the wall (figure 9-18). To begin construction, cement mortar is placed on the foundation in two parallel bands, one on the inside and the other on the outside. Appropriately sized pieces of cordwood are placed on the mortar, then worked into

Figure 9-17. Papercrete blocks, one of the newest "alternative" building materials, can be used to build exterior walls of homes.

Source: Gordon Solberg

Figure 9-18.
Workshop participants work on a cordwood wall. The cavity between the inner and outer layers of mortar will be filled with insulation, usually sawdust. Logs are worked into the mortar to create an airtight wall.

Source: Richard Flatau

it. After the logs are secured, insulation (usually sawdust) is poured into the space between the inner and outer mortar joints. After the first course is laid, the second is laid down in similar fashion, starting with mortar. When it is completed, another layer is added and so on until the wall is complete.

The first cordwood homes were built in the United States in the mid-1800s. Where the technology came from is anyone's guess. In recent years, cordwood has gained popularity among owner-builders in large part due to the efforts of Rob and Jaki Roy of northern New York State and Richard Flatau of Canada, among others.

Pros and Cons of Cordwood Construction. Besides using a locally available material, cordwood construction is easy to learn and is a practical and economical building option for do-it-yourselfers. Cordwood walls are strong and durable. Cordwood structures built a hundred years ago are still standing in good condition.

Cordwood offers the rustic look of stone masonry and can be quite pleasing to the eye (figure 9-19). Cordwood walls offer other benefits as well, specifically, mass and insulation. Thermal mass is provided by the cement mortar and the logs themselves. The logs and the sawdust between the inner and outer mortar joints provide insulation. The R-value of a cordwood wall depends on its thickness and the type of wood that's used. A 16-inch wall has an R-value ranging from 16 to 20, plus some mass for holding heat and maintaining thermal stability inside a home.

Cordwood homes are being built primarily in heavily wooded areas with cold, wet climates, like upper New York State and southeastern Canada. Here this unusual building material performs admirably, working well with passive solar design.

Figure 9-19.
The cordwood home of Rob
and Jaki Roy, in upstate New
York. Two of America's pioneering cordwood builders,
they offer workshops on
cordwood building.

Source: Rob Roy

Although it is a suitable building technique, cordwood construction is fairly labor intensive and not well known outside a few areas in North America, which can make it difficult to acquire building department approval.

Log Homes

Log homes are by far the most popular natural homes, with approximately 70,000 built each year in the United States, often as prefabricated "kits" delivered and assembled on a site. Like other forms of natural buildings, log homes have been around for centuries. Early American settlers built log cabins, barns, and homes. Log home building, however, is not an indigenous technology. Early settlers brought this technique from their native homelands, especially Finland, Sweden, and Germany.

To build a home, logs are laid on a foundation, often made from stone or concrete. In the early days of log home building, logs were laid down one at a time with corners joined by notches. After the walls were erected, the

Figure 9-20 a & b.
In log construction, spaces
between logs (a) are packed
with mud (b) to prevent cold
air from entering, as in this
log cabin in the mountains of
northern New Mexico.

Source: Dan Chiras

spaces between the logs were filled with a material referred to as *chinking* (figure 9-20). A mud mortar was typically used. These walls looked nice, but they performed poorly from a thermal standpoint. Chinking cracked, and these cracks often let cold air enter. Chinking also permitted heat to be conducted to the outside.

To address this problem, the Swedes invented chinkless log construction. In this technique, logs are notched along their length so they fit tightly on top of one another, eliminating the gap and the need to apply chinking between logs (figure 9-21). This innovative design has resulted in a more airtight and energy-efficient wall and a more comfortable home. Although notching grooves in logs requires more time up front, it does reduce maintenance as well—notably, periodic chinking needed to repair cracks.

Pros and Cons of Log Homes. Log homes offer many benefits. One of the most noteworthy is their exquisite beauty (figure 9-22). Part of their charm arises from their rustic, back-to-the-earth look—an aesthetic that appeals to people who've grown tired of the rectilinear world of conventional stick-frame construction.

Log home construction is practiced widely throughout North America. If you're unfortunate enough not to have access to a competent builder, you can always build one yourself, cutting logs yourself or building one from a log-home kit. Kits are widely available at relatively affordable prices.

Unfortunately, log homes can be one of the least sustainable building options. Although logs are natural and require far less processing than the 2 x 4s and 2 x 6s used in conventional construction, many log homes are built from trees felled in distant forests. The logs are debarked and then transported hundreds of miles to the job site. And many of the logs used in kits are obtained from companies practicing unsustainable logging. If your heart is set on a log home, you will want to obtain locally harvested trees that are sustainably grown and harvested.

On the plus side, log walls offer thermal mass as well as insulation and perform as well or slightly better than conventionally insulated framed walls in respect to heat loss. When the U.S. National Bureau of Standards performed tests on log walls, they found that a 6-inch-thick log wall equaled or exceeded the energy performance of any other type of exterior wall during all seasons tested except in the dead of winter. In the winter, conventionally insulated wood-frame walls won by a small margin.

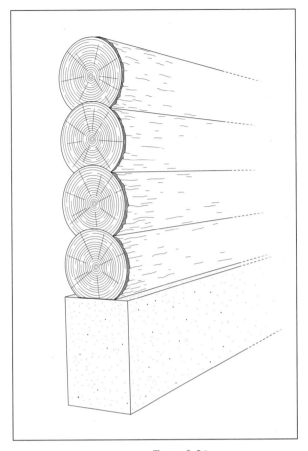

Figure 9-21.
Today, log builders carve out logs so they nest, leaving little room for air infiltration, as shown here.

Source: Lineworks

Figure 9-22.
Log homes are some of the most beautiful and most popular natural homes. Be sure the logs are harvested locally and in a sustainable manner.

Source: Lodge Log Homes

Jim Cooper, author of *Log Homes Made Easy* and numerous articles on log homes in popular magazines, notes that "Too many people see the tag 'energy efficient' as a license to ignore sensible energy conservation measures. They run amuck with energy-inefficient cathedral ceilings, glass in the wrong quantity and the wrong place, and disregard for the role of house siting in maintaining energy efficiency." Don't fall into that trap with this or any building technology.

What Is the Best Natural Building Material?

Every natural building enthusiast has a favorite material and technique. For years, proponents of various technologies have vigorously defended their favorite materials while denigrating others. In time, though, many proponents have learned the value of other materials and even begun to use them when building with their preferred approach. It is not uncommon for a natural house builder to use two or more natural building options in combination.

If you are interested in natural building, I recommend that you be careful not to be swayed too early in your exploration. Take a long, hard look at all of your options. Visit projects or completed homes. Enroll in workshops. Read books. Study each natural building material carefully before deciding.

My book, *The Natural House,* provides a comprehensive and unbiased examination of each of the natural building options, listing the pros and cons of each. It supplies sufficient detail to help you narrow down your choices and also contains an extensive resource guide. If you have specific questions on natural building, you can log on to www.greenbuilding.com and click on "Ask the Experts." Your question will be routed to an expert in the field who will respond to your query. I respond to questions on passive solar design and general queries on natural building. This web site also contains a wealth of information on other aspects of green building. For a list of books that provide more detailed information on individual natural building techniques, be sure to see the resource guide at the end of this book.

Natural Building: Back to the Future

Despite their offbeat status, many natural home building techniques have a history much longer than that of the new, supposedly improved building practices we rely on today. And many natural buildings could outlast contemporary buildings—perhaps enduring for centuries—further adding to their appeal.

Earthen and straw walls are energy efficient and ideal for passive solar heating and passive cooling, a topic we'll explore in more detail in chapters 11 and 12. Combined with proper siting and other essential design features, including good windows, airtight construction, and good ceiling insulation, they can provide a lifetime of comfort at little cost.

Natural building is not a panacea. And natural building techniques have their shortcomings, as acknowledged in this chapter. In addition, natural building may seem odd at first—but visit a few homes and you'll see they're often elegant and comfortable, and, except for the newest techniques and materials, which haven't stood the test of time, many promise economical, durable, long-lasting shelter. As time goes on, more and more builders are devoting their lives to creating homes from natural materials. The number of architects and engineers involved in the movement is growing, too, and there's a wealth of information on natural building, from videos to books to articles. But don't expect to look up natural builders in the yellow pages. Chances are there are some in your area, but you will have to find them by asking around or contacting the growing number of organizations dedicated to straw bale building and other forms of natural construction (see the resource guide for a listing).

Over the years, I've been gratified to learn that many building departments are now looking favorably on natural materials and techniques, especially when plans are stamped by a state-licensed structural engineer or architect. I've been equally gratified to learn about builders who are creating entire state-of-the-art developments from straw bale homes.

As our environmental problems deepen natural building could play a larger and larger role in the housing market. Who knows— maybe someday you will live in a subdivision of straw bale homes built by one of the nation's leading home builders. Remember: Natural building may seem like a case of going out on a limb, but as Will Rogers once asked, "Why not go out on the limb? That's where the fruit are!"

CHAPTER 10

Earth-Sheltered Architecture

On January 5, 2002, I returned from a ten-day trip to southern Florida with my two boys to visit their grandparents. While we were away, the nighttime temperature at our home nestled in the foothills of the Rockies, 8,000 feet above sea level, frequently hovered around -20°F. Even so, the interior temperature never dropped below 52°F . . . and get this . . . there was no heating system running while we were gone! How could this be?

This remarkable feat results from the fact that our house is earth sheltered: built into a hillside with much of the roof covered by a thick layer of dirt, with a dense carpet of vegetation growing on top. In the spring and summer, wildflowers blossom in profusion on our roof. At times I can hear elk grazing overhead, nibbling at the luxuriant grasses.

Also contributing to this house's exceptional performance is its reliance on solar energy for heat (chapter 11). South-facing windows allow the low-angled winter sun to enter my home, warming it during the day. Much of that warmth lingers through the night, helping maintain a comfortable interior temperature.

Appropriate for a wide range of climates and a variety of building sites, even flat sites, passive solar earth-sheltered homes cost very little more to build than conventional homes—and, in some cases, they cost less.

Earth sheltering provides superior year-round comfort, keeping those of us who venture slightly underground warm in the winter and cool and comfortable in the summer, and with far less energy than is required in a conventional home. When teamed with passive solar design, an earth-sheltered home can bring you as close as possible to full self-sufficiency in heating and cooling and can save you tens of thousands of dollars in fuel bills over your lifetime.

Properly designed earth-sheltered homes are also great places to live in. They're bright and cheery inside. And they blend naturally with the landscape. Even though there are no windows on the north side of my house, it's difficult to get used to the idea that you are nearly underground. Most visitors are shocked to learn this fact. "You just wouldn't know it," one visitor remarked while touring my home during the 2001 National Tour of Solar Homes. In fact, there's so much sunlight that my boys and I rarely turn on lights during the day.

Earth sheltering represents "a gentler kind of architecture," says author-architect Malcolm Wells, who specializes in earth-sheltered design. Earth-sheltered homes not only are comfortable and affordable, but also can help

Properly designed earth-sheltered homes are great places to live in. They're bright and cheery inside. And they blend naturally with the landscape.

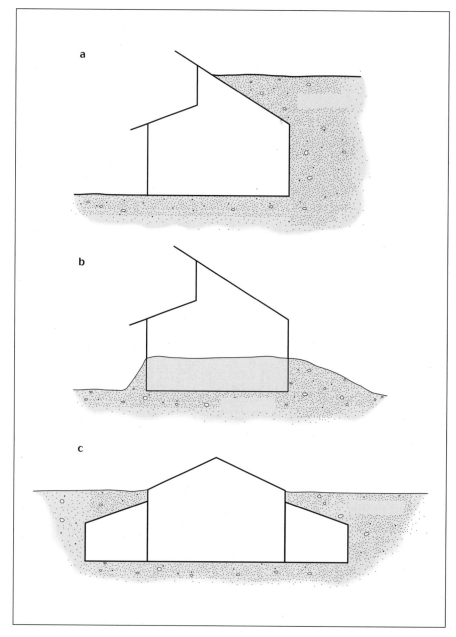

restore the biosphere, the life-support system of our planet. If widely uti-
lized, this technology could help reverse the rampant land destruction so
prevalent today.

Anatomy of Underground Existence

An earth-sheltered home is one that is partially or completely covered with
earth. They generally fit into three basic categories, as shown in figure 10-1.
    First is the partially underground home. Typically built into a hillside, all

LIVING ROOFS

Most roofs in new homes are finished with steel roofing or shingles manufactured from wood, asphalt, or one of the new, more environmentally friendly materials such as recycled plastic or rubber. Those interested in returning more of their land to nature, however, may want to consider a living roof.

A living roof is a rooftop ecosystem consisting of soil and plants over a waterproofed wood or concrete roof. The three most important considerations when designing a living roof are providing adequate support, waterproofing, and drainage.

Living roof construction is a traditional method that's been used for centuries. Living roofs can be nearly flat, domed, or sloped. To ensure that the substructure—the decking and roof framing in wooden roofs or the concrete in concrete roofs—is kept dry, a waterproof membrane must be applied.

Dirt is carefully placed over the roof in a layer from 6 inches to 9 feet in depth, depending on the strength of the substructure. The soil can then be planted with seed. Native plants are highly recommended: Well adapted to local soil and weather conditions, they are more likely to survive the vagaries of the local climate. (Note that the roof may need to be watered a bit during the first year to ensure good seed germination and growth.) Planting wildflowers on the roof will greatly add to its beauty and make your roof the focus of conversation in the neighborhood. Low-growing grasses reduce the need to mow the roof.

Better yet is a "sod planting." Sod can be removed from the ground before the excavation of the foundation or grading of a driveway and replanted on the roof when construction of the structure is complete. Sod plantings greatly accelerate the establishment of healthy living cover on your roof.

Although living roofs are beautiful and functional, you may want to consider designing your roof so that part of it is uncovered to support solar panels for domestic hot water or photovoltaic modules to produce electricity (see chapter 13). I used a clerestory design, which left a portion of my roof available for these valuable functions and also to collect rainwater (see figure 10-2).

of its walls, except the south-facing facade, are embraced by the earth. In such designs, dirt may also be placed on the rooftop, then seeded with grasses, wildflowers, or other native plants. Some folks even grow gardens on their "living roofs."

The second design is the bermed home. Ideal for flat or slightly sloped terrain, the bermed home is built largely aboveground. In this design, dirt is bulldozed up against waterproofed walls, usually to just below the bottoms of the windows. Doorways are left unbermed for obvious reasons. To provide additional earth sheltering, some builders install a living roof.

The third, and least widely built, design lies entirely underground. This type of home is built around an atrium with rooms looking out onto this sunken courtyard to ensure adequate lighting. Visitors and occupants enter via a stairway that descends into the atrium. Skylights, known as *lightwells,* may be installed to deliver light deep into the interior of adjacent rooms for more uniform lighting. Most atria are paved or filled with decorative rocks. Plants may be grown in the atrium, but because so little direct sunlight falls on them, plants tend to fare poorly.

Figure 10-2 a, b, c. In the author's home, shown here, the north side of the house (a) is largely earth sheltered to protect against cold winter winds and to cool the home in the summer. The south-facing roof (b) is finished with metal roofing and is used to collect rainwater and to house solar electric panels. This flower-adorned roof (c) on a nursery and garden supply center in Silverthorne, Colorado, provides incredible beauty.

Source: Dan Chiras

Table 10-1
Comparison of Underground Shelters

	Bermed	Partially Underground	Completely Underground
Visual disruption	Least	Intermediate	Most
Potential for passive solar heating	Excellent	Excellent	Poor
Thermal stability	Good, but least of the three options	Intermediate	Greatest
Protection from wind	Good, but least of the three options	Intermediate	Greatest
Outside views	Excellent in all directions	Excellent, but in one direction only	Very little, principally open sky
Protection from noise	Good, but least of the three options	Intermediate	Greatest
Potential for natural lighting	Excellent	Good to excellent, if done correctly	Least of the three options
Cost	Least expensive	Intermediate	Most expensive

Each design comes with its own set of pros and cons, as outlined in table 10-1. Which design you choose depends on personal taste and a host of other practical factors.

Why Build an Earth-Sheltered Home?

According to Jerry Hickok, president of Earth Sheltered Technology in Mankato, Minnesota, "There are many good reasons why earth-sheltered living may be right for you . . . and the planet." Here are ten of them:

1. **Efficient and comfortable.** Earth-sheltered homes provide year-round comfort, as noted above, but not because dirt is a good insulator. It's not. Soil has an R-value of only about 0.25 per inch, fourteen times less than that of wet-blown cellulose insulation and twenty times less than that of certain types of rigid foam insulation.

    The secret of earth-sheltered homes actually lies in the stability of the Earth's temperature. In colder climates, the subsoil beneath your feet remains a fairly constant 50°F, plus or minus a little, depending on the location. Homes nestled in the Earth benefit from this constant temperature year-round. During the winter, for instance, the heat from the Earth moves into a house. If it is unheated, inside temperatures in an earth-sheltered home in a cold climate such as mine hover around 50°F to 55°F. Although this is too cool for comfort, it's quite easy to boost the temperature to a satisfactory level, say to around 70°F. The modest 20-degree temperature hike can easily be provided by solar heat, as in the case of my house. A conventional aboveground home in a cold climate,

Earth-sheltered homes provide year-round comfort, but not because dirt is a good insulator. Soil has an R-value of only about 0.25 per inch, fourteen times less than that of wet-blown cellulose insulation and twenty times less than that of certain types of rigid foam insulation. The secret of earth-sheltered homes actually lies in the constancy of the Earth's temperature.

on the other hand, will need much more supplemental heat to maintain a comfortable interior temperature. If outdoor temperatures hover around -20°F, an aboveground home will need a temperature infusion of nearly 90 degrees.

Although some sources claim that earth-sheltered homes work best in cold climates, not all builders agree. One of them is Jay Scafe, owner of Terra-Dome, a company located in Green Valley, Missouri. His company has built over five hundred earth-sheltered homes. "In desert climates," Scafe notes, "ground temperatures remain around 70°F, benefitting occupants year-round."

The constancy of the temperature of the ground means that earth-sheltered homeowners never have to worry about water lines freezing, even in the coldest weather. The Earth's thermal stability also translates into exceptional comfort and low heating and cooling bills in the summer.

Although it may be a blistering 95°F outside, earth-sheltered homes remain in the low 70s—as cool as any airconditioned home, and without a noisy, energy-guzzling air conditioner, uncomfortable air currents, and astronomical electric bills.

2. **Economical to build and operate.** The energy efficiency and comfort of an earth-sheltered home are enhanced by the reduced area available for air infiltration. Together, the earth's constant temperature and the reduction in air leakage cut heating and cooling bills by 80 to 90 percent. "Even when passive solar is not incorporated," notes Scafe, "fuel bills are frequently 50 percent lower than for a conventional, aboveground home."

    Not only are they economical to operate, earth-sheltered homes can be quite affordable. Scafe's homes, for instance, run the same as custom-built homes in his area, around $75 to $100 per square foot.

3. **Bright and cheery.** Earth-sheltered homes emit abundant natural light, if thoughtfully designed. Bright interiors create more productive and healthier work space and living area and are pleasant to be in. Discard any notions you might have of a cavelike existence.

4. **Beautiful.** Contemporary builders are creating earth-sheltered homes that are functional and extraordinarily attractive. You'll very likely find a house that suits your taste.

5. **Durable and long-lasting.** Because they are typically built from concrete, rammed earth tires, or cement blocks, earth-sheltered homes resist termites, rodents, rot, and earthquakes. Because they are sheltered by the Earth, they also resist wind, hail, and violent storms, including hurricanes and tornados, all of which can raise havoc with conventional stick-frame buildings. An earth-sheltered home could last well over a hundred years.

6. **Low maintenance.** Because much of the structure is underground and because there is so little exposed surface, earth-sheltered houses require very little maintenance. There's no scraping and painting of siding or periodic reroofing. You won't be cleaning gutters each spring, either.

7. **Fire safe and burglar resistant.** With so little exterior wall exposed to the elements and the use of concrete and natural plasters or cement stucco on exposed surfaces, earth-sheltered homes are practically fireproof. In addition, with fewer points of entry, earth-sheltered homes present a greater challenge for thieves, reducing the threat of break-ins.

8. **Clean and healthy.** Earth-sheltered homes are nearly dust free, thanks in large part to the fact that there is very little exposed surface for air infiltration. Decreased infiltration dramatically reduces pollen levels, making it much easier for those who suffer from asthma to breathe. Because they tend to be a bit more humid than standard housing, earth-sheltered homes are ideal for sinuses, especially in arid climates. My sinuses, which in the past caused me extreme discomfort in the dry Colorado climate, haven't acted up since I moved into my earth-sheltered home in 1996.

9. **Quiet.** Earth sheltering is a natural soundproofing method that reduces unwanted racket from noisy neighbors, dogs, cars, trucks, trains, and jets. You'll be amazed at how quiet it is to live in an earth-sheltered home.

10. **Environmentally sound.** Earth-sheltered homes are good for the environment for many reasons. In addition to reducing fossil fuel use, they return large amounts of the Earth's surface to natural vegetation, helping reduce habitat and species loss. Because they require less maintenance and could outlast standard housing by decades, they also require fewer resources over their lifetime.

Earth-sheltered homes are nearly dust free because of reduced air infiltration. Decreased infiltration dramatically reduces pollen levels, making it much easier for those who suffer from asthma to breathe.

Proceed with Caution

As with any form of home construction, building an earth-sheltered home has some disadvantages. One of the main drawbacks is the lack of suitably experienced architects and builders in many parts of the country. Don't be dismayed, however. Some builders, like Jay Scafe of Terra-Dome, will travel to your location. Scafe's company has built earth-sheltered homes in every state in the country, including Alaska, as well as the Virgin Islands. In addition, several companies, such as Davis Caves and Terra-Dome, sell plans that can be modified to meet your needs. If these options don't work for you,

you may be able to locate an appropriate building professional in your area through the American Underground Construction Association, listed in the resource guide at the end of the book. The box on this page lists some additional on-line resources that may be helpful.

Earth-sheltered construction is trickier in wetter climates. Special precautions are needed to protect walls from moisture and to reduce interior moisture levels. Terra-Dome, for instance, installs ventilation systems controlled by humidistats that bring in and mix fresh air with the more moist interior air to achieve the right humidity level.

The public's lack of familiarity with earth-sheltered construction can present another problem. As with any alternative home, marketing an earth-sheltered house may require greater diligence than with a conventional home. In fact, because uneducated buyers may be unaware of the true benefits of earth-sheltered design, sellers often don't highlight this feature, and in some cases they don't even list earth sheltering in their promotional literature. In addition, in the past, buyers encountered resistance when applying for mortgages for earth-sheltered homes. Today, however, loans for earth-sheltered homes are available through the Veterans' Administration, the Federal Housing Administration, and progressive local lenders.

On the other hand, locating an earth-sheltered home to purchase may also be problematic, according to Scafe. Of all the earth-sheltered homes he's built, only ten have been put on the market in the past eight to nine years. Most of the people for whom he builds, he remarked, "like them so much that they stay for life."

Tips on Successful Earth-Sheltered Design and Construction

Creating an effective and suitable earth-sheltered design depends on many factors, among them soil type, topography, proximity to groundwater, and aesthetics.

The completely underground home built around an atrium is ideal for those who seek ways to minimize land disturbance and blend unobtrusively with the landscape. This design is best suited for flatter land with permeable, well-drained soils and no threat of groundwater intrusion resulting from a high water table.

The partially buried earth-sheltered home is suitable for those who want to blend with the environment and take advantage of passive solar heating. This type of design is ideal for hilly or mountainous country. Because water tends to drain downhill toward the building, and off the roof toward the back of the building, it is advisable to build in soils that are highly water permeable and to install a drainage system around the perimeter of the buried walls (as discussed on page 187).

For a list of earth shelter architects, builders, books, and plans in the United States, Canada, Europe, and New Zealand, log on to www.earth-house.com.

For general information, photos, building plans, government information, and guidelines, log on to architecture.about.com/cs/earthsheltered/

Earth-sheltered homes may have gotten a bad reputation from early experiments with this design. Many early models looked more like bunkers and were much too dark and humid indoors to be comfortable. However, today's builders are cognizant of the problems of the past and are working diligently to produce strong, waterproof, well-lighted, and visually appealing homes.

For those who are less concerned about blending in on flat terrain or for whom high water tables or water-impermeable soils pose a problem, the aboveground bermed structure may fit the bill.

Whichever type of design you choose, the following procedures are crucial:

1. **Choose a site with good natural drainage.** Avoid areas in which water could collect during rainstorms or as snow melts. Also, avoid building in the natural surface drainage, that is, areas where water runoff tends to flow; be sure to channel natural water flows around the building site.

   A good site with natural drainage requires the grade to be sloped away from the house and also requires permeable soils. The most permeable soils are granular in nature, for example, consisting of a high proportion of sand or gravel, which allows moisture to pass through with ease. The least suitable soils are those that have a high clay content. Clay soils expand and contract with varying moisture content, which can cause structural damage, and are relatively impermeable to water, so moisture tends to pool on the surface. Be sure to hire a soil engineer to test the subsoil on all prospective sites.

2. **Install drainage systems** around the house in wetter climates to remove water that may accumulate next to the house. A French drain, consisting of a porous 4-inch pipe covered with filter cloth and located in a bed of ¾-inch crushed rock along the perimeter of the building at the base of the walls, works well (see figure 2-3 on page 37).

3. **Waterproof the walls and the roof** before they are covered with earth. Water is a relentless force of nature that will find a way to enter a building if you've left an entry point. Even a tiny hole in a waterproofing material can permit moisture to gain entrance, potentially causing significant damage over time. More often than not, by the time you notice a leak, the damage is quite extensive. Finding and repairing a leak can be difficult and costly, requiring removal of large quantities of dirt.

   Earth Sheltered Technology installs a triple-layered waterproofing system consisting of a layer of dry bentonite clay, which repels water, and a layer of heavy polyethylene sheeting covered with a heavy, oversized pool liner. They're so confident in this system and have had such good results with it that they offer a lifetime warranty against leakage. Terra-Dome paints butyl rubber onto concrete dome roofs, then applies a product called Paraseal (bentonite clay applied to a plastic sheet) over areas that are most likely to leak. On my wooden roof, I used bituthene sheeting covered by a ½-inch-thick layer of foam to protect the waterproofing during backfilling.

> Water is a relentless force of nature that will find a way to enter a building if you've left an entry point. Even a tiny hole in a waterproofing material can permit moisture to gain entrance, potentially causing significant damage over time. More often than not, by the time you notice a leak, the damage is quite extensive.

4. **Insulate the structure.** Although the performance of an earth-sheltered home depends on the thermal constancy of the surrounding dirt, good insulation (usually a rigid foam) between the exterior walls and the ground helps ensure greater year-round comfort. In the winter, for instance, insulation helps reduce heat migration out of a home into the cooler earth surrounding it. Insulation also keeps walls warmer and prevents condensation that can lead to mold and mildew. Their spores can contaminate indoor air, causing health problems.

Earth Sheltered Technology installs 3 inches of rigid polystyrene foam insulation over vertical walls and 6 inches over the roof, covered with 3 or more feet of dirt to ensure comfort. Terra-Dome installs 2 inches of rigid polystyrene foam over the top (which is typically 3 to 7 feet below grade), 1 inch over back walls, and 2 to 3 inches over exposed concrete, which is then typically stuccoed or covered with brick.

In addition to insulation on the earth-sheltered roof and walls of my home, I installed wing insulation— that is, rigid foam insulation extending 2 to 4 feet horizontally from the walls, 18 inches below the surface (figure 10-3). Wing insulation traps heat around the structure, reducing heat loss. In my home, insulation was placed over a 6-inch layer of crushed granite to help keep the area around the walls drier. I also installed drainage ditches lined with 6 mil plastic sheeting and filled with rock either on the surface or slightly below the surface to remove water percolating down from the surface as an added precaution. The drier the soil, the lower the heat loss from the adjacent structure.

Figure 10-3.
Water is a primary concern when building an earth-sheltered home in almost any climate. Be sure to insulate and waterproof walls to prevent moisture from entering the home.

Source: Dan Chiras

5. **Check radon levels** in the soil before you build. Use an open land radon test kit from Air Chek (at www.radon.com) or a regular indoor radon test kit. The latter can be placed on a couple bricks on the ground, then covered with a bucket and removed after an appropriate period. Although radon can be easily dealt with, in order to incorporate abatement measures you must know that this harmful form of radiation is present. With advance warning, you can install a simple, effective radon removal system (see chapter 3).

6. **Build above the water table,** that is, the upper limit of the groundwater. Building below the water table, while possible, increases the likelihood of leakage.

7. **Incorporate passive solar design** for optimal winter comfort. Orient your home toward the south to take advantage of the low-angled winter sun (see chapter 11). In an earth-sheltered home, passive solar gain could provide virtually all of the heat you need, although your home may require a small backup heating system for extended periods of cold, cloudy weather. Be sure not to oversize the system, for reasons explained in chapter 11.

   If you can't orient the house toward the south for one reason or another, rest assured that an earth-sheltered home will still outperform a standard aboveground home. Scafe of Terra-Dome builds homes that face in any direction, and even north-facing ones enjoy "in excess of 50 percent energy savings in the winter and considerably more in hot summer months."

8. **Choose a south-facing slope.** South-facing slopes are naturally warmer than north-facing slopes and are ideal for passive solar earth-sheltered homes.

9. **When backfilling** against the walls and placing dirt on the roof, be sure to do so carefully to avoid damaging insulation or waterproofing materials. Compact the backfill against vertical walls—a little at a time in successive layers—so that the soil doesn't settle later and rip insulation from the walls or damage buried pipes. Note that heavy loads dropped suddenly—or a tractor driven onto a roof—can exceed the building's structural strength, causing damage or collapse. Work slowly and carefully at this stage.

10. **Hire qualified professionals** to design and build your home or, if you are planning to go it on your own, to consult with you, . Experienced earth sheltering professionals will help ensure the best results. They can assist you with building-code compliance to meet fire safety requirements, roof specifications, and insulation standards. They may also be able to help you secure financing.

Restoring the Earth

As Malcolm Wells aptly observes in his book, *How to Build an Underground House,* "Every square foot of this planet's surface—land and sea—is supposed to be robustly alive. It is not supposed to be shopping centered, parking lotted, asphalted, concreted, condo'd, housed, mowed, polluted, poisoned, trampled, or in any other way strangled in order that we—just one of millions of species—can keep on making the same mistakes."

Wells admits that "all construction causes land damage." Fortunately, he says, "Underground architecture can heal the wounds, and in many cases can improve the health of the land."

"Four hundred years ago," Wells remarks, "this land of ours was in perfect health. One hundred years from now it can be well on its way to recovery," that is, if we choose earth-sheltered building.

PART THREE

SUSTAINABLE SYSTEMS

CHAPTER 11

Passive Solar Heating

Marc Rosenbaum has dedicated over two decades of his professional life as an engineer and home designer to creating energy-efficient homes in the northeastern United States. Relying on superinsulation, airtight design, energy-efficient lighting, and sunlight to reduce fossil-fuel consumption in the buildings he's designed, Rosenbaum, like other environmentally conscious designers, is saving his clients substantial sums of money while helping to build a more sustainable future.

In 1994, Rosenbaum was approached by a couple who wanted to build a super-energy-efficient home in Hanover, New Hampshire. Although Hanover is far from the nation's sunbelt, Rosenbaum jumped at the chance. Months later, he finished the design for a remarkable home—one that would consume 95 percent less heat than an energy-efficient home built to Model Energy Code standards (figure 11-1). How could he do this?

In addition to designing an airtight home with extremely high levels of insulation, Rosenbaum incorporated measures to heat the home with sunlight, a technique known as *passive solar heating*. A passive solar home permits low-angled winter sunlight to enter during the winter months through south-facing windows. Inside the building, the sun's energy is converted to heat, providing warmth.

Figure 11-1.
This solar home in chilly New Hampshire, designed by Marc Rosenbaum, uses 95 percent less energy for heating than a typical energy-efficient home, thanks to active and passive solar systems. It has won several awards for energy efficiency.

Source: Marc Rosenbaum

In Rosenbaum's design, the passive solar features were complemented by an active solar system, a set of panels on the roof that absorb the sun's energy.

Together, these passive and active solar technologies provide an astonishing 95 percent of the annual heating requirement for Rosenbaum's clients—a remarkable feat in this seasonally cold and often cloudy region of the country. The cost savings were even more amazing. According to a computer analysis, a standard home, built without provisions for harvesting solar gain, would cost about $2,400 a year to heat, compared to a meager cost of $145 per year for Rosenbaum's house. This would save the homeowners $2,255 a year on their fuel bills. Over a few decades, the savings from solar heat, if wisely invested in modest financial instruments, could easily pay the college expenses of a child or two.

Passive solar heating can be incorporated in any style of home (see figure 11-2), so long as the south side of the house receives full sun for most of the day (this is where solar energy is most intense). Many who are first learning about passive solar are amazed to learn that most areas of the United States actually provide sufficient sunlight to fulfill at least half of a household's annual heat demand. Even in the cloudy "gloom belt" of the northeastern United States, passive solar can supply a substantial portion of a home's heat.

When correctly designed, solar homes provide unrivaled comfort through all kinds of weather, from searing summer heat to brutal winter cold. Passive solar homes also afford generous views of the surrounding countryside and offer bright and sunny interiors that will lift anyone's spirits. Moreover, many passive solar homes are designed with open floor plans, yielding a feeling of spaciousness in which many revel.

Passive solar offers many benefits to homeowners, you say, but surely it must cost a lot more. I posed this question to Ron Judkoff, director of the Buildings and Thermal Systems Center at the National Renewable Energy Laboratory (NREL), a government agency that spearheads much of our nation's research on passive solar heating. Judkoff observed that passive solar design and construction increase the cost of building a new home anywhere from zero to about 3 percent. On a $200,000 home, for example, the maximum additional cost of incorporating passive solar heating may be only another $6,000. Why so little?

As Judkoff explained, many building codes now require much more energy-efficient windows, walls, ceilings and foundations than in the past. As a result, building a superefficient passive solar home adds insignificantly, if at all, to the initial cost. In addition, Judkoff reminded me, passive solar homes often require much smaller backup heat sources (furnaces or boilers). The lower initial cost and subsequent savings here can be used to offset higher costs elsewhere.

Judkoff based his cost estimates in part on a series of case studies sponsored by NREL and the American Solar Energy Society (ASES). Data was collected on passive solar homes from a variety of locations, including Arizona, Indiana, Maine, Massachusetts, North Carolina, and Wisconsin

> Passive solar heating can be incorporated in any style of home, so long as the south side of the house receives full sun for most of the day. It can also be used in almost any part of the country to provide a significant portion of a home's annual heating requirement.

Figure 11-2.
Passive solar energy can be designed into any style home, as shown here.

Source: Dan Chiras

Table 11-1

Energy Costs and Savings Resulting from Passive Solar Design

Location (date construction was completed)	Additional Cost	Reduction in Heating and Cooling Cost	Annual Savings in Fuel Bill	30-year Savings (Based on current fuel costs)	30-year Savings (Based on projected 5% and 10% increase in fuel costs per year)
Jonesport, Maine (1988)	$1,000	70%	$300	$9,000	$19,900–$48,700
Falmouth, Massachusetts (1995)	$3,500	82%	$1,260 with PV system for electricity	$37,800	$76,894–$207,750
Burlington, North Carolina (1990)	$5,000	64%	$840	$25,200	$55,339–$138,160
Naperville, Illinois (1984)	$3,000	72%	$550	$16,500	$36,540–$90,430
Stevens Point, Wisconsin (1995)	$6,000	70%	$600	$18,000	$42,179–$98,690
Hanover, New Hampshire (1994)	0	95%	$2,255	$67,650	$141,390–$403,976
Andover, Connecticut (1981)	0	58%	$958	$28,740	$63,650–$157,675
Santa Fe, New Mexico (1985)	0	81%	$220	$6,600	$14,614–$36,196

(table 11-1). Analysis of these buildings showed that initial cost ranged from zero to 3 percent higher than that of standard homes while savings in fuel bills resulting from solar heat ranged from a modest $220 a year in Santa Fe, New Mexico, to an impressive $2,255 a year in New Hampshire (the home designed by Marc Rosenbaum and described in the introduction of this chapter). Depending on the location and other factors, over a thirty-year period these monthly savings could save the homeowners $7,000 to $67,000, even if fuel costs remain the same as today.

Because energy prices will very likely increase over the next few decades, I decided to run some calculations to assess potential savings in the homes in the study. Assuming a 5 percent yearly increase in natural gas prices, I found that savings at the low end ($220 per year) would increase to over $900 per year by 2032. At the high end, the current annual savings of $2,255 would increase to over $8,800 by 2032. Total savings over the thirty-year period at the low end would be an astounding $14,600. At the high end, the thirty-year savings would be an even more impressive $141,400. Quite a return on investment.

If energy prices increase even more, say 10 percent per year, the $220 per year savings would increase to nearly $3,500 per year, reaping a homeowner a thirty-year net savings of $36,200. The $2,255 per year savings would jump to nearly $40,000 per year, with a total thirty-year savings of nearly $404,000!

No one knows for sure how fast fuel prices will increase in the coming decades, but if the summers of 2001 and 2003 are any indication, Americans could be in for a big shock in coming years as natural gas supplies dwindle.

Even if price increases don't take place, passive solar makes good financial sense. "You'd be fool not to incorporate solar into your home," says a friend of mine. "You wouldn't turn down a free furnace or a lifetime of free energy bills, would you?"

In addition to free heat, passive solar provides a wide range of environmental benefits, among them reduced air pollution. Debbie Rucker Coleman of Sun Plans, Inc. (www.sunplans.com), estimates that, compared to the average home in Alabama, an energy-efficient solar home in Georgia, which averages $12 per month to heat and cool, will save 574,410 pounds of carbon dioxide emissions over thirty years.

So what is required to design and build a passive solar home?

Keys to Successful Passive Solar Design

Passive solar design, whether for a new home or a retrofit, requires a working knowledge of ten key principles. I will highlight them here with the friendly advice that those seriously interested in utilizing passive solar explore each one in much more detail *before* they start pounding nails or hefting straw bales onto a foundation. My book *The Solar House* offers extensive information on this topic.

1. The first requirement for passive solar heating is **a site with unobstructed solar access,** one that is bright and sunny from around 9 A.M. to 3 P.M. during the heating season (figure 11-3). In colder regions, such as Vermont and Wisconsin, the heating season typically extends from mid-to-late fall, through the winter, and well into the spring. In warmer areas, such as North Carolina, the heating season may be a comparatively brief two-month period in the "dead of winter."

   Choose your site carefully. Avoid wooded lots, or, if you can't, remove trees on the south side of the home to open up to the sun. Whatever site you choose, be on the lookout for large obstructions, such as evergreen trees, hills, or nearby buildings, that could block the low-angled winter sun. When in doubt, visit the site on December 21 (when the sun is lowest in the sky) to be certain the home will receive full sun from around 9 A.M. to 3 P.M. Some solar designers use a handy little device

Figure 11-3.
For best performance, a solar home requires access to the sun from 9 A.M. to 3 P.M. during the heating season. This means keeping the south side free from major obstructions such as trees at least 60 degrees from each corner.

Source: David Smith

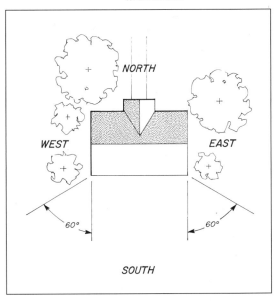

BUILDING NOTE

In subdivisions, proper orientation of homes for maximum solar gain can be quite a challenge. But if streets are oriented on an east-west axis, builders are able to orient homes correctly for solar gain. Outside of Chicago, Bigelow Homes built a subdivision with 1,100 new homes. By orienting the streets on an east-west axis and minimizing windows on the east and west sides of homes, the developer was able to ensure homeowners a lifetime of low energy bills.

known as the Solar Pathfinder (www.solarpathfinder.com), which allows them to fully assess the year-round solar potential of a site in about the time it takes to drink a cup of tea.

If you're choosing a small lot to build on, select one that is as deep as possible from north to south to ensure good solar access. That way, someone will be less likely to build in your way and block the sun. In rural settings, locating the septic drainage field within the solar access zone is a good strategy to maintain good access, since that area will need to be kept clear of trees that could obstruct the southern sun.

2. **For optimal solar gain, orient your house so that its long axis lies on an east-west axis** (figure 11-4). Be sure to orient the long axis of the home within 10 degrees (east or west) of true south. True north and south form an invisible line stretching from the North Pole to the South Pole (parallel to longitude lines). True north and south are not the same as magnetic north and south, which are determined by magnetic fields created by iron-containing minerals in the Earth's core. Magnetic lines rarely run true north and south. In fact, in some places, such as northern California, they run nearly east to west. In others, true north and south vary by 10 to 15 degrees from magnetic north and south.

When designing a passive solar home, be sure to determine magnetic north and south, then adjust compass readings to determine true north and south. Contact a local surveyor to find out how many degrees to adjust from magnetic south. Then orient the house on the lot within 10 degrees east or west of true south. Either the front, the back, or the side of the house can serve as the "solar face."

Deviating from this orientation will cost you in effectiveness. That's because departing from true-south orientation reduces wintertime heat gain and increases summertime heat gain, which can lead to serious overheating in some locations.

Figure 11-4.
Orient the long axis of your home to the south and you'll be amply rewarded in the winter and summer.

Source: David Smith

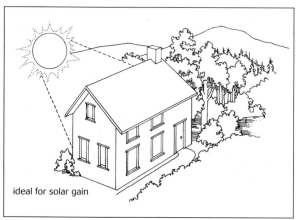

ideal for solar gain

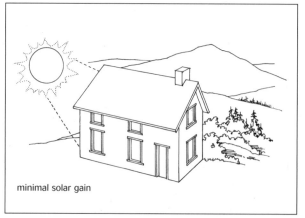

minimal solar gain

Figure 11-5.
South-facing glass is key to passive solar design. When the sun cuts a low arc across the sky during the winter, south-facing glass allows ample sunlight into a home to provide heat. During the summer, the sun cuts a high arc through the sky, and little sunlight penetrates the windows, provided there's adequate overhang and the house is oriented as close to true south as possible.

Source: Dan Chiras

3. **When designing a passive solar home, concentrate windows on the south side of the house**—the "solar face" just mentioned (figure 11-5). Simply shifting a few windows to the south side of a house increases solar gain enough to cut annual heating bills by about 15 to 30 percent. The result is a *sun-tempered house.*

   To achieve even greater results, place more windows on the south side and minimize north, east, and west glazing (see sidebar). As a rule, the more rooms that have south-facing windows, the better. This will eliminate the need for fans or ducts to move warm air from one area to another. But don't go wild on windows! Too many south-facing windows can result in overheating and can also result in excess heat loss during evenings and long, cold cloudy periods.

   As a rule, south-facing glass in passive solar homes should fall within the range of 7 to 12 percent of the total square footage of heated space. For a 2,000-square-foot home, that means the solar glazing should occupy 140 to 240 square feet of wall area for optimal performance. Generally, the colder the climate, the more glass you'll want, provided you install low-E glass and window shades to prevent nighttime heat loss.

   The type of glass you install is also very important. I've discussed rudiments of energy-efficient windows in chapter 6. Because this subject is so complicated, you should also refer to more comprehensive treatments.

4. **Include overhangs, especially on the south side of a house.** Overhangs perform two vital functions that contribute mightily to year-round comfort in a solar home. During the summer,

---

Window Allocations in Direct-Gain Systems

South-facing glass—7 to 12%*

North-facing glass—no more than 4%

East-facing glass—no more than 4%

West-facing glass—no more than 2%

*Percentages are based on total square footage of a home. Window space is glass area (total window space minus frame).

**Figure 11-6.
Overhangs control solar
gain and protect a home
from overheating during
the summer months.**

Source: Dan Chiras

Thermal mass—solid materials, such as tile and concrete inside a passive solar house, that absorb solar heat and help stabilize internal temperatures.

overhangs shade the windows and exterior walls from the sun (figure 11-6). Shading these areas from intense sunlight bearing down from the high-angled summer sun helps keep a home cool.

Overhangs also regulate wintertime solar gain, helping determine when solar heating begins and ends each year (figure 11-7). In other words, the overhang regulates heat input. In most northern locations, a 2-foot overhang works well. It shades an 8- to 9-foot wall well, yet permits solar heating to commence in the fall when heat is required. In southern locations, less heat is required and a longer overhang is advisable. Generally, the warmer and sunnier the climate, the more advisable it is to have generous overhangs. Log on to the website www.susdesign.com and click on the Design Tools button to access a window overhang design program that can help you design overhangs for your region.

While we are on the subject of overhangs, be sure to choose a home design with a few well-placed projections and porches on the south. These features shade adjacent windows, preventing the summer sun from entering the home and therefore reducing solar heat gain. Porches on the east and west sides of a house are also often beneficial as they shade windows and exterior walls from the hot summer sun.

5. **For maximum comfort, include an adequate amount of thermal mass in your design.** Thermal mass refers to solid materials, such as tile and concrete inside a house, that absorb solar heat. Some homes use water stored in rigid columns as thermal mass, but masonry materials are more common. When sunlight strikes the thermal mass or when the temperature of the thermal mass is lower than the air temperature inside a home, heat is absorbed into the material and stored for later release—that is, when room temperature drops below the surface temperature of the mass. When this occurs, heat is slowly relinquished. The result is relatively stable and comfortable indoor temperatures, despite dramatic oscillations in outside temperatures.

For sun-tempered homes—that is, homes with less than 7 percent solar glazing—mass in floors, framing, wallboard, and furniture is generally sufficient to accommodate the heat produced by solar energy. When south-facing glass falls within the 7 to 12 percent range to capture more sunlight energy, however, additional mass is required to

prevent a house and its occupants from overheating. Concrete slabs, adobe floors, masonry
walls, and planters all work well.

How much mass does a house need to perform well? Passive solar energy books and software programs, such as Energy-10 (discussed
shortly), can help a designer or builder determine the amount of mass a solar home
requires. For optimal performance, mass should
be located in direct contact with incoming
solar radiation throughout at least part of the
day, and it should have as much surface area as
possible distributed throughout the house for
even heat.

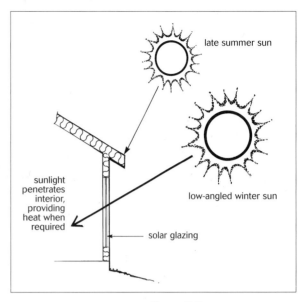

Figure 11-7.
**A solar home operates auto-
matically, with overhangs
preventing overheating during the summer, and south-
facing windows allowing
sunlight to enter during the
winter, when it's needed
most for heating.**

Source: David Smith

6. **All solar design relies on energy efficiency.**
When a home has been designed and constructed to maximize energy
efficiency, the building will be easier to heat with solar energy.
Remember, ordinary levels of insulation are insufficient in passive solar
homes. A passive solar home requires superior levels of insulation in
ceilings, walls, floors, and foundations to conserve solar heat and buffer
against outdoor temperature swings. In nearly all cases, the better insulated a home is, the better it will perform thermally, in summer as well
as in winter. NREL's Ron Judkoff recommends insulating at least to
the level prescribed by the International Energy Conservation Code
for residential structures or ASHRAE 90.2 for commercial buildings—
but preferably beyond them. (These guidelines offer region-specific
recommendations for envelope elements and mechanical systems.)

The EPA and DOE's Energy Star program, discussed in chapter 6, also
specifies exceptional levels of insulation. Even so, many energy-conscious
builders prefer to insulate at levels that exceed these recommendations.

General insulation guidelines for efficient passive solar homes are
R-30 walls and R-60 roofs in temperate climates and R-40 walls and
R-80 roofs in extremely hot or cold climates, according to Ken Olson
and Joe Schwartz, authors of "Home Sweet Solar Home: A Passive
Solar Design Primer," an article published in *Home Power* magazine.

If you can't design in as much south-facing glass (solar glazing) as
you would like, adding extra insulation may offset the lower heat gain
by lowering heat loss. Insulation must be installed correctly. Careful
attention to detail is essential. To be effective, insulation should not be
compressed, and air should not be able to leak in around it. Some
builders are fond of the liquid foam insulation products, which are
sprayed into wall and ceiling cavities and then expand as they dry,
because they create an airtight seal and are resistant to water, which

BUILDING NOTE

If you can't include as
much south-facing glass in
a home as you'd like,
adding extra insulation may
offset the lower heat gain
by lowering heat loss.

renders other forms of insulation such as fiberglass and cellulose much less effective.

Energy-efficient windows are vital to passive solar heating as well. To make a home even more energy efficient, be sure to provide a means of covering windows at night—for example, insulated shades or thermal shutters—and seal all cracks in the building envelope around windows and doors and elsewhere to reduce air infiltration and exfiltration. Sealable entryways or airlocks separated from the main living space by an inner door are especially helpful in cold climates in preventing cold air from rushing in when the outside door is opened.

As noted in chapter 3, a good airtight design allows 0.35 to 0.5 total air changes per hour to ensure healthy indoor air. While sealing a home is important, indeed essential, you need to take extra steps to ensure adequate ventilation. Installing a whole-house ventilation system with a heat-recovery ventilator will ensure proper fresh air turnover and help maintain indoor air quality.

7. **Protect insulation from moisture.** Most commonly used forms of insulation lose their ability to restrict heat flow even when only slightly moistened. To prevent this problem, builders frequently install a vapor barrier on the warm side of the wall. Additional protection may be provided by installing a plastic housewrap on the exterior sheathing.

8. **Design your house so that most, if not all, of the rooms are heated directly by incoming sunlight.** Most passive solar homes are rectangular rather than square. This shape results in the largest possible amount of exterior wall space and window area on the south side of the building and permits sunlight to penetrate more deeply into the interior of the structure than in a square design. It also ensures that each room is heated independently, eliminating the need for fans and ducts to move warm air from one part of the house to another. If a rectangular design isn't possible, place rooms that require less heat, for example, workshops, bedrooms, and kitchens, on the north side of the house and locate frequently used rooms on the south side, where they can be steadily warmed by incoming solar radiation.

Rectangular floor plans also minimize the exposure of east and west walls to summer sun, which is especially important in hot climates. East- and west-facing windows can increase heat gain during the cooling season, resulting in mild to severe overheating.

9. **Create sun-free zones in your home** (figure 11-8). In the past thirty years as I have toured passive solar homes, it has become painfully

Figure 11-8.
Although passive solar heating is a valuable and economic source of heat, too much sun in living space can create considerable discomfort. Be sure to create sun-free areas in homes to protect occupants from overheating.

Source: Dan Chiras

Create Sun-Free Zones

Although passive solar homes are designed to let the sun in to produce heat, sunlight can render rooms useless. In family rooms and home offices, glare on televisions and computer screens is a major nuisance. Design your home so that sun can heat spaces without frying you and your family and bleaching furniture and carpeting in the process.

clear to me that their comfort and utility depends on the presence of sun-free zones for computer work, watching television, or just plain lounging.

Unfortunately, many passive solar designs in the 1970s and 1980s were little more than boxes with south-facing windows that permitted maximum sunlight penetration. During the day, many homes were bathed in blinding sunlight. Baseball caps and sunglasses became standard equipment for occupants during the daytime.

Although the purpose of passive solar design is to allow sunlight in to generate heat, too much sunlight in the wrong place can make a house extremely hot and uncomfortable. It can also render rooms useless for various activities. In sun-drenched family rooms and home offices, glare on televisions and computer screens can be a major nuisance and can greatly reduce the habitability of a home.

Design your home so that the sun can heat the interior without frying you, blasting you and your family with sunlight, and bleaching furniture and carpeting. As figure 11-9 shows, there are many creative ways to achieve this goal.

10. **Install an efficient, properly sized, environmentally responsible backup heating system.** Most passive solar homes require some form of backup heat. However, because most of the heat in a passive solar home comes from the sun, and because passive solar homes are so well insulated, solar builders are frequently able to install smaller than average heating systems—for example, a woodstove or wall-mounted space heater rather than a large furnace or a costly radiant-floor heating system.

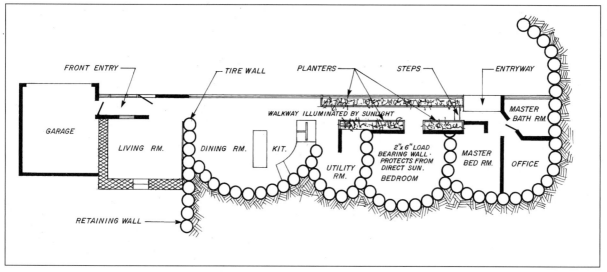

FRONT ENTRY — TIRE WALL   PLANTERS —   STEPS —   ENTRYWAY

WALKWAY ILLUMINATED BY SUNLIGHT   MASTER BATH RM.

GARAGE   LIVING RM.   DINING RM.   KIT.   UTILITY RM.   2"x 6" LOAD BEARING WALL · PROTECTS FROM DIRECT SUN.   MASTER BED RM.   OFFICE   BEDROOM

RETAINING WALL —

Figure 11-9.
A floor plan of the author's home in Colorado shows an extensive buffer zone to prevent sun drenching.

Source: David Smith

Installing a smaller system saves money and often offsets the added costs of passive solar construction, if any.

Choosing a backup heating system that meets your needs but is efficient and nonpolluting helps make a passive solar home even more environmentally friendly. Solar hot-water systems, masonry heaters, clean-burning woodstoves, and heat pumps are great examples of technologies that complement passive solar, providing backup heat efficiently without contradicting the environmental ethos of solar design. These options are discussed in chapter 6. For more detail, you may want to refer to the chapter on backup heating systems in *The Solar House*. If you want to study the subject in even greater depth, have a look at *Natural Home Heating* by Greg Pahl.

Integrated design views the house as an integrated whole. It recognizes that decisions about design and materials in one part of a home affect others. Because the number of possible interactions is great and the potential for error is large, many professional solar designers use powerful new software when designing homes.

Helpful Design Tools

As noted in chapter 6, solar design requires a holistic approach that views the house as an integrated whole, recognizing that decisions about design and materials in one part of a home affect others. Because the number of possible interactions is great and the potential for error is large, many professional solar designers use powerful new computer software such as Energy-10, developed under the leadership of Doug Balcomb of the Building and Thermal Systems Center of the National Renewable Energy Laboratory. This program is available from the Sustainable Buildings Industry Council, which is listed in the resource guide.

Energy-10 allows designers to predict the energy performance of various building design strategies, permitting a fine-tuning of designs to achieve optimal comfort, performance, and economy. Architect Debbie Coleman prefers a simpler approach and uses *Passive Solar Design Strategies: Guidelines for Home Building,* by the Sustainable Buildings Industry Council (see resource guide). This book comes with software—*Builder Guide for Windows*—that is easy to use. "Most homeowners would be able to follow these guidelines, as the notebook is put together very well. I like the program because it tells you the percentage of your heating needs that the sun is providing, and it lets you compare strategies such as adding insulation or increasing south glass. You can complete the worksheets manually or with the computer software version."

Passive solar design is a rattling good idea. Like many good ideas, it is not new but, rather, draws upon approaches practiced for thousands of years by our ancestors, including the ancient Greeks and the Anasazi Indians of the desert Southwest. Today, as fossil fuel resources, especially natural gas, become depleted and as concern for global climate change mounts, passive solar design is more important than ever. Be careful, however. Read more. Lots more. Design holistically or hire a designer who practices integrated design. Follow the guidelines I've outlined in this chapter. Use the sophisticated new tools such as Energy-10 to analyze the performance of your design and to tweak it to achieve the best performance at the lowest cost.

Then let the sun shine in!

Passive solar design is not a new idea but, rather, an ancient one that's been neglected far too long. Practiced thousands of years ago by cultures as disparate as the ancient Greeks and the Anasazi Indians of the desert Southwest, passive solar design is more important than ever as fossil fuel resources, especially natural gas, become depleted and as concern for global climate change mounts.

CHAPTER 12

Passive Cooling

A few months after I moved into my earth-sheltered passive solar home, two middle-aged women appeared at my front door. With their clipboards and stern, no-nonsense expressions, I knew I was about to get an official visit from some government agency. When I opened the door, the taller woman introduced herself as a representative of the county tax assessor's office. They were there, she explained, to measure my home and note any features that might add to its value, so they could assign appropriate property tax.

I let the women in so they could begin their work. Stepping out of the 95-degree temperature into the naturally cool interior of my home, both women sighed in relief, then went right to work. The leader of the duo looked at me, then said, "You must have central air conditioning."

"No," I politely responded. "This house stays cool naturally."

She frowned and narrowed her eyes suspiciously.

"Really," I said. "It's superinsulated and earth sheltered. It stays cool naturally." Another skeptical look shot my way.

But before I could explain, her companion chimed in, "It's true. I've been in other earth-sheltered homes and they don't need air conditioners."

With that exchange, the two women left me to my writing and set about their work, the skeptic no doubt on the lookout for telltale signs of an air-conditioning system—ducts, vents, and a compressor—that she'd never find.

My house, like hundreds of others throughout the world, was built to cool itself naturally . . . without any outside help. Just as a home can be heated without expensive, energy-intensive, polluting mechanical equipment, so too can it be cooled. The process is called *passive cooling*.

To cool a home passively requires knowledge of the local climate and an understanding of some simple but highly effective measures. To help you understand how these measures work, I've organized them into four categories: (1) those that reduce internal heat gain, (2) those that reduce external heat gain, (3) those that purge built-up heat from buildings, and (4) those that are designed to cool people directly.

Reducing Internal Heat Gain

As in so much of life, prevention is the best medicine. In keeping a house cool, one of the easiest and least expensive methods is to reduce internal

Incandescent light bulbs convert only about 5 to 10 percent of the electricity that flows through them into light. The rest is converted to heat. It would therefore be more accurate to call them "heat bulbs"!

Figure 12-1.
Compact fluorescent light bulbs save energy and money—considerable amounts of both—and outlast conventional bulbs by nearly a decade! Sunwave lightbulbs provide a full-spectrum lighting—like the light at noontime—and are ideal for reading. Although they're a bit bluish, I find them wonderful to read and work under. I use one in my office by the computer.

Source: American Environmental Products

heat sources during the cooling season—that is, reduce heat generated inside homes, known as *internal heat gain*.

Heat comes from a variety of internal sources such as people, lights, stoves, and appliances. Let's take a look at the major culprits.

Lighting

Incandescent light bulbs are one of the most common sources of internal heat gain. Widely used throughout the world, incandescent bulbs are actually rather inefficient sources of light. They convert only about 5 to 10 percent of the electricity that flows through them into light. The rest is converted to heat. It would be more accurate to call them "heat bulbs"!

Incandescent light bulbs can be replaced by much more efficient compact fluorescent lights (CFLs), screw-in bulbs that fit into standard incandescent sockets. CFLs convert electrical energy into visible light much more efficiently than incandescent bulbs, using 75 percent less energy to produce the same amount of light. Not only do they consume less electrical power and save you money, they also emit less heat—about 90 percent less heat than a similarly rated incandescent light. They are also color adjusted to produce a pleasing light, not the cold bluish light of a standard fluorescent bulb.

By substituting compact fluorescent light bulbs for incandescent bulbs, especially in areas where lights are on for long periods, a homeowner can achieve marked reductions in electrical use and also reduce internal heat gain, which means a cooler, more comfortable interior in the summer.

Another way of reducing internal heat gain is through task lighting, providing light to high-use areas or work zones within a room. This technique allows us to selectively increase the level of light in portions of a room where it's most needed, rather than bathing an entire room in bright light when using only a portion of the space.

Yet another successful means of reducing internal heat gain is daylighting, the use of natural light instead of artificial light. In passive solar homes, south-facing windows often contribute significantly to daylighting, as do skylights and windows strategically placed throughout the house. My passive solar home with generous south-facing windows is a testament to the effectiveness of daylighting, as we rarely require supplemental daytime lighting.

To provide daylighting without increasing cooling loads, windows need to be positioned so that they permit light to enter during the cooling season without causing excess heat gain from the outside. That means they need to be placed primarily along a south-facing wall that is protected by overhangs (see chapter 11). Skylights tend to lead to overheating during the summer, although they may be okay on the north side of a home where they bring in daylight but not direct sunlight.

Appliances

Household appliances such as dishwashers, ovens, clothes dryers, and washing machines also contribute to internal heat gain. When possible, use heat-generating appliances in the morning or late evening, when a little extra heat can be tolerated. Shifting certain activities to locations outside the house also helps minimize internal heat gain. For example, drying clothes on an outdoor line or cooking outdoors during the summer reduces internal heat gain. Designers and builders of new homes can help make this possible by providing areas for these activities, for example, patios and decks. Shade trees and other vegetation planted around patios and porches ensure privacy and can help keep these areas cooler. A small fountain may provide additional cooling as well.

Using appliances that are less energy intensive also helps reduce internal heat gain. Microwave ovens, for example, produce much less waste heat than regular stoves and ovens. Newer, more energy-efficient appliances also help keep a house cooler because they use less energy and generate less waste heat than less efficient models. When buying a new home or remodeling, be sure to look for the most energy-efficient models, now easily identified by comparative rating and Energy Star labels.

Moreover, in many homes it is possible to seal off internal heat producers. Laundry rooms and water heaters can be isolated from the rest of the house with the close of a door. Waste heat can be vented to the outside during the cooling season. Be sure to isolate heat producers when designing a new home or remodeling an existing one.

Reducing External Heat Gain

Important as it is to reduce internal heat gain, the primary source of unwanted heat in a home during the cooling season is the great outdoors. External heat gain comes from several sources, such as sunlight shining on or around a building, and warm air surrounding it. Heat from these sources is transferred to roofs, walls, windows, and skylights, then migrates into the interior. Hot air can also enter through openings in the building envelope via infiltration. Because external heat gain contributes so significantly to cooling loads, finding ways to reduce or eliminate it are vital to passive cooling.

Orientation of a Home

One of the most effective means of preventing external heat gain is proper orientation. As noted in chapter 11, orienting the east-west (long) axis of a passive solar home in the northern hemisphere as perpendicular as possible to true south maximizes desirable solar gain in the winter and minimizes unwanted solar gain during the summer. As the orientation of a home deviates from true south, summertime solar gain increases, as does the cooling

BUILDING NOTE

A far better option than a traditional skylight is a solar tube skylight, which consists of a flexible tube that extends from the ceiling through the rafters to the roof. Equipped with a light-gathering lens mounted on the roof and an internal reflective surface, this device transmits large amounts of diffuse light from a small surface area. The result: enormous light transmission with minimal heat loss. Solartubes, Sun Tunnels, and other such products take up a fraction of the space of conventional skylights. Inside, they look like ceiling lights. And they don't require any special framing.

BUILDING NOTE

When designing landscaping, it is wise to use plants native to the area. They're adapted to local conditions and thus are hardier than many exotic species. They'll also require less care. Once they're established, many native species can survive on rainfall alone.

load. The warmer the climate, the greater the penalty and the more difficult it is to passively cool a home.

Avoiding Two-Story Glass and Skylights

External heat gain during the cooling season may also result from too much glass. Two-story walls of glass and traditional skylights are particularly deleterious to summertime comfort (figure 12-2). Both are difficult to shade and can result in overheating. So, be careful not to overdo windows, especially on the east and west sides of a house. Instead of conventional skylights, consider installing solar tube skylights.

Biological Shading: Trees and Other Vegetation

Vegetation, such as trees and vines, or built structures, such as overhangs, provide shade that helps reduce external heat gain and make a home considerably more comfortable. Trees and other vegetation also cool the air surrounding a home by evaporation. Water evaporating from leaves, a process known as *transpiration*, draws enormous amounts of heat out of the surrounding air. Don't underestimate the effectiveness of a tree as a means of cooling a house. As Anne S. Moffat and Marc Schiler point out in their book, *Energy-Efficient and Environmental Landscaping*, a single mature tree gets rid of as much heat as five 10,000-Btu air conditioners, and without adding pollution to the atmosphere. When it comes to providing shade and evaporative cooling for a house, deciduous trees are a great investment. Trees, vines, and grass can lower the temperature of the air surrounding a home by as much as 9 degrees F. Grass also helps cool the air around a home, by as much as 10 degrees F, when compared to bare dirt, because grass absorbs less sunlight than bare dirt and loses moisture by transpiration.

Shade can be provided by planting evergreens, such as pines and spruces. Be careful, however, where you place them. Because they retain their leaves (needles) year-round, evergreens planted along the south side of a passively conditioned home can block wintertime sun and reduce solar gain. Along the north, south, and east sides, evergreens generally work fine, although they may block breezes that contribute to passive cooling.

Mechanical Shading Devices

Window shades and awnings also help cut down on unwanted heat gain. The most effective shading is provided by external devices: awnings, external louvers, rigid shutters, roller shades, and solar screens.

Interior shades such as drapes or curtains, conventional pleated shades, and blinds also reduce heat gain. Tightly woven, light-colored, opaque fabrics work best, as they reflect more incoming sunlight than dark or open fabrics. Double-layered drapes and curtains block even more sunlight and improve the insulation value, reducing heat gain during the summer and heat loss in the winter. The closer a curtain or drape is to the window, the better it works.

Although interior shades offer some significant advantages over most exterior shading devices—for example, they are easier to operate—they are not as effective at blocking heat gain. That's because exterior shades block sunlight before it penetrates a window. Interior shading devices do not. As a result, considerable heat can build up between a window shade and the glass. Much of this heat will enter the room, increasing the cooling load.

House and Roof Color

Further reductions in heat gain can be achieved by painting a house a light color. Light-colored walls reflect sunlight, rather than absorbing it, which reduces external heat gain. As an added benefit, light-colored walls increase the life span of siding, especially on the south, west, and east facades of a home, which receive the most direct sunlight.

Light-colored roof shingles or roof tiles also help reduce external heat gain and can have a significant effect on comfort levels. That's because most summer heat gain (up to two-thirds) occurs from the rooftop. When combined with other measures outlined in this chapter, a light-colored roof and light-colored walls can make a big difference in summer cooling loads.

The type of roofing material and method of installation also influence heat gain. For example, metal roofing, when installed over sleepers, as shown in figure 12-4, creates an air space that retards heat migration into a building. Rigid

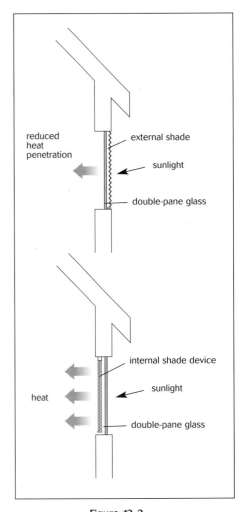

Figure 12-3.
External shades (top) provide better protection against the summer sun. Internal shades (bottom) permit sunlight to penetrate the glass, which allows heat to enter the interior.

Source: David Smith, from *The Solar House* by D. Chiras

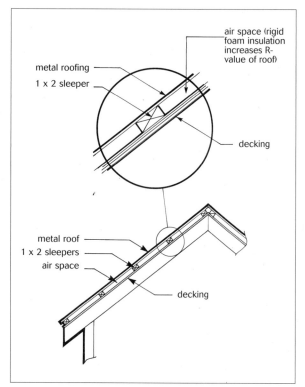

Although white shingles absorb 70 percent of the sun's radiation striking them, it makes sense to install them or other light-colored materials on roofs in hot, sunny climates. Why? They permit less heat gain than darker materials. And the roof is a major source of heat gain. When combined with other measures such as light-colored walls, light-colored roofing makes a notable contribution to summer-time cooling.

foam insulation placed on the decking between the sleepers reduces heat migration even more.

Tile roofs are popular in many hotter areas such as Florida, and for good reason. Spanish tile, for instance, creates a sizable air space over the roof decking that retards heat migration into a house (figure 12-5).

Radiant Barriers in Roofs

External heat gain through roofs can be reduced by installing radiant barriers, as described in chapter 6. A radiant barrier consists of aluminum foil stapled to roof rafters, as shown in figure 6-6 on page 116, or a thin film of aluminum applied to oriented strand board that is then used as roof decking. Radiant barriers block heat penetrating through roofs on hot summer days and are most effective in hot climates.

Good Quality, Low–E Windows

According to the U.S. Department of Energy, about 40 percent of the unwanted summer heat that enters a house comes through the windows. This is another reason to choose windows wisely. Low–E windows are especially helpful in reducing heat gain; they are discussed in chapter 6.

Insulating Your Home

Insulation in walls and ceilings and on foundations plays a valuable role in reducing external heat gain, especially when used in conjunction with other measures described in this chapter. Because most external heat gain occurs through the ceiling, pay special attention to roof and attic insulation. Wall insulation is not as important as ceiling insulation for cooling (or heating), but don't skimp when it comes to insulating your walls. Floor insulation generally has little effect on cooling, but it is important during the cold months of the year in many regions.

Reducing Air Infiltration

Hot air entering a home through cracks in the envelope—around poorly sealed doors and windows, for example—can also contribute to external heat gain. To prevent this problem, homes should be designed and built airtight. Vapor barriers, house wraps, and certain building materials, such as insulated concrete forms and structural insulated panels (discussed in chapter 4), can be used. They all help reduce air infiltration. Be sure that all remaining cracks in the building envelope are sealed with caulk, foam, or weatherstripping.

Figure 12-5.
Spanish tile, because of its barrel shape, produces an air space between the tile and the roof, reducing heat gain through the roof, a major source of overheating in warmer climates.

Source: Stan Chiras Sr.

Removing Built-Up Heat

Reducing internal and external heat gain can lower cooling loads significantly, and in some climates, these strategies may be all that is required to passively cool a home. In others, heat buildup may still lead to uncomfortable interior temperatures. Measures may be required to remove this heat to achieve comfort. Several options are available.

Natural Ventilation

Natural breezes have been used for centuries in all parts of the world to cool people's homes, even in hot, muggy climates. On tropical islands, for instance, homes of bamboo were built on stilts to capture breezes. Bamboo walls permitted good air circulation through the house.

Whenever possible, a house should be sited and designed to take advantage of prevailing winds to provide natural ventilation and cooling. In a well-designed home, a few strategically placed windows often allow the purging of internal heat by creating good cross-ventilation, air flowing naturally through the interior from one side to another. If air is drawn from a cool side, for example, a shady backyard, the result is quite spectacular. Stairwells and open design also promote natural ventilation. So can careful landscaping, discussed in more detail in chapter 15. For example, trees and shrubs, when planted correctly, can funnel breezes toward a house.

When breezes are insufficient, window fans may be required to increase ventilation. Fans can be directed outward to force air out of a house or inward to draw air in. A combination of the two may be useful as well, with incoming air drawn from the cool side of a house and outgoing warm air vented through windows on the warm side.

BUILDING NOTE

"As a general guideline for natural ventilation openable windows should equal 12 to 15 percent of the total floor area if insect screens are used or 6 to 7.5 percent if screens are not used."

**Sustainable Buildings
Industry Council**
Green Building Guidelines

Natural ventilation can also be achieved by opening upper- and lower-story windows to take advantage of the stack or chimney effect—the natural tendency for heated air to rise. Because of this natural phenomenon, opening high-placed windows in a two-story home flushes hot air near the ceiling, replacing it with cool air siphoned in through a cool basement or through first-story windows on the shaded side of a home. Cupolas can also exhaust hot air from a house.

Attic and Whole-House Fans

Heat often builds up in the attic of a house on hot summer days. Because it migrates through ceilings into the living space, removing the heat from the attic can help ensure a cool, comfortable interior.

Attics can be ventilated actively or passively, that is, with or without fans. Passive ventilation occurs through roof or gable vents that allow hot air to escape naturally. Small fans in roof vents increase the removal of heat and use relatively little power. Even better are solar-powered roof vents, which operate during the hottest times of day, when sunlight falls on a dedicated photovoltaic module that powers an electric motor in the device.

A more effective approach is the whole-house fan, which is relatively easy to install and fairly inexpensive. As shown in figure 12-6, a whole-house fan draws cool outdoor air into a home through open windows and exhausts hot room air through the attic to the outside, purging heat from the house.

Whole-house fans are typically run at night to remove heat that has built

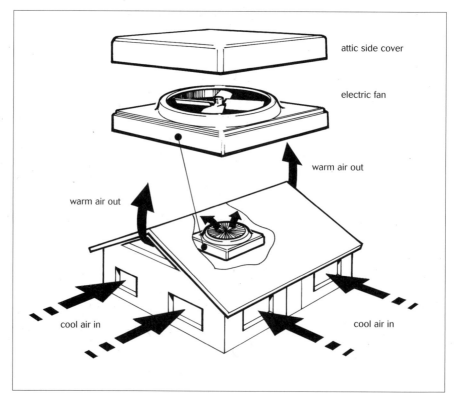

Figure 12-6.
Whole-house fans help provide supplemental cooling in well-built homes and may be all the backup cooling required for summertime comfort.

Source: David Smith

attic side cover

electric fan

warm air out

warm air out

cool air in

cool air in

up during the day. They are much less expensive than central air-conditioning systems. Unlike air conditioners, though, whole-house fans can only cool a home to the outside temperature.

Thermal Mass

Thermal mass, described in chapter 11, can contribute significantly to passive cooling in some locales. Thermal mass in a house absorbs and radiates heat in response to temperature differences. When the air inside a house is warmer than the mass, heat is absorbed by the mass. When the indoor air is cooler than the mass, heat is given off.

During the cooling season, thermal mass inside a home acts as a heat sink, absorbing heat building up in indoor air. The heat absorbed in this way can then be purged from the house at night, for example, by natural ventilation or by running a whole-house fan. However, purging works only if the outside air temperature falls significantly at night.

Interior thermal mass is especially useful for passive cooling in hot, arid climates because nighttime temperatures fall so dramatically, even during the hottest months of the year, due to the lack of heat-trapping moisture in the atmosphere. At night, opening a few windows allows the cool desert air to flow through a house, drawing off heat from walls, floors, and other forms of mass. By morning, the mass has cooled down and will begin absorbing heat from various internal and external sources.

In desert climates, massive exterior walls also help cool a home. During the day, the walls absorb heat, which then begins to migrate toward the cooler interior. If the walls are thick enough, the heat won't make it all the way to the interior before sunset, at which time the outdoor temperature drops and the heat reverses direction (figure 12-7). It's for this reason that thick-walled earthen buildings—like the pueblos of the desert Southwest—have been built in many hot, desert areas throughout the world.

And yet, in hot, humid climates massive exterior walls may be a detriment. In these areas, evening temperatures often remain too high for nighttime ventilation to purge heat from the walls. However, internal thermal mass can be used in conjunction with active cooling systems, such as air conditioners. Rather than open windows at night, a homeowner can turn on an air conditioner for a few hours to remove heat absorbed by the mass during the day. Operating the air conditioner at night saves money because the units do not need to

Figure 12-7.
Thick exterior mass walls help to passively cool homes in hot, arid climates.

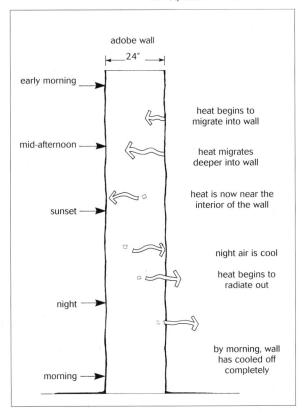

adobe wall

← 24" →

early morning →

heat begins to migrate into wall

mid-afternoon →

heat migrates deeper into wall

heat is now near the interior of the wall

sunset →

night air is cool

heat begins to radiate out

night →

by morning, wall has cooled off completely

morning →

work as hard as during the day, and because nighttime electricity rates are sometimes lower than they are during the day.

Cooling People

Reducing internal and external heat gain and purging heat from homes all help us passively cool our homes. However, there is one more approach that pays huge dividends in comfort: cooling people directly.

Ceiling Fans

One of the most effective and least expensive means of cooling the occupants of a home is an ordinary fan. Although fans can be used to promote ventilation, which removes heat from a house, they can also be used to move air over and around people, drawing off heat and cooling them.

A fan's cooling effect results from the movement of air against the body, which strips heat from the boundary layer, a thin region of warm air lying next to the skin. Removing this warmer air makes us feel cooler. This effect is especially important in hot, muggy climates where evaporation from the skin is reduced by high moisture levels in the air surrounding our bodies. (In the winter, wind strips heat from the boundary layer, resulting in "wind chill"—that is, a feeling that outside temperatures are colder than they really are.)

One of the most popular types of fan is the ceiling fan. Ceiling fans can be adjusted either to draw air up from the floor for cooling in the summer or to push warm air down from the ceiling in the winter. Although portable fans and ceiling fans do not change the internal temperature of a home, they do make us feel cooler in the summer. Studies show that a ceiling fan has the same effect as lowering air temperature by about 4°F.

Ceiling fans are effective in all climates. Although they do require the use of outside energy, fan motors consume very little electricity compared to air conditioners. Ceiling fans can also be used in conjunction with air-conditioning. Researchers at the Florida Solar Energy Center found that using ceiling fans in conjunction with air conditioners in hot, humid climates allows homeowners to set the thermostat 2 to 6°F higher. According to Scott Sklar and Kenneth Sheinkopf, authors of *Consumer Guide to Solar Energy*, each degree of increase saves about 8 percent on cooling costs. It also reduces noise from air conditioners and the environmental pollution caused by the production of electricity.

The Last Resort

If the measures outlined so far do not provide sufficient comfort, it may be necessary to consider other, more conventional methods of cooling, such as evaporative coolers and air conditioners.

Studies show that a ceiling fan has the same effect as lowering air temperature by about 4 degrees F.

An evaporative cooler is a mechanical device that draws outside air through a moistened mat and then pumps it into the house. The cool, moist air is either transported to a central location, where it is released through a single vent, or distributed throughout a house through ducts and ceiling vents. Evaporative coolers consume much less energy than conventional air conditioners, but because they add moisture to the air, they are of little value in hot, humid climates. They also require a fair amount of water.

If you can't use an evaporative cooler, you may need to install an air conditioner or a heat pump. Of the two, heat pumps are more efficient. Heat pumps draw heat from the air inside a house and dump it outside, thus cooling the interior. (See chapter 6 for more details.)

If you want to install an air conditioner or are considering the purchase of a home with an air conditioner, be sure the builder has selected the most efficient model possible. Also, be sure air conditioners are placed in locations that are well shaded. Air conditioners function most efficiently when protected from hot summer sun. If your home includes many of the passive cooling features outlined here, air conditioner run time should be minimal. For more details on mechanical cooling systems and listings of performance data, read the *Consumer Guide to Home Energy Saving,* by Alex Wilson, Jennifer Thorne, and John Morrill.

Combining Measures

Providing comfort in the summer requires an understanding of climate, diligence, and faith—diligence to "design in" a variety of small measures and faith in the cumulative effect of these sensible, cost-effective strategies. Reductions in internal heat gain through the use of compact fluorescent lights, task lighting, daylighting, and energy-efficient appliances may be small. However, when combined with techniques that reduce external heat gain, such as proper solar orientation and window placement, shading, light-colored exterior surfaces, insulation, radiant barriers, and controls on air infiltration, the total effect can be quite significant. Additional gains are achieved by using natural means to eliminate heat that builds up inside a house, for example, by smart use of thermal mass and ventilation. Moving air inside a home with strategically placed fans further advances the goals of natural cooling.

Passive cooling is feasible in any well-built home, whether a standard wood-frame or masonry structure, no matter where it is located. However, many natural building systems, such as adobe and straw bale, are even more conducive to passive conditioning. The thick walls of a straw bale home provide a superbly insulated shell that reduces heat gain during the summer, while the plastered surface provides a considerable amount of thermal mass well distributed throughout the house. Earthen walls made from adobe, cob,

Providing comfort in the summer requires an understanding of climate, diligence, and faith—diligence to "design in" a variety of small measures and faith in the cumulative effect of these sensible, cost-effective strategies.

BUILDING NOTE

Good builders insulate hard-to-reach places during construction, a process called pre-insulation. That way, they're sure these areas don't get missed. One study of new homes in Fort Collins, Colorado, showed that builders who missed these spots ended up installing air conditioners over 200 percent bigger than would have been needed had they done a better job of insulating.

rammed earth, and rammed earth tires provide excellent thermal mass, and are ideal for warmer and drier climates. In colder climates, however, exterior mass walls lose too much heat unless insulated externally.

Laziness vs. Intelligent Design

When I was in graduate school, I lived in a number of poorly constructed apartments. Because these apartments were terribly energy inefficient, their heating systems had to run incessantly to maintain some degree of comfort. Since then, I've been in dozens of houses that suffer the same malady. Why is this so?

Over the years, builders have become cost conscious. Cutting corners to maintain or increase profits, many of them have built inexpensive and extremely inefficient homes. To compensate for poor design and construction, they installed gigantic heating and cooling systems. The results of this unfortunate situation are excessive fossil-fuel consumption, high fuel bills, and unnecessary environmental damage. As you have seen in this chapter, however, attention to detail in the design and construction of a home can virtually eliminate energy-intensive heating and cooling systems, creating greater levels of comfort, huge savings to homeowners, and enormous advantages to the air we breathe.

CHAPTER 13

Green Power: Electricity from the Sun and Wind

California reeled under a sudden, widespread shortage of electricity in 2001. Although growing evidence suggests that the shortage was the result of mismanagement of plentiful electrical resources—and perhaps even some criminal wrongdoing—the shortage dealt a crippling blow to West Coast homeowners. In peak periods, electricity was running as high as $1.30 per kilowatt-hour (kWh) compared with about 8 to 10 cents per kWh in most parts of the country.

In January 2001, before these energy troubles began to emerge, Shea Homes, the tenth largest production builder in the United States, was building solar electric homes in San Diego. Working with AstroPower, a leading manufacturer of solar electric modules, Shea was looking for a way to distinguish itself in a field teeming with competition. "Solar electric power offers value, comfort, and security to home buyers that few building products can provide," writes Colleen Gourley, public relations specialist with AstroPower. Jonathan Done, Shea Homes's director of partnering, adds, "We were already building energy efficiency into our homes, so this was truly a logical step."

Shea is not the only production builder to take this leap into twenty-first

Figure 13-1.
Premier Homes in California is adding solar electric systems to some of their new homes.

Source: AstroPower

century power. Clarum Homes, based in Palo Alto, California, just south of San Francisco, is also building homes that come equipped with solar electric systems. What is remarkable about Clarum Homes is that they're incorporating solar electricity in affordable starter homes intended for first-time buyers. Premier Homes, another production builder, is building homes with rooftop solar electric systems in a new subdivision in Roseville, California.

California is a logical place for solar electricity. With skyrocketing electric rates caused by deregulation, poor management, and possibly seedy manipulation of the system by power brokers, consumers are now offered generous rebates for tapping into the sun's energy to produce electricity. Further sweetening the deal for homeowners, many are able to sell back to their local utilities any excess electricity their homes generate during the day while they're at work.

California's efforts are part of a national movement for a responsible energy future. One big player in this movement is the U.S. Department of Energy, which has mounted a nationwide effort to promote net zero-energy homes, that is, homes that produce more clean, renewable energy than they consume. Someday you may be part of the nation's electrical generating system, too.

In this chapter, we will examine what can be done to turn an existing or new home into a net energy producer. We will begin by looking at solar electricity. We will then turn our attention to wind power and end with a discussion of the green power—clean, renewable energy—now being offered by many utilities the world over.

Solar Electricity

A solar, or photovoltaic, cell is a solid-state device designed to convert energy in sunlight into electricity. Most solar cells in use today consist of very thin layers of silicon, a semiconductor that conducts electricity. When exposed to sunlight, solar cells produce direct-current electricity.

Solar electricity originated over a century ago in the 1870s, according to John Perlin, author of *A Golden Thread: 2500 Years of Solar Architecture and Technology*. It was not until the 1950s, however, that solar electricity began to be used. Silicon solar cells, the kind used today, made their debut flight in the satellite Vanguard I in 1958. Since then, nearly all U.S. satellites have been powered by solar cells made from one of the world's most abundant raw materials, sand. Today, solar cells are being used in calculators, watches, remote signs, and buoys, in remote villages in less developed countries, and also, in recent years, on more and more homes and businesses. Working quietly without producing clouds of pollution, solar cells produce clean electricity.

What Is a Solar Cell?
A solar, or photovoltaic (PV), cell is a solid-state device designed to convert energy in sunlight into electricity. Most solar cells in use today consist of

very thin layers of silicon, called a *semiconductor* because of its ability to conduct electricity. When exposed to sunlight, solar cells produce direct-current electricity. A module is comprised of many cells that each produce a tiny amount of electricity.

The module is framed in metal and glazed with a protective layer, such as glass or weather-resistant plastic, that lets light through but prevents damage from hail and other natural forces.

Each solar electrical module typically produces between 50 and 125 watts of electricity. Because a single module is insufficient for most homes, most solar electric systems consist of anywhere from ten to thirty modules, mounted on racks on the roof of a house, on the ground, or occasionally on poles. Some modules are installed on a rack with a tracking device that follows the sun in its path across the sky, thereby increasing the efficiency of the system.

Figure 13-2.
Solar cells come in several varieties. The most popular today is the polycrystalline solar cell. Single cells above are assembled to form modules. Most homes have a dozen or more modules.

Source: BP Solar

Although a solar electric system mounted on a roof is one of the most beautiful sights I've ever seen (I have a love affair with solar electricity that blinds me at times), not everyone likes their appearance. In recent years, solar manufacturers have begun to produce less conspicuous solar electric materials, for instance *thin-film solar,* which will be familiar to most readers since this is the type of photovoltaic cell used in calculators.

Thin-film materials are used to produce a noncrystalline sun-absorbing layer using a fraction of the semiconductor material of their predecessors. The most widely used thin-film material, amorphous silicon, is less expensive and easier to produce than earlier types. However, amorphous silicon is only about 5 percent efficient.

Figure 13-3.
The newest solar commercial product, PV roofing, does double duty: It protects our homes from the elements like normal roofing, but it also generates electricity. Many homeowners like this product because it is not as visually apparent as standard solar electric modules.

Source: Uni-Solar

Amorphous silicon has other significant disadvantages. It operates best in low light and is actually damaged by bright sunlight. When first developed, its use was restricted to consumer products, such as hand-held calculators, which are usually used indoors, not in direct sunlight. To address these shortfalls, researchers developed layering techniques that improve the efficiency and durability of thin-film solar cells. Multiple layers of amorphous silicon, for example, are about 8 percent efficient and are now being used in solar electric products exposed to direct sunlight, such as the solar roofing materials shown in figure 13-3. Thin coats of this material are also applied to glass, allowing windows and skylights to produce electricity. Both roofing materials and solar glass are known as *building-integrated photovoltaics.*

SINGLE CRYSTAL AND POLYCRYSTALLINE SOLAR CELLS

The first solar cells produced commercially are known as *single crystal silicon* (SCS) solar cells because they were manufactured from a single, pure crystal of semiconductor-grade silicon. Carefully sliced from pure silicon rods, single crystal silicon solar cells are expensive to produce and capable of converting 15 percent of the sunlight striking them into electricity.

To reduce costs, manufacturers began making *polycrystalline silicon* (PCS) solar cells made from a less pure silicon. Consisting of many small crystals, polycrystalline silicon solar cells are considerably less expensive to produce but slightly less efficient—only about 12 percent.

Figure 13-4.
This sleek inverter made by Xantrex converts the direct-current electricity produced by solar modules to alternating current and boosts the voltage from 24 or 48 volts to 120, the voltage required to run virtually all household appliances.

Source: Xantrex

As a rule, if you are building a house more than half a mile from an electric line, it is cost-effective to install a stand-alone solar electric system.

What Is a Solar Electric System?

Solar cells, solar roofing materials, and solar windows are the power source of a solar electric or photovoltaic system. They produce direct-current (DC) electricity, the same type produced by batteries. Direct current results from the unidirectional flow of electrons in wires. Low-voltage DC electricity produced by solar cells and other solar electric devices is drawn off by an array of wires attached to their surface.

For most homes, direct-current electricity is of no value. Virtually everything in our world runs on alternating current (AC). Rather than a continuous one-way flow of electrons in a wire, alternating current results from the rapid flow of electrons back and forth along a wire, a remarkable sixty times per second.

In most solar electric homes, DC electricity from the system is converted to AC electricity by a device known as an *inverter* (figure 13-4). Besides converting DC to AC, the inverter also boosts the voltage from 24 or 48 (typical voltages of solar electric systems) to 120 volts, the voltage commonly required by lights and household appliances. Direct-current electricity also can be stored in batteries, then converted to AC for use in the house.

Solar electric systems come in two basic varieties, stand-alone and utility intertie. A stand-alone system consists of solar modules or solar roofing, an inverter, batteries, and a back-up generator. In this system, solar energy is used to produce DC electricity, which is stored in batteries and then converted to AC current when needed. Houses equipped with such systems are autonomous, which is to say they generate all of their electricity from solar energy, although other devices, such as wind generators, may be added for additional power.

Stand-alone systems are typically installed in remote areas where the cost of bringing electricity to a home via poles or underground wires can be quite costly. As a rule, if you are building more than half a mile from an electric line, it is cost-effective to install a stand-alone solar electric system. Even so, many people like myself who live close to electric lines have opted to build off-grid homes to free ourselves from utility power and help promote a more sustainable future.

A utility-intertie system, on the other hand, typically consists of solar modules or solar roofing and an inverter. There are no batteries or backup generators. DC current produced by the system is converted to AC current by the inverter, then used in the home. Excess electricity is sent into the utility grid—that is, your community's electrical wires—and distributed to others.

In most utility-intertie systems, the electrical grid acts as the "batteries," storing electricity when the system is using more than is required by the household. At night, when the solar cells cease electrical production, the home draws power from the grid, just as it would from a bank of batteries.

Thanks to federal law, utilities must buy excess power from homeowners and other small generators. Although the utilities are only required to purchase power at their production rates, usually about a third of what the consumer pays, some power companies simply let the meter run backward when power is being diverted onto the grid, since it is easier and less costly than installing a separate meter to monitor electricity entering the grid. As a result, power companies are reimbursing the homeowner for power at the full retail cost. If a solar electric home produces as much electricity as it requires in a month, the electric bill will be zero. (Although there will very likely be a fee for reading the meter.) If the system produces more electricity, you, the homeowner, could (theoretically) receive a check from the utility company paying for the power they've purchased from you. (They're more likely to give you a credit on your next bill or to keep the excess without paying you.) If a solar electric home produces less electricity than is needed, the homeowner will be charged for the difference between production and consumption. So why don't more people install solar electricity?

Economics of Solar Electricity

Although the cost of solar electricity has fallen impressively in the past three decades, plummeting 80 percent, and its efficiency has doubled, in most locations solar electricity costs two and a half to four times more than electricity produced from conventional sources. When the cost of a solar electric system, including installation, is amortized over the lifetime of the system, the electricity costs about 24 to 27 cents per kilowatt-hour. In most places, electricity from conventional sources such as coal-fired power plants and nuclear power plants costs residential customers of utilities 8 to 10 cents per kilowatt-hour. So how can builders like those mentioned in the introduction of this chapter afford to install solar electricity?

The builders described earlier are in California, where electricity during peak usage periods now retails for about 50 cents per kilowatt-hour.

By being efficient in their use of electricity, many homeowners can get by with a small PV system, making it economical to install a stand-alone solar electric system if they're as close as 0.2 mile from a utility line. Contact your local utility and ask them how much they will charge to hook your new home up to the grid, then compare this to the cost of a solar electric system to meet your needs.

Figure 13-5.
Solar electricity is the second fastest growing source of electricity in the world thanks in part to the efforts of BP, British Petroleum, the third largest oil company in the world. Besides manufacturing inverters and solar panels, BP is installing solar electric canopies over many of its gas stations.

Source: BP

How Much Solar Is Enough and What Will it Cost?

One of the most common questions I am asked is "How much do I need to spend for a solar electric system?" Of course, the answer depends on how much electricity a home requires. On average, a typical U.S. home will require two watts of PV power per square foot of floor area, according to solar designer Stephen Heckenroth. Thus, a 2,000-square-foot home will require a 4,000-watt or 4-kilowatt system. By making the home energy efficient, a family could easily get by with a system half as big, that is, a 2,000-watt system. (Incidentally, I've been able to live well on a system half that size.) A grid-tied system would cost about $16,000 to $24,000. If your state or utility offers rebates, the cost could be cut in half, that is, to $8,000 to $12,000 to meet your lifetime needs.

Generous subsidies from the California Energy Commission and a 15 percent tax credit from the state also help by underwriting the cost of a solar electric system—paying for half the bill! The high cost of conventional electricity and financial incentives make solar electricity much more affordable, even profitable, in this state.

But California is not alone in its quest to promote greater reliance on the sun and other renewable energy sources. Other states, such as Florida, also offer generous incentives for solar electricity. New York and New Jersey currently offer some of the best rebates in the United States.

As noted earlier, solar electricity also makes sense when a new home is being built more than a half mile from a utility line. Running electric lines this far can cost a homeowner around $50,000, much more than even a large solar electric system. If a home is designed and equipped for energy-efficient operation, solar can save even more money.

Solar electricity makes sense from a number of other perspectives. It helps create personal security by reducing reliance on infrastructure that is vulnerable to a variety of threats and problems. It contributes to national security by creating more homegrown energy. During the first year of operation of their solar electric system in Washington, D.C., Alden and Carol Hathaway reported that the neighbors suffered eight power outages. As is typical, these solar homeowners didn't know about the power outages until after the fact. Their home kept on working without faltering. In my area, we have two or three power outages every spring, some that last for many hours. I typically find out about them the next day or when I show up at the hardware store to find the clerks searching for items with flashlights.

Solar electricity, like other renewable technologies, reduces the carbon dioxide emissions that are largely responsible for global warming. Solar electricity helps put power in the hands of people, creating a wonderful sense of independence. I haven't paid an electric bill since June of 1996! And buying a solar system helps stimulate the solar electric industry. For those who install solar roofing, remember that you're not only getting solar electricity, you're getting a long-lasting roof in the deal!

In the 1990s, solar electricity was the second fastest growing source of energy in the world, and its future is looking brighter every day. Even British Petroleum, the third largest oil company in the world, has climbed enthusiastically onto the solar bandwagon, having purchased Solarex, a leading producer of solar cells. British Petroleum is also equipping many of its gas stations with new canopies fitted with solar electric glass to produce electricity to power the station (figure 13-5). Pretty ironic, isn't it? Using solar electricity to pump gasoline, the fossil fuel on the verge of extinction.

Wind Energy and You

Solar electricity is a highly practical alternative for homes in cities, towns, and rural areas throughout the world. It fits on the roofs of buildings and quietly goes about its business making electricity from a free fuel source. Another option is the wind turbine.

Wind-energy technology can be used to pump water or grind grain; however, most modern wind machines generate electricity (figure 13-6). They go by the names *wind generators, wind turbines, wind machines,* or *wind plants.*

In the first part of the twentieth century, people began to use windmills to generate electricity. In fact, in the United States in the 1930s thousands of small farms, most of them located on the Great Plains, used windmills not just to pump water for livestock, but to produce electrical power. As America industrialized, however, rural windmills began to disappear, going the way of the American bison, because of an ambitious, large-scale rural electrification program sponsored by the federal government and the electric utility industry. Slowly but surely, wind machines were replaced by grid power—electricity delivered from electric lines strung up across the nation and supplied by centralized power plants.

Interest in wind energy was revived in the 1970s as a result of two crippling oil embargos. Thanks to private

Rebates for Renewables

States and utilities offer a variety of rebates and incentives for using solar and other renewable energy resources, including tax credits, property tax exemptions, low-interest loans, and rebate programs. To find out if rebates and other incentives are offered in your area, call a local solar or wind dealer or call your local state energy office (listed in Appendix D) or log on to the Database of State Incentives for Renewable Energy at www.dsireusa.org.

Figure 13-6.
This relatively small wind generator manufactured by Bergey is designed to power entire households.

Source: Bergey

Farmers across the nation are being paid $2,000 a year to lease one-eighth-acre sites on which stand large wind generators mounted on 200-foot-tall towers.

investment and funding from the federal government, by the early 1980s more than thirty companies were involved in manufacturing wind turbines, largely for electrical generation. Several utility companies erected California wind farms containing more than a hundred large turbines.

Unfortunately, wind power didn't fare well early on. Many early models were inefficient and plagued with mechanical problems. Thankfully, the picture has changed. After more than two decades of research and development, the efficiency and reliability of wind generators have been dramatically increased. Since the 1980s, the efficiency of large, commercial wind generators has doubled and tripled. The reliability of modern wind generators, measured as the percent of time a unit is available for operation, now falls in the range of 97 to 99 percent. Today, commercial wind generators are often quite competitive with electricity produced from coal-fired power plants. Wind energy is frequently cheaper than electricity produced from nuclear power plants.

Modern Wind Machines

Modern wind generators vary in size from tiny units designed to power remote signs or ranger stations to large commercial units designed to provide electrical power to many homes, even entire communities.

The smallest wind generators are known as *microturbines*. They are used to provide electricity to sailboats, electric fence chargers, and other low-power applications. Microturbines produce about 300 kilowatt-hours of electricity per year when installed at locations experiencing average wind speeds of approximately 12 miles per hour or 5.5 meters per second, similar to those found on the Great Plains of North America, according to Paul Gipe, author of *Wind Energy Basics*. They're not sufficient for homes, however, in which monthly electrical consumption may be as high as 900 kilowatt-hours.

The next larger size is the *miniturbine,* which is typically used to provide electricity for remote vacation cabins. However, miniturbines can also be used in conjunction with solar electric systems, providing electricity when storms move in to block the sun or at night. Miniturbines generate 1,000 to 2,000 kilowatt-hours of electricity over a year's time in areas with average wind speeds of 12 miles per hour (5.5 meters per second), according to Gipe. They're still too small for most homes.

The next larger unit is household-size wind machines. These produce 1,000 to 20,000 kilowatt-hours per year in conditions similar to those noted above, depending on the size of the generator.

The largest of all wind machines are the commercial generators (figure 13-7). Each of these units produces thirty to fifty times more energy than large household-size units. Although they can be installed singly in remote areas to power entire towns, most are found in large wind farms, where they generate electricity for the nation's power grid.

Figure 13-7.
Large commercial wind generators assembled in giant wind farms provide power to meet the demands of many homes in urban and rural settings.

Source: National Renewable Energy Laboratory

Wind Energy Systems

Like solar electric systems, wind systems can be stand-alone or tied into the grid. Stand-alone systems consist of a power source, the wind generator, an inverter to convert DC electricity to AC electricity, and a battery bank to store electricity for nonwindy periods. They also contain various meters and control devices to prevent the batteries from overcharging or being severely depleted. Grid-connected systems lack batteries, although you may opt to have them just in case grid power fails.

Combining wind and solar electricity is commonplace, as many renewable energy enthusiasts like myself have found that wind and sunlight in combination are a more reliable source of electricity than either alone. In many locations winds tend to peak in the winter. Wind machines often provide a large percentage of the electricity at this time. In the summer, when winds are calmer and the sun is a more common sight, PVs supply the bulk of one's electrical demand.

Is Wind Energy Right for You?

Wind energy is not appropriate for all homes and all locations. For example, it typically doesn't work well in urban and suburban homes. Even though wind generators are usually installed on poles or towers 30 to 100

BUILDING NOTE

Wind generators should be installed in areas away from trees or other large objects like barns. For optimal performance, they should be mounted 20 to 30 feet higher than and 400 to 500 feet away from the closest major obstruction

feet above the ground, trees and buildings may block the wind or greatly reduce its velocity, diminishing the potential for wind power. In addition, local regulations may prohibit wind generators for aesthetic reasons. In such instances, however, a homeowner may be able to purchase wind-generated electricity from a local utility (see below).

Wind generators are more practical in rural settings where access to winds is greater. Turbines also fit better into the rural landscape and can be less obtrusive than in more highly populated areas. Poles, towers, and turbines may be less likely to offend neighbors. Any noise they produce is swallowed up by the wide spaces between homes. However, don't forget that the noise of some high-speed machines can disrupt your own peace of mind. Whatever you do, be sure not mount a wind machine on your roof or against a wall. Vibrations from the unit will be transmitted into the house, disrupting your peace of mind and even your sleep.

Before installing a wind generator, it is essential that you assess wind resources and potential obstructions (see sidebar). Although the cost of a small system, including installation, runs in the range of $1,000 to $3,000, larger systems (including towers) may cost up to $20,000 to $30,000. You don't want to invest that kind of money without assurances that it will pay off. For directions on how to assess the wind potential of a site, consult Paul Gipe's book listed in the resource guide. I've also covered wind power in more detail in my book *The Natural House.*

Be sure to review local building codes, as there may be height restrictions. Research different models and talk to experts as well as people using wind power to find out which machines perform best. For an analysis of noise from wind machines, I strongly recommend Mick Sagrillo's wind generator comparison in *Home Power* magazine (listed in the resource guide).

If your site proves sufficient for wind power, you will be happy to know that many states and utilities offer financial incentives that help offset your initial investment, cutting costs up to 60 percent in some instances (for example, in Illinois and California).

Relying on the Wind?

"The idea of relying on the wind as an energy source may strike you as risky," Michael Hackleman and Claire Anderson observe in a story published in *Mother Earth News,* "since wind seems to be so variable from day to day." They go on to say, "But wind actually acts in fairly predictable ways. Analysis of more than a half-century's recorded data, from thousands of sites, shows distinct patterns in both wind direction and speed through the seasons." Hackleman and Anderson note that there are two distinct kinds of wind: prevalent winds that blow frequently and reliably, and energy winds produced by storms or gusts that piggyback on the prevalent winds. "On average," they write, "there are seven days of prevalent winds and three days of energy winds in any two-week period." Although that may not seem like much wind, it turns out that while energy winds blow only 20 percent of the time, they contain approximately 70 percent of the energy that can be harvested from a wind generator. With batteries to store the energy for use during nonwindy periods, a homeowner can do quite well.

Even in less windy areas, you may live in a microclimate (for example, on top of a ridge or hill) that experiences strong winds that are significantly greater than reported average local wind speeds. Mounting a wind generator on a higher pole can also dramatically increase its electrical production, as higher elevations experience more reliable and stronger winds. Because electrical production increases three-fold for every one mile per hour increase in wind speed, mounting a wind machine higher can double electrical output. As Hackleman and Anderson note, "Most wind speed measurements are taken at a height of 6 feet because that's where wind affects us. But the wind speed is always greater higher above the ground. . . . A wind of 8 mph measured at 6 feet indicates a speed of 11.4 mph on a 36-foot tower and 13.9 mph for a 96-foot tower." This modest increase of 2.5 miles per hour achieved by mounting a wind machine 60 feet higher results in a 100 percent increase in electrical output!

Wind's Bright Future

The potential for wind power is enormous, and although it presently represents a tiny fraction of total global electrical production, wind was the fastest-growing source of energy in the world in the 1990s. It continues to gain in popularity, with Germany (the world's leader in wind-generating capacity), the United States, Denmark, and Spain leading the way. In the United States, Iowa, Minnesota, and California are the major wind states, but many others have great potential. Winds in North and South Dakota, for instance, are sufficient to meet two-thirds of the total U.S. electrical demand.

Even though most wind development is occurring on a grand scale, homeowners can often tap into generous supplies of wind. Be sure to study your site carefully first.

Green Power from Your Local Utility

Fortunately, you don't have to install solar electric panels or a wind generator to tap into the Earth's generous supply of renewable energy. Many utilities are now offering their customers green power, that is, electricity generated from wind and other renewable sources. Where I live, Excel Energy offers customers an option to purchase blocks of wind energy from their ever-expanding wind farm in northern Colorado. Customers pay an additional $2.50 per 200 kilowatt-hours and can opt to buy as little or as much of their monthly electricity as they want from wind farms.

There are two distinct kinds of wind, prevalent winds that blow frequently and reliably, and energy winds produced by storms or gusts that piggyback on the prevalent winds. "On average, there are seven days of prevalent winds and three days of energy winds in any two-week period," according to Michael Hackleman and Claire Anderson ("Harvest the Wind," *Mother Earth News*). Although that may not seem like much wind, it turns out that while energy winds blow only one-third of the time, they contain approximately 70 percent of the energy that can be harvested from a wind generator. With batteries to store the energy for use during nonwindy periods, a homeowner can do quite well.

Winds in North and South Dakota are sufficient to meet two-thirds of the total U.S. electrical demand.

Even if your provider does not have its own wind farms or PVs, it may be purchasing power from renewable producers and can offer you green energy just the same. At this writing, May 2003, eighty utilities in twenty-eight states offer some form of green power. More than a third of all U.S. households could choose some type of green power directly from their local utility or through the competitive marketplace, according to Blair Swezy and Lori Bird, authors of "Businesses Lead the Green Power Charge," published in *Solar Today*. California and Pennsylvania have been the most active markets for green power.

So, if PVs or a wind generator aren't quite right for you, call your local utilities and inquire about the availability of green power through them. As the markets for electricity deregulate, you may be able to obtain green power from their competitors. Even if it costs a few bucks more each month, and it usually will, it's well worth the added expense.

If no utilities in your area provide green power, you can still play an active role in this growing industry by purchasing a green tag. A green tag is a small subsidy to power companies producing green power. It supports their green power programs by paying the additional cost incurred when producing green power. You won't actually receive the electricity. Someone else will, but you will help make it happen.

Green tags are sold by a number of companies and go by different names. The Los Angeles Department of Water and Power calls theirs Green Power Certificates. Pacific Gas and Electric sells Pure Wind Certificates. Waverly Light and Power in Iowa sells Iowa Energy Tags.

With city governments—such as Chicago, San Francisco, and Salem, Oregon—and federal agencies—such as the U.S. Environmental Protection Agency, U.S. Department of Energy, and U.S. Postal Service—making substantial commitments to buying green power, homeowners are very likely going to have even greater opportunities to purchase electricity generated from clean, renewable sources in the near future. At the present time, electricity from these sources costs more, but as I like to remind people, good planets are hard to come by. Spending a little money for environmentally friendly energy is a small price to pay to protect the Earth's life-support systems and create a sustainable way of life.

CHAPTER 14

Water and Waste: Sustainable Approaches

Michael Reynolds is one of America's most innovative green builders. Not only does he build passive solar/solar electric homes, he makes them largely out of an abundant waste product, used automobile tires. Moreover, he has spent much of his adult life creating and improving highly effective systems to capture nutrients from household waste, notably water from sinks, showers, and toilets. Instead of being flushed down the drain to end up in a septic tank or sewage treatment plant, waterborne wastes are used to nourish plants that provide beauty, humidity, and food (figure 14–1).

Reynold's Earthships also capture rainwater off the roof, storing it in large cisterns. The water flows into the house when needed and is purified by an array of filters, providing 100 percent of a family's annual needs in desert-like Taos, New Mexico.

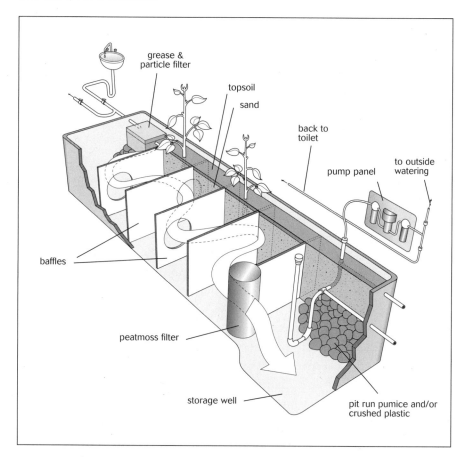

Figure 14-1.
This indoor planter serves triple duty. It provides food and flowers, beautifies the home, and treats gray water. Graywater from sinks and showers enters the lined planter, then makes its way through a thick layer of volcanic pumice, where bacteria and other microbes gobble up nutrients. Plant roots absorb the water.

Source: Lineworks

Although the idea of capturing domestic water off a roof and using waste-water for growing plants may sound mildly old-fashioned, out of step with contemporary building practices, these two techniques could help people in industrialized nations meet future needs while protecting groundwater and surface-water supplies from the twin curses of modern society: resource depletion and pollution. In this chapter, we'll explore these and a host of other ways to obtain fresh water and to recycle used water from sinks, showers, washing machines, and toilets.

Supplying Water Sustainably

Water for human consumption and use comes from two principal sources: surface waters (lakes, streams, and rivers) and groundwater (aquifers). Unfortunately, in many parts of the world surface and groundwater supplies are being depleted by the continually growing human population. There's hardly a country in the world that isn't facing water shortages—or won't face them in the near future. When drought hits, as it does with increasing frequency in these days of global climate change, providing water to thirsty cities and towns is proving to be especially troublesome. So how do we meet our needs for water without further depleting valuable lakes, streams, rivers, and aquifers?

In many parts of the world, surface and groundwater supplies are being depleted by the continually growing human population. There's hardly a country in the world that isn't facing water shortages—or won't face them in the near future.

Conservation: Use What You Need and Use It Efficiently

The most important potential "source" of fresh water is wasted water. Many municipalities lose tremendous amounts of water due to leaks in water mains. Fixing them could supply enormous amounts of this precious liquid. But there are many other effective ways to end the flagrant waste of water. Installing water-efficient fixtures such as showerheads, toilets, and faucets and water-efficient appliances in apartments and homes can make significant inroads into daily household water usage (figure 14-2).

Because water consumption outside a house, such as lawn watering, accounts for an even larger percentage of total demand in many parts of the country (especially in arid and semiarid environments), conservation measures here are especially important in a strategy to reduce water use. Fortunately, much can be done to decrease water consumption outside the house, as described in chapter 15. Even more fortunate, as with energy, water conservation is almost always far cheaper than developing new supplies—for example, damming up rivers to expand a municipality's water supply. Water conservation in and around the house therefore helps protect rivers for wildlife and recreation—and saves money!

So, whether you are buying a new home or building new, be sure to incorporate all of the indoor and outdoor water-saving techniques and

Figure 14-2.
Efficient showerheads help dramatically reduce water use with no loss of comfort or functionality.

Source: Dan Chiras

technologies you can. Not only will they reduce water demand and help reduce your monthly water bill, they will make it easier to rely on an alternative source, such as rainwater and snowmelt.

Catchwater: Capturing Rainwater and Snowmelt

The U.S. Virgin Islands, like other islands in the world, are surrounded by water, billions upon billions of crystal-clear aquamarine water—a vast resource that is, regrettably, too salty to be consumed by humans. Rather than remove the salt from this massive reservoir to produce freshwater, however, residents of this and most other island nations turn to the skies, seeking rain to supply water for domestic uses. Water that falls on their roofs is diverted into large tanks, called *cisterns,* where it is stored for later use.

Cisterns were used in many parts of the continental United States before well-drilling equipment became widely available. Even today some new homes are being built with cisterns to capture rainwater and snowmelt to meet some or all of a family's annual demand.

For best results, water needs to be collected off a clean roof surface, for example, a metal roof—one that won't release potentially toxic chemicals into the water like typical asphalt shingles will. Water flows off the roof, runs down gutters and downspouts, then empties into pipes that deliver it to the cistern (figure 14-3). Cisterns can be installed aboveground or underground, next to the house, or in basements. They are typically made from plastic, fiberglass, metal, or ferrocement (steel-reinforced cement). (Be sure all tanks

Because water consumption outside a house, such as lawn watering, accounts for a large percentage of total demand in many parts of the country (especially in arid and semiarid environments), conservation measures here are especially important in a strategy to reduce water use.

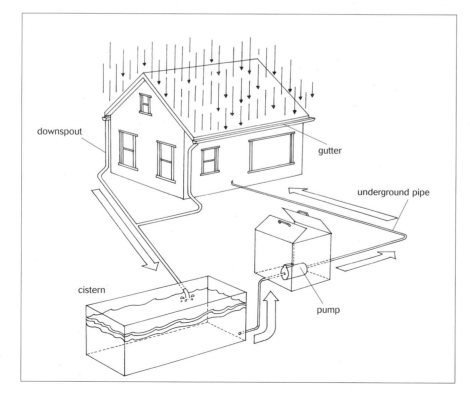

Figure 14-3.
Rainwater and snowmelt captured from a roof, are stored in a cistern, filtered, then used for a variety of indoor functions like bathing, flushing toilets, washing dishes, cooking, and drinking.

Source: Michael Middleton

downspout

gutter

underground pipe

cistern

pump

are rated for drinking water, and that tanks to be buried are rated for burial.) When needed, water is drawn from the cistern, filtered, and used.

Many catchwater homes incorporate a roof-wash system that diverts the first water coming off the roof away from cistern. This prevents dust, dirt, bird droppings, and other potential pollutants from contaminating the cistern.

Unless a catchwater system is installed solely to water plants, in most places it requires filters to remove pollutants, such as particulates or acids, scrubbed from the sky by rain and snow. Many homeowners rely on a combination of filters to remove bacteria, organic pollutants (such as pesticides), acids, and heavy metals. My recommendation is to install filters with recyclable cartridges, so you don't add mountains of disposable filters to our already significant waste stream. Better yet, consider installing systems that contain filter media that can be periodically purged of waste and thus restored to full use.

Although ensuring the purity of a home's water supply is vital, you must address other matters *before* you decide to install a catchwater system, notably how much water you can capture each year and if this will be sufficient to meet your needs. As a rule of thumb, each inch of rainwater will provide about 0.55 gallons of water per square foot of roof. As an example, if your roof is 2,000 square feet and your area receives 20 inches of rain per year, you can expect to acquire 22,000 gallons of water, provided the rain doesn't come all at once and overflow your cisterns. Don't measure actual square footage of your roof but, rather, the square footage of protection it provides, as demonstrated in figure 14-4.

Cistern size should be large enough to get you and your family through the driest part of the year; I generally recommend oversizing them. Cisterns should be designed so they can be filled from a water truck in case you run out, and they should also be equipped with an overflow in case the tank fills to capacity and more rain is on its way. Be sure to install clean-out valves, which allow you to drain the tanks, if necessary, and be sure tanks are accessible so you can clean them, if necessary. In cold climates, you may need to insulate tanks, even underground tanks, to prevent freezing. Aboveground tanks should be protected from sunlight; for example, fiberglass tanks should be coated with gel coat.

Catchwater systems not only help reduce pressure on surface and groundwater supplies, they also save considerable amounts of energy over their lifetime because water does not have to be pumped from a deep well or a domestic water supply. Don't expect huge economic savings, however, when it comes to installing a catchwater system. The costs are often comparable to the costs of installing a well. A catchwater system will probably

---

As a rule of thumb, each inch of rainwater will provide about 0.55 gallons of water per square foot of roof.

Figure 14-4.
Calculate the square footage of a roof very carefully to estimate annual yield. Notice that it is not the total square footage of roof but the total square footage of "rain intercept" (a x b) that determines how much water a roof will capture in any given climate.

Source: Michael Middleton

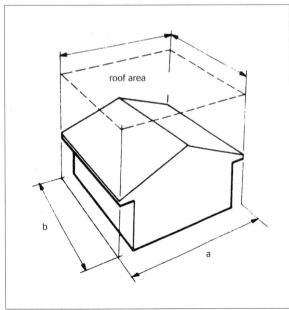

roof area

b

a

cost more than hooking up to a domestic supply (a city or town water system), although once a system is up and running it quickly begins to generate savings, as you won't be paying a monthly water bill. And don't fret about running out of drinking water. In most areas, commercial water haulers can deliver clean, fresh water at a reasonable cost, should drought strike.

If you are not keen on the idea of installing a catchwater system to provide water for domestic uses, you may want to consider installing a system for collecting water to supply gardens, trees, and lawns, which could supplement city water, well water, or a graywater system (discussed below). In addition, your cistern could provide a measure of safety against fire. For further details on catchwater systems, you may want to consult chapter 13 of my book, *The Natural House,* or read other books specifically on catchwater systems, such as Suzy Banks and Richard Heinichen's *Rainwater Collection for the Mechanically Challenged.*

Capturing Nutrients from Waste

Most of the water that enters our houses leaves shortly thereafter, contaminated with a wide assortment of wastes. Which is to say, we consume very little of the water that we "use" and pollute large quantities as we bathe, shower, wash dishes and clothes, and rid our bodies of waste. In most homes, the wastewater from various sources is comingled, then piped out of the house to a sewage treatment plant or to a septic tank.

In a sewage treatment plant, most of the wastes are promptly extracted or broken down. The remaining water is typically chlorinated and then dumped into nearby surface waters—rivers, lakes, bays, or oceans. Organic waste extracted from sewage, known as *sludge,* is buried in landfills or sometimes used to fertilize agricultural land.

In a septic system, the waste decomposes in a tank while the liquid portion drains into a series of pipes buried beneath the ground, known as the *leach field* (figure 14-5). The solid waste that accumulates in the tank is periodically pumped out of the tank and delivered to local sewage treatment plants.

Like so many aspects of our society, household wastewater is part of a linear system—a one-way flow of nutrients from source, say farm fields, to dumps. In nature, however, all waste is recycled, as the wastes from one organism become food for others. Recycling of waste ensures the continuation of life and also prevents unsupportable levels of toxic substances from building up in the environment.

We would be wise to take instruction from natural systems. In waste management, that means creating a more cyclic system in which waste is turned into useful products. Fortunately, there are a variety of ways to follow nature's lead at the household level.

Graywater Systems. Graywater is wastewater from bathroom and utility room sinks, showers, tubs, and washing machines; it gets its name from the color it

Most of the water that enters our houses leaves shortly thereafter, contaminated with a wide assortment of wastes. Which is to say, we consume very little of the water that we "use" and pollute large quantities as we bathe, shower, wash dishes and clothes, and rid our bodies of waste.

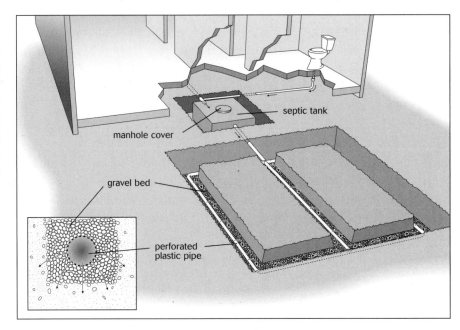

Figure 14-5.
A septic system typically found in rural areas consists of a septic tank that receives graywater and blackwater. Solids settle in the tank, while liquids drain into a series of underground pipes in the leach field.

Source: Lineworks

takes on. Comprising about 80 percent of a household's wastewater, graywater is actually a rather valuable asset, as it contains much-needed water and a host of nutrients useful to plants and soil microorganisms.

Graywater systems capture and recycle these wastes, and they range in complexity. The simplest—and one you could implement immediately—is a plastic tub placed in a kitchen sink for rinsing dishes. The rinse water can be used to water plants. Another simple system is a hose connected to the washing machine (figure 14-6). It can be run out of a window to outdoor plants. More complex systems consist of a tank to receive graywater from washing machines and sinks and a series of hoses or pipes to deliver it to vegetation (figure 14-7)

For best results, graywater should be used immediately. If it sits for a while, even a short period, in a storage tank, it turns smelly as the organic matter in the solution begins to decompose. Although there are no recorded incidents of disease resulting from human contact with graywater, most building codes that allow graywater systems require it to be deposited beneath the surface (near the roots of plants) or onto highly porous material, such as bark or sand, located around outdoor plants (figure 14-8). These techniques prevent graywater from pooling on the surface of the ground where people, pets, or wildlife might come in contact with it.

Graywater can be used to irrigate vegetable gardens and fruit trees, but most experts recommend its use only for plants that bear fruit aboveground, such as tomatoes and squash, and not for root

Figure 14-6.
The simplest graywater system consists of a hose that drains a washing machine outdoors. Be sure not to use bleaches or conventional detergents when watering plants. Use only biodegrable, or, better yet, biocompatible detergents.

Source: Michael Middleton

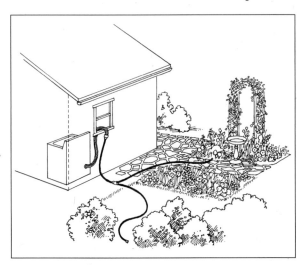

crops, such as potatoes and carrots. When watering gardens, be sure the graywater is applied below the surface.

One innovative way of applying graywater below the surface is the Watson-wick filter, invented by Tom Watson of southern New Mexico. This system, shown in figure 14-9, consists of a plastic device, called an *infiltrator,* placed in a pit dug in the ground near the house. Pumice, a light, porous volcanic rock, is shoveled into the pit and covered with soil. Fruit trees and vegetables can be planted in the overlying soil. Graywater empties into the infiltrator, then migrates outward into the pumice bed, where it is decomposed by bacteria and other microorganisms colonizing the surface of the pumice. As the plants grow, they send roots through the soil and into the pumice, further soaking up nutrients and water, putting your household waste to good use.

In most locations, graywater is used to irrigate outdoor plants. However, graywater can be used indoors, for example, to water decorative plants or vegetables in indoor planters. Here again, there are many options. Michael Reynolds has devised an indoor planter, shown in figure 14-1. Graywater empties into the lined planter, flows through a filter to remove lint and such, then flows by gravity into an underlying bed of rocks. From here it travels through a bed of pumice overlain by soil and plants. As the graywater travels through the planter, nutrients are broken down by bacteria and other microorganisms in the pumice layer. As in the Watson-wick filter, plant roots extend into the pumice, soaking up water and nutrients, growing luxuriantly in their indoor planter.

Blackwater Systems. The remaining 20 percent of household wastewater is blackwater derived from toilets and kitchen sinks, containing blood from meat and vegetable scraps ground up in garbage disposals. (It is more brown in color.) Blackwater is potentially more dangerous than graywater, as it contains fecal matter that may contain pathogenic (disease-causing) microorganisms. Therefore, blackwater is handled even more carefully than graywater.

Blackwater can be delivered to Watson-wick filters, replacing costly septic tanks and leach fields. Reynolds, however, delivers blackwater to solar septic tanks that accelerate the decay of organic matter (figure 14-10). The

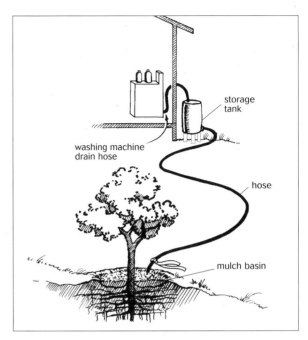

Figure 14-7
A storage tank can be used to accept graywater from washing machines. Water is then released more slowly to plants.

Source: Art Ludwig

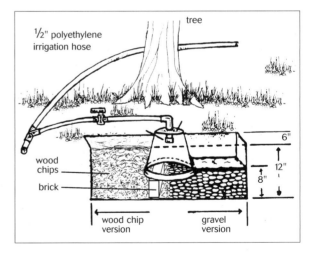

Figure 14-8.
For best results, and to please code officials, graywater should be released belowground in minileach fields as shown here to prevent water from pooling on the surface.

Source: Art Ludwig

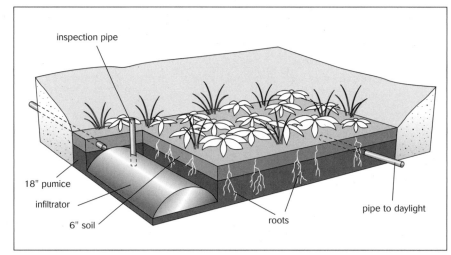

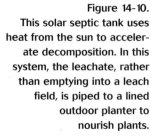

For best results, graywater should be used immediately. If it sits for a while, even a short period, in a storage tank, it turns smelly as the organic matter in the solution begins to decompose.

leachate, that is, the liquid draining from the tank, then enters outdoor lined planters similar to the graywater planters just described. Here the remaining wastes are taken up by plants and the water is purified. Early results from experimental installations show that these systems are highly effective. In one field test, for example, the wastewater treated by the system contained 0.5 ppm nitrate. It was so clean, in fact, that local officials were skeptical. The best their local sewage treatment plants could achieve was 10 ppm nitrate.

Another option is the constructed wetland. One of the safest is the submerged wetland, consisting of a lined depression filled with crushed rock or pumice, then covered with dirt. Sewage from the house empties into the system underground, eliminating any possible contact with humans, pets, or wildlife. Organic matter is decomposed by the bacteria in the rock or pumice. Plants growing in the soil absorb moisture and nutrients from the sewage.

Figure 14-10.
This solar septic tank uses heat from the sun to accelerate decomposition. In this system, the leachate, rather than emptying into a leach field, is piped to a lined outdoor planter to nourish plants.

Source: Earthship Biotecture

Surface wetlands are used by homeowners but more commonly by municipalities and businesses. In these systems, wastewater empties into a lined pond with abundant vegetation growing around its edges and on its surface. These plants and microorganisms remove wastes. Wastewater may flow into secondary ponds where it is further purified.

Composting Toilets. Another option for recycling human waste is the composting toilet, not to be confused with those smelly outhouses you find in parks and national forests. A composting toilet consists of a throne and a waste repository, a chamber below the seat where fecal matter and urine are deposited and broken down or composted by a diverse community of aerobic (oxygen-requiring) microorganisms. Liquid from the waste is evaporated through a vent pipe.

In a properly functioning composting toilet, the organic matter in human wastes breaks down quickly to form a dry, fluffy, odorless material suitable for enriching the soil in flower gardens and around trees.

Composting toilet systems may be purchased from manufacturers or built on-site from standard building materials. One of the best composting toilets on the market is produced by Sun-Mar. In this system, waste is deposited into a rotating drum, where composting occurs. Turning the drum from time to time helps oxygenate the organic material, which accelerates decomposition. Oxygen enters through holes in the front of the unit, which establishes an airflow into the system. Any odors are carried away from the unit by a vent pipe. When the drum fills to two-thirds of its capacity, a turn of the crank causes partially composted waste to spill into a drawer in the bottom of the unit. Here the organic matter undergoes its final decomposition and dessication. When the process reaches completion, the tray is removed and emptied, usually one to four times a year, depending on the level of use, into the soil of flower gardens.

Sun-Mar composting toilets convert human waste, toilet paper, and kitchen scraps into compost in a few weeks at the proper temperature, working one hundred times faster than a septic system. Because they have been certified by the National Sanitation Foundation and approved by the Canadian government, it is relatively easy to obtain a permit to install them in both the United States and Canada, though not all building departments will cooperate.

Although some readers may be concerned about the potential for catching a disease from a composting toilet, there appears to be little to worry about. "According to current scientific evidence, a few months retention time in just about any composting toilet will result in the death of nearly all human pathogens," says Joseph Jenkins, author of *The Humanure Handbook*. Why? Human pathogens don't last very long outside the body. In addition, acidity and bacteria in composting toilets destroy pathogens, as do the high temperatures present in these enclosures.

Jenkins goes on to suggest that, when properly composted, human feces

Composting toilets can be grouped into two categories. The first is a centralized or remote composting toilet system. It consists of a throne in the bathroom and a receptacle located outside the bathroom, usually in the basement just below the bathroom. The toilet itself may be one of three basic types: waterless, ultra-low-flush (microflush), or urine-diverting.

The second option is known as a self-contained unit. It consists of a toilet seat and waste receptacle, which are all part of one attractive assembly.

In a properly functioning composting toilet, the organic matter in human wastes breaks down quickly to form a dry, fluffy, odorless material suitable for enriching the soil in flower gardens and around trees.

can be used to fertilize fruit trees, flower beds, and even vegetable gardens. He adds, "It is well known that humanure contains the *potential* to harbor disease-causing microorganisms (pathogens). The potential is directly related to the state of health in the population which is producing the excrement. If a family is composting its own humanure, for example, and it is a healthy family, the danger in the production and use of compost will be very low."

If you have heard stories about bad composting toilets, chances are they originated from experiences with early systems that did not perform as originally intended. Over the years, however, composting toilet manufacturers have made significant improvements in their designs. "The performance of many composting toilets is more commensurate with system designers' and manufacturers' claims. And more and more composting toilets are appearing in mainstream bathrooms," write David Del Porto and Carol Steinfeld, authors of *The Composting Toilet System Book*. Today, composting toilets are satisfactorily used in thousands of homes in the United States. They're available through Gaim Real Goods and other suppliers.

Satisfaction and Service

Creating sustainable housing requires a wide variety of measures to reduce the impact of our homes during construction and during years of operation. Alternative water systems are part of the list and well worth the investment. Whatever you do, research these topics, especially graywater systems, much more thoroughly, or hire experts to design and install your systems or provide advice. Well-designed systems can provide years upon years of satisfactory service with little impact on the environment.

CHAPTER 15

Environmental Landscaping

George Mitchell is a Texas developer. Unlike the majority of his colleagues, Mitchell chooses to design with nature, rather than waging war on it. In The Woodlands, a subdivision near Houston, Mitchell put his ideas to work, building roads and houses in the heavily wooded area on high ground, leaving natural drainage intact. The result is an attractive open space that removes water and serves as a sanctuary for wildlife. Careful placement of roads and houses also protect aquifer recharge zones, areas where surface water percolates into groundwater, which in this case supplies nearby Houston.

But there are other reasons for environmental landscaping. In The Woodlands, relying on natural drainage made it unnecessary to install a storm sewer, saving $14 million! Attesting to the effectiveness of Mitchell's approach, in 1979, shortly after the development was completed, heavy rains fell on the area, causing streams to swell by 55 percent. In nearby subdivisions built without regard for preserving natural drainage, water flows increased by 180 percent and resulted in extensive flood damage.

This example illustrates one of the most important but frequently overlooked aspects of landscaping: the need to protect natural drainage. Far more commonly, landscaping is used to block unsightly views, to provide shade, and, of course, to beautify the exterior of a house. But today, more and more landscapers are pursuing a wide range of environmental goals, from restoring land disturbed during construction with native plants to promoting the health of native wildlife populations to enhancing the energy performance and comfort of homes. Like so many ideas we've been exploring in this book, environmental landscaping saves money and helps reduce pollution and environmental destruction. It creates a win–win–win situation. That is to say, it is good for people, their pocketbooks, and the environment.

In this chapter, I'll begin by briefly reviewing energy-efficient landscaping. We'll then examine natural drainage, ways to reduce water use in landscaping, and landscaping to preserve wildlife populations. Then we'll turn to the importance of using native species in landscaping and the concepts of edible landscaping and permaculture, ways of turning the land into a garden that produces food, fiber, and other basic necessities.

Today, more and more landscapers are pursuing a wide range of environmental goals, from restoring land disturbed during construction with native plants to promoting the health of native wildlife populations to enhancing the energy performance and comfort of homes.

Energy-Efficient Landscaping

As you may recall from chapters 11 and 12, vegetation and landscaping are key elements of energy-efficient design. They help keep a home warmer in the winter and cooler in the summer. Tall trees, for instance, provide shade in the summer and create a cooler environment through evaporation. Both shade and evaporative cooling promote natural cooling and can make significant inroads into fuel consumption. But, as important as shading is to summertime comfort, remember to place your trees carefully. Shading the south side of a home can reduce passive solar heating dramatically. Even deciduous trees, which lose their leaves in the fall, will obstruct the sun, reducing solar gain. Also, do not shade the roof if solar hot-water or solar electric panels are to be incorporated. A small shadow across a solar electric panel can dramatically reduce the output of the entire array.

Smaller trees and bushes should be planted on the east and west sides of a home to protect against the low-angled summer sun. Trees and shrubs can also be strategically placed to funnel breezes toward a house, providing significant natural cooling in the summer.

Trees can be used to enhance thermal comfort in the winter, too. Planted upwind from the house, they can create a windbreak, protecting a house from cold winter winds. In cold climates, evergreen shrubs such as junipers can be planted alongside a house to create a thermal break, a slight buffer against the cold, which reduces heat loss.

Bear in mind, however, that landscaping strategies vary from one region to another and must be devised with an intimate knowledge of heating and cooling requirements and climatic peculiarities. To learn more, you may want to consult my books *The Natural House* and *The Solar House* and *Energy-Efficient and Environmental Landscaping* by Moffat and Schiler.

Landscaping for Natural Drainage

Poorly drained home sites can be a nightmare. Poor drainage may result in damp basements, causing valuable possessions stored there to take on a musty odor or become damaged. Damp basements also pose a potential health risk from mold.

Moisture accumulating around a house can also cause structural damage, especially in cold climates subject to freezing temperatures. When moisture in the soil under a foundation freezes, it expands. Expansion of the soil may cause the foundation to shift and crack, and cracks in the foundation can cause cracks in the walls of a house, which can lead to costly repair. If severe enough, cracking in the foundation and overlying walls can lead to structural failure.

As the story of George Mitchell illustrates, there are ways of providing drainage naturally. Natural drainage strategies can be inexpensive and effective, and they are far less disruptive to the environment than conventional means.

Natural drainage takes advantage of the existing terrain to remove water from rainstorms and melting snow. Above all, it requires that we site our homes carefully, out of natural water flows. Build your home on higher ground, out of natural drainage pathways, as Mitchell did in The Woodlands. It is also imperative to avoid building in areas where moisture naturally accumulates, for example, in depressions. Preserving the existing topography of a site can promote natural drainage.

Natural drainage also requires efforts to preserve vegetation on building sites by minimizing land disturbance. A well-vegetated site acts like a sponge, soaking up rainwater and snowmelt. The more water that percolates into the soil, the less water runs over the surface, causing erosion, damage to homes, and flooding in nearby streams and rivers. When surface flows are diminished, additional water percolating into the ground nourishes plants and recharges vital groundwater supplies. The more moisture that seeps into the ground, the less time and money a homeowner will need to spend on watering the landscape.

Surface runoff can be decreased by minimizing impervious surface areas such as driveways and patios. Builders can install porous pavers, wherever possible, in place of concrete or asphalt (figure 15-2). These are open blocks laid down to create parking places, driveways, walkways, and patios. Dirt and grass or crushed rock typically fill the spaces. Porous pavers, if properly installed, provide a fairly solid surface for vehicles or walking, while permitting

A well-vegetated site acts like a sponge, soaking up rainwater and snowmelt, because vegetation increases the percolation of water into soil, effectively reducing surface flows that can lead to flooding.

Figure 15-2.
Porous pavers or patio blocks are installed in place of concrete and asphalt. They provide stability, yet permit moisture to percolate into the ground, where it replenishes groundwater. They also reduce flooding.

Source: Nonpoint Education of Municipal Officials, University of Connecticut

virtually all of the water that falls on the "paved" surface to soak into the ground.

Another method builders use to reduce surface runoff is to place turf or gardens between impervious surfaces, for example, between a sidewalk and a street. This allows water to flow from the sidewalk into the growing area, rather than down the gutter. This not only reduces the amount of water entering sewers, it waters lawns and other vegetation.

When building a new home, be sure to explore the surrounding area. Denuded areas uphill from a building site caused by poor land management, such as overgrazing, can greatly increase surface runoff, dramatically increasing soil erosion and sediment deposition in streams, rivers, and lakes. To protect your home and nearby surface waters, you will probably need to revegetate denuded upslope areas. If you don't own the property, you may have to volunteer to replant the land with the owner—and pay the expense of doing so. The investment could be well worth your efforts.

Working with nature—that is, properly siting a home out of trouble spots, taking advantage of the existing terrain to promote natural drainage, protecting vegetation, and replanting disturbed land—can save a lot of money and hassle. In the long run, we'll all be further ahead by cooperating with nature. She knows her business far better than we do!

Water-Efficient Landscaping

In chapter 14, I discussed the importance of conserving water indoors, by using efficient showerheads, fixtures, and such. I noted that much of the water used in many households, especially in arid regions, goes into lawn watering. In such areas, water conservation makes good sense, though in principle it's important everywhere. What can be done to reduce water use outside the home?

The most important measure you can take is to ensure that the soil on the property you are landscaping is in good shape. That means the topsoil must be thick and rich in both organic matter and inorganic nutrients, such as nitrogen and phosphate. A thick layer of nutrient-rich topsoil nourishes grass and other vegetation. Having sufficient organic matter in the soil provides a favorable environment for the soil bacteria needed to maintain healthy vegetation. Organic matter also helps conserve water by acting like a sponge in the soil, so water falling on a garden or lawn does not evaporate or percolate into deeper layers of the soil where it is inaccessible to many plants.

If the soil is nutrient poor, be sure to build it up before planting. Till compost and fertilizer into the soil. Buy additional topsoil if yours is too thin (less than 8 to 12 inches deep).

Once the soil is in good shape, further gains can be made by planting vegetation that requires little, if any, additional watering—that is, plants that can survive on rainfall. In many arid parts of North America, more and more landscapers practice the colorful art of xeriscaping, a landscaping strategy that minimizes water demand. Don't make the common mistake of thinking that xeriscaping means planting cacti in your front lawn. Over the years, I've discovered that there are many delightfully beautiful and colorful xeric flowers, shrubs, and trees that can brighten a landscape.

Xeriscaping requires intelligent planning and design, for example, placing plants that require more water than others near sidewalks, driveways, or downspouts. In all three locations, water-hungry plants are treated to the additional water they require. Besides grouping plants with similar water requirements, skilled gardeners also group them by sun and maintenance needs.

Xeriscapers reduce the amount of yard devoted to grass, especially areas that are small or hard to water or mow. Because common lawn grasses require a lot of water, in part because they lose so much from transpiration, many xeriscapers plant native, drought-resistant grass species, such as buffalo grass, blue grama, Bermuda grass, and fairway wheat grass. Slower-growing species require less mowing, which can dramatically reduce air pollution (see sidebar). I planted my backyard in Denver with a low-water grass and was able to get by with a handful of waterings a year while neighbors watered every three days all summer long. Planting shade trees and shrubs on lawns helps conserve water by reducing evaporation from grassy lawns.

Watering early or late in the day reduces water consumption by 60 percent or more. If you don't want to be hauling hoses and sprinklers around, consider installing an automatic sprinkler system. However, be sure to set the timer so that the water comes on in the morning a few hours before sunrise or in the evening after sunset. Watering during the heat of the day results in massive losses due to evaporation. Also remember that it is best to

BUILDING NOTE

Planting native grasses, trees, and shrubs can reduce outdoor water use by as much as 80 percent.

The most important measure you can take is to ensure that the soil on the property you are landscaping is in good shape. The topsoil must be thick and rich in both organic matter and inorganic nutrients, such as nitrogen and phosphate.

BUILDING NOTE

Studies have shown that automated sprinkler systems typically use 20 to 30 percent more water than manual hosing. This is mainly due to overwatering and leaks. Homeowners who insist on a sprinkler should consider a drip system.

water deeply and less frequently. Watering deeply ensures much deeper root growth and plants that are able to withstand dry spells. Also be sure that automated sprinkler systems do not spray much water on sidewalks, streets, and driveways.

Unfortunately, conventional sprinkler systems lose 30 to 70 percent of their water to evaporation and runoff. A better option is a root-zone irrigation system, which waters grass through pipes buried beneath the surface. Drip irrigation systems are ideal for trees, shrubs, and flowers. They provide water through tiny pipes that slowly drip water near the base of the plants. Both root-zone and drip irrigation systems help reduce outdoor water use and will save homeowners a considerable amount of time. For maximum benefit, be sure that these systems are free of leaks.

Reducing outside water use also requires mulching, the application of bark, grass clippings, or compost on the surface of the soil in flower and vegetable gardens or around trees. Mulches cut down on the evaporation of water from the soil and reduce watering requirements.

For more ideas on the subject, I recommend contacting your local water department or a nursery. In dry areas, many water departments provide free guides to outdoor water conservation. State agricultural extension agents may also be of some assistance, especially when it comes to selecting suitable vegetation for landscaping. You can also log on to www.waterwise.org for ideas on outdoor water conservation. A good book on the subject is *The Resource Guide to Sustainable Landscapes and Gardens* by Environmental Resources, Inc.

Planting Native Species

Planting native species is essential to conserving water, as noted earlier, but also to supporting local wildlife. Native vegetation can survive extremes of weather, such as drought, often faring much better than the exotic species frequently sold in local nurseries. Once established, native plants often require less care and attention than exotics. For example, they are typically resistant to local insects and, because they're adapted to local conditions, frequently require little, if any, additional water and fertilizer. So, when landscaping your home, be sure to select native species. Many nurseries in my area are beginning to stock native grasses, flowers, shrubs, and trees. You may also be able to purchase native plant species through mail-order suppliers. For advice on which species work best, contact your local extension office. They should have a list of appropriate species and may provide some advice on where you can purchase them. Local water departments may also provide information on native species for xeric plantings, including local suppliers, if water conservation is part of their strategy for ensuring a reliable water supply.

Contacting Your State Agricultural Extension Office

The U.S. Department of Agriculture funds state agricultural extension offices, which can provide information on the types of vegetation that are suited to your climate. They also offer information on drip irrigation and organic fertilizers. For a contact list, log on to www.reeusda.gov/ and click on the "State Partners" button.

Because typical lawns provide few benefits to most types of wildlife, reducing the size of the lawn, and replacing it with a variety of native plant species, will often make a yard more suitable for wild creatures. Reducing or eliminating herbicide and insecticide use will also help by protecting insects, the food source of many species.

NATURAL SWIMMING POOLS: THE WAVE OF THE FUTURE?

Swimming pools are popular in many parts of the country. They offer summertime fun, relaxation, and exercise for the young and old—and everyone in between. But pools consume lots of water; use a great deal of energy, especially in areas where they must be heated; require constant vigilance to ensure chemical balance; use a potentially dangerous chemical, chlorine, and acids; and are expensive to build.

There is an alternative, however. It's the natural swimming pool.

Popular in Europe, "where nude beaches are common . . . it was only a matter of time before pools went *au naturel*," writes Michele Taute in *Natural Home* magazine. Natural swimming pools are gaining acceptance slowly in Europe but are viewed with skepticism in the United States. Pioneered by Biotop, an Austrian company, which began working on the idea in 1985, natural swimming pools are being built in Germany, Switzerland, the Netherlands, and England. The company has installed more than 1,500.

Natural swimming pools do not need to be drained at the end of each season, operate with little outside energy, rely on plants for natural filtration and chemical balance, and may be far cheaper to build than conventional swimming pools. A natural swimming pool is made by excavating an appropriate size "pond" in the backyard. Its edges slope gently toward the deeper middle; concrete or cement blocks are not necessary. The pond is then lined with bentonite clay or a synthetic liner, which is covered with 5 inches of clean gravel.

Gravel provides a habitat for bacteria that help biodegrade leaves and other natural materials that accumulate on the bottom of the pond, say Douglas Buege and Vicky Uhland in their detailed story on natural swimming pools in *Mother Earth News*. Additional water purification is provided by a shallow plant zone, as shown in figure 15-3, extending partway or nearly all the way around the pond. It houses lily pads, sedges, reeds, and other aquatic plants. Bacteria associated with the roots of the plants purify the water, removing contaminants and excess nutrients. According to Taute, these microorganisms "keep natural pools clean enough to comply with the European Union's strict water quality standards."

The plant zone is separated from the swimming area by a well-concealed wall or rim an inch or two below the water's surface. This allows water to flow freely in and out of the plant zone. It also allows water heated in the filtration zone by the sun to gently warm the swimming area. In addition, the plant zone serves as a home to frogs that feed on insect larvae, especially mosquitoes. As a result, mosquitoes are rarely a problem.

Although some natural pools require aeration and additional filtration—most often an inexpensive skimmer to remove leaves and seed pods that might fall in the water from nearby trees—the natural swimming pool is relatively simple, inexpensive to operate, and effective.

To learn more about natural swimming pools, contact Biotop and consult the articles and Tim Matson's book (*Earth Ponds Sourcebook*) listed in the resource guide.

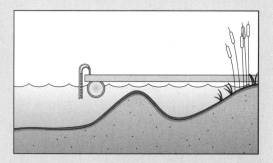

Figure 15-3.
This shallow shelf around the perimeter of a natural swimming pool houses a variety of aquatic plants and microbes that purify the water, creating a healthy, clean pool for swimmers.

Edible Landscapes and Permaculture

Grass is pretty, there's no denying it, but it is otherwise pretty useless. People can't eat it and they can't convert it to fuel. Although it's fun to roll around on with the kids and beats dirt hands-down on aesthetics, grass is a pretty pointless adornment. But what could it be replaced with? Why not an edible

LANDSCAPING FOR WILDLIFE

Several years ago, while working on another book, I got up from my desk to stretch. There on the front lawn was a red fox—this one, a rare black morph. As I watched, the petite and graceful animal lunged forward, snatching an unsuspecting vole from its subterranean tunnel. He then proceeded to skin and devour his prey while I watched in amazement.

Watching this wonderful creature feed on the uncut meadow grass kept wild because I love the long, thick grasses and the wildflowers that sparkle like diamonds in the sea of grass reminded me of the importance of landscaping to provide food and shelter for our fellow travelers, the wildlife.

Wildlife is an important element of the rural landscape but also often coexists with humans in suburbs, towns, and even cities. (Friends have spied coyotes in Los Angeles!) In the ever-growing suburbs of Denver, it is not uncommon to see foxes, squirrels, skunks, and raccoons in yards in search of food. They're accompanied by a delightful variety of songbirds and insects, including those delicate marvels, butterflies.

Because human dwellings have an enormous impact on native species, many people find the idea of landscaping to provide for displaced creatures quite appealing. I'm not suggesting you create bear habitat on your property, or even raccoon habitat. They can be pretty pesty. But there are ways to promote songbirds, butterflies, and other desirable species. Contact your state's or province's division of wildlife for advice. They should be able to help you determine what species live nearby and how to create habitat for them. The National Wildlife Federation has a program that provides information to homeowners who are interested in converting some portion of their yard to wildlife sanctuary. For information on their Backyard Wildlife Habitat Program, contact them at the address listed in the resource guide. You will also find that there are numerous magazine articles on the subject. Once again, don't forget that protecting habitat during construction is the easiest and most cost-effective approach. Protective measures are outlined in chapter 2.

Figure 15-4.
Small pools provide habitat and water for wild species.

landscape? Why not dig up your grass and plant raspberries and fruit trees and carrots and tomatoes? And while you're at it, why not plant some trees for firewood?

The idea of creating a landscape around a home—urban or rural—that provides food, fiber, fuel, and other vital resources is being popularized by the permaculture movement. Coined by two Australian ecologists, Bill Mollison and David Holmgren, the term *permaculture* is a contraction of two other terms, *permanent agriculture* and *permanent culture*.

Permaculture relies on ecological principles of design. According to Brad Lancaster, a teacher at the Permaculture Drylands Institute, "By learning to mirror the patterns found in healthy natural environments, you can build profitable, productive, sustainable cultivated ecosystems" that "have the diversity, stability, and resilience of natural ecosystems" in your own back-yard . . . and front yard as well

You won't be alone, however, because permaculture is inching its way around the planet as people strive for more environmentally benign and self-sufficient lifestyles.

Permaculture is guided by dual goals of sustainable development: care of people and care for the Earth. Although the emphasis is on cultivating plants that provide vegetables, fruit, fiber, and other resources, many permacultur-alists also raise chickens and livestock for food and clothing. Animals help foster a productive ecosystem. Chickens, for example, may feed on insect pests in the garden while fertilizing the land with their droppings.

Principles of permaculture can be applied in the city, on farms, and in remote wilderness locations. Although the idea may seem absurdly impractical to urban and suburban builders, the landscape around a house can often be made to provide for you and your loved ones.

Some suburban home sites have been converted to create a significant amount of staples. Kimberly Reynolds and her hus-band and four children live in the suburbs on a small piece of property that supplies many of their needs. In their tiny gar-den—measuring only 20 feet x 20 feet—the family grows enough corn and tomatoes to last all winter. Surpluses are frozen, stored in the freezer, then popped into the microwave at mealtimes. They also freeze zucchini, green peppers, and hot peppers. Cucumbers are made into relish and bread-and-butter pickles. Beets are pickled as well. Kitchen scraps go into the compost bin and, when they've turned to rich organic matter, are added to the garden. Their home is heated by wood from trees cut down by neighbors. Says Kimberly, "I don't live on a farm or a remote mountainside—nowhere near the 'boonies,' as my mother calls the countryside. I am just another suburbanite with a half-acre plot and a brick ranch house in

Permaculturalists recog-nize that, without a perma-nent agricultural base, no culture can hope to be permanent.

On-line Assistance: Permaculture and Environmental Landscaping

Introduction to Permaculture: Concepts and Resources. An extremely valuable web site with a wealth of information on newsletters, journals, and organizations through-out the world.
www.attra.org/attra-pub/perma.html

Permaculture and Sustainable Living and Livelihood. This web site offers numerous links to other sites covering a wide range of permaculture topics.
csf.colorado.edu/perma/

Sustainable Building Sourcebook. See the section on landscaping for energy savings.
http://www.greenbuilder.com/sourcebook/

"At home, my brother and I harvest over 100,000 gallons of rainwater a year on a 1/8th-acre urban lot and adjoining right-of-way. This harvested water is then turned into living air conditioners of food-bearing shade trees, abundant gardens, and a thriving landscape, incorporating wildlife habitat, beauty, edible and medicinal plants, and more. Such sheltering landscapes cool buildings by 20°F, reduce water and energy bills, and require little more than rainwater to thrive. Outside the home, I have helped others do the same and enabled clients to create ephemeral springs, raise the level of water wells, and shade and beautify neighborhood streets by harvesting their street runoff in adjacent tree wells."

Brad Lancaster
www.harvestingrainwater.com

the middle of it. Our yard is modest, but we've found even on this small amount of land, a family can go a long way toward self-sufficiency."

Tucson permaculturalist Brad Lancaster has taken the idea of living within an urban environment even further. Brad and his brother purchased a dilapidated home in a run-down neighborhood. Over the years, the two young men have fixed up the property, making improvements in the land and the house, turning it into a self-sufficient oasis. They painted the house white to reduce heat gain. They increased the length of the overhang, planted vines (actually vegetables) that creep up wires to shade the house during the summer, and planted trees and vegetable gardens, fed by rainwater collected from their roof. They even cook outdoors in a solar oven during the summer to prevent overheating. At night, they open windows to let in cool desert air to remove heat that may have accumulated inside the house during the day.

During the winter, they heat their home with solar energy that streams in through south-facing windows. A small woodstove provides backup heat and is fueled with wood from trees grown on the premises. All year round, hot water comes from a solar water heater that requires no backup.

Electricity comes from one solar electric panel with two batteries. Among other things, the PV panel supplies electricity to ceiling fans they built out of old bike parts. "As they are powered by the panel, we literally cool the house with the sun," remarks Lancaster. Slowly but surely, they've become more and more self-sufficient—all within the confines of a city.

The indefatigable Lancaster has also organized an annual neighborhood tree-planting program. Since 1996, he and his neighbors have planted more than eight hundred low-water native trees. "The neighborhood has been beautified and cooled, and folks are investing in their community. Neighbors are getting to know one another, people are regaining familiarity with native vegetation of the Sonoran Desert, and native bird populations and diversity are soaring within the inner city," notes Lancaster. He has also organized the creation of an organic community garden, mini nature park, and an orchard.

If you are interested in following in Lancaster's footsteps, take a look at my newest book, *Superbia!* Written with Dave Wann, this book offers a wealth of advice on ways to live more self-sufficiently and economically within suburbs—changing Suburbia to Superbia! We group ideas in the book by degree of difficulty. (This book and several articles we've published on the topic are listed in the resource guide.)

A vegetable garden is a great place to start your edible landscape. A few fruit trees will further expand your culinary repertoire. If you have a woodstove or a masonry heater, starting your own forest might help meet your demand for fuelwood. Fast-growing species such as cottonwoods are an ideal choice. From these beginnings, you can pursue many other options.

Much more than a place to relax in the sun on a sunny afternoon, a drain on resources, and a source of pollution, your yard can be a productive element of your life, making your home more comfortable year-round, absorbing rain and snowmelt while preventing flooding, conserving limited water supplies, providing habitat for butterflies and songbirds, and even supplying you and your family with a host of edible fruits and vegetables.

Making Your Dream Come True

In 1995, I set out to build a state-of-the-art environmental home. Armed with what I thought was a substantial amount of knowledge on important aspects of green building, such as passive solar heating and cooling and natural building materials, but having very little building experience, I consulted with a builder from Colorado Springs. He had some experience building the type of home I was interested in, and I liked his ideas and the sketches he provided. So, I hired him to draw up a set of blueprints. Over the next few months, I began issuing checks to him. We spoke a few times on the phone, then one cold day in March we met in his apartment to go over plans. I was disappointed to find that he'd made very little progress.

Shortly after this meeting, he disappeared, taking $2,000 of my hard-earned cash and the partially completed plans with him. That was the last I heard from him, except for a rumor several years later that he was bragging about designing my home.

Needless to say, this was a rude welcome to the sometimes unscrupulous world of home building.

A few months after this fiasco, I hired two knowledgeable and reputable builders and began the process all over. This time, our collaboration was far more fruitful. These two mavericks produced a great set of plans in a timely fashion, and in short order we were on our way to creating my home in Evergreen, Colorado. Unschooled in construction, I soon found myself on a learning curve steeper than the north slope of Mount Everest. In fact, it was this experience that convinced me to begin writing building books, so that others might find the path to sustainable shelter far less daunting. In this section, I'll share some more ideas to make your life easier.

Buying a Green Home

Although buying an already-built green home is simpler and less demanding than building your own from scratch—or, as is more common, having someone build one for you—there are some things you should know when setting out on this course. First, how do you locate a green-built home in your area?

The first step is to check the local newspaper for ads. You should definitely contact your state's or city's green building program, if one exists. The staff there can probably supply a list of builders and developers who are active in green building in your area. Local architects and builders may also be able to steer you in the right direction. Or attend a local green building conference. Chances are, if you ask enough people, you will find a few builders in your area who are building high-quality environmental homes

When you have identified local green builders, read their promotional material, interview them, and talk to people living in their homes. Compare the features of their homes, paying close attention to energy efficiency, renewable energy systems, indoor air quality, the use of green building materials, water efficiency, efficient appliances, landscaping, optimum value engineering, and recycling of construction waste. Then compare your lists to the checklists I've provided in appendices A, B, and C, or to the green building requirements set forth by a local green building program. If the lists reveal substantial shortcomings, you may want to consider having a home built specially for you or building one yourself.

If you are considering buying an existing home, it's a good idea to check its actual energy consumption by contacting a local utility. They'll usually provide the information free of charge. If it is a new home, ask for energy performance projections the builder or developer may have made. If these aren't available, you can hire an energy specialist to make projections using sophisticated computer software, such as Energy-10, which will allow you to estimate the energy demands for heating and cooling a home in your area.

Once you locate a home you would like to purchase, be sure to test indoor air quality, including radon, and check into air infiltration, *before* you finalize the transaction. In rural settings, you may want to have well water tested as well. I recommend adding a clause to the contract that stipulates successful performance of the home on these tests as a final condition of the sale. If the house does not measure up, you can back out of the contract with a full refund of your earnest money.

Be sure to look into special Energy Star mortgages for houses that have been built to use energy efficiently. These are discussed in chapter 6. Remember that even though a green home may cost a little more, lower interest rates and other incentives could offset the additional costs. Substantially lower energy bills could translate into immediate savings as well.

If all of this seems daunting, you may want to hire a green building consultant to work with you. He or she can assess the environmental aspects of a home and can compare several homes, offering advice on the best buy. While you are at it, don't forget to assess the neighborhood and other vital aspects of a community, such as fire and police protection, access to schools, noise level, and air quality.

Building a Green Home or Remodeling
Far more involved than buying a new home is building anew or remodeling an existing home. When pursuing these paths, you have two basic options: You can do the work yourself or you can hire a contractor to do the work for you.

When building yourself, the easiest course is to assume the role of the general contractor. This requires no special training or license. As a general contractor, you file for building permits and hire various subcontractors, from framers to plumbers to roofers, to complete the construction. Be sure to write detailed contracts with all subcontractors so misunderstandings are avoided.

Although assuming the role of the general contractor can be difficult, especially if you are not familiar with home construction, this option can save 10 to 25 percent of the cost by eliminating the overhead and profits of a builder you would hire. If you are interested in this path, be sure to talk to others who have done it. Books such as Carl Heldmann's *Be Your Own House Contractor* offer much useful information. Hiring a green building consultant is not a bad idea either. That person may be able to recommend reliable and competent subcontractors and communicate with them and inspect their work better than you can.

A far more demanding way of building your own home is to do the construction yourself. Even in this scenario, you may want to hire specialists for tasks such as drawing up blueprints and installing plumbing or electrical wiring.

Building your own home is hard work and extremely time-consuming. If you are not a skilled builder, the process can be difficult and enormously frustrating. It can take forever, too, and can cost far more than you ever imagined. Lest we forget, it has destroyed many a marriage.

Because natural building materials are often easy to master, many owner-builders are using them to construct their homes. Even so, a structural engineer's stamp may be required by the local building department for approval of your plans.

For guidance on building a home yourself, read the last chapter of my book *The Natural House*. It describes the stages involved in home construction as well as the permits that are typically required by local building departments. It also contains indispensable and practical information that could save you thousands of dollars.

If you have specific questions on green building, you can log on to www.greenhomebuilding.com and click on "Ask the Experts." Your question will be routed to an expert in the field who will respond to your query. Incidentally, I respond to all questions on passive solar design and general queries on natural building.

If constructing a house doesn't appeal to you, you can always hire a builder. In this scenario, it is even more important to interview candidates, to make sure you find someone whose level of commitment matches yours, and with whom you can work successfully over an extended period of time. Beware of those who only dabble in green building, especially if you are interested in pushing the envelope. I've known a few people who have hired such builders and have been disappointed by the final product. Whatever

Although assuming the role of the general contractor can be difficult, especially if you are not familiar with home construction, this option can save 10 to 25 percent of the cost by eliminating the overhead and profits of a builder you would hire.

Building your own home is hard work and extremely time-consuming. If you are not a skilled builder, the process can be difficult and enormously frustrating. It has destroyed many a marriage.

you do, be sure to write detailed contracts with builders, stipulating materials to be used and other important aspects. The more you have written down on paper, the less room there is for confusion and potential cheating.

If your choices are limited and you cannot find a builder who meets your expectations, you can hire a green architect or building consultant to work with your builder to ensure that your home is as environmentally sound and healthy as you would like. Be sure your builder is open to working with "an outsider" and is not going to resist the consultant's suggestions. Many builders are comfortable with practices and materials they are familiar with using and can be quite reluctant to try something new. A few meetings up front, that is, before you sign a contract, will give you a feeling for your contractor's openness to a consultant's ideas and how well the two people will get along.

Whatever path you take—building a home yourself or hiring a contractor to do the job for you—remember that an integrated approach, discussed in chapter 6, is vital to the success of your project. Extensive thought and planning before construction begins is vital. Although hiring a consultant to look over plans or help develop them may consume valuable time and money, in the long run preplanning can save enormous amounts of money and can greatly increase the quality of a home. One way preplanning saves money is by eliminating change orders, special amendments to your contract that call for changes in construction from the original agreed-upon plans. Change orders cost money, sometimes a lot of money, so the better job you do of planning your home, the less likely it is that you will have to make changes along the way.

Whether you build a home yourself or hire someone to do it for you, be prepared to make lots of decisions. When in doubt, take time to research options. Suppliers and manufacturers of green building materials can be called to help with the installation of materials if a builder is unfamiliar with them.

The Long Term

Green building promotes human health and comfort. Because a house represents a long-term contract with the environment, however, green building is also a high-impact way of reducing damage to the environment. Its benefits will ripple silently through human and natural systems for decades, perhaps even centuries, after a home is built. No matter what path you take—buying a green-built home or building one yourself—homes that are good for people, the planet, and our economy provide enormous personal rewards, among them a pride of ownership unrivaled in the modern world. And you don't have to be a dyed-in-the-wool environmentalist to build green, as the accompanying story about George W. Bush's home in Texas proves.

Although hiring a consultant to look over plans or help develop them may consume valuable time and money, in the long run preplanning can save enormous amounts of money and can greatly increase the quality of a home.

Green building promotes human health and comfort. Because a house represents a long-term contract with the environment, however, green building is also a high-impact way of reducing damage to the environment. Its benefits will ripple silently through human and natural systems for decades, perhaps even centuries, after a home is built.

THE GREEN TEXAS WHITE HOUSE

In April 2001, *USA Today* published an article on George and Laura Bush's Crawford, Texas, home, located halfway between Dallas and Austin. A sprawling 4,000-square-foot home on 1,600 acres of grassland graced by live oak trees, this newly built home has become the president's retreat, his home away from the White House. Much to the surprise of many, the home also features numerous green building features.

Although a 4,000-square-foot home hardly constitutes a small building, it's a far cry from the 16,000-to-20,000-square-foot mansions most people of the Bushes' wealth and power are having built. With exterior walls made from local limestone, waste from a local quarry, the house was designed to blend into the Texas landscape. It's large by most people's standards, but thanks to its low profile, the home is not obtrusive.

Moreover, the house was situated to take advantage of breezes for passive cooling. It's one-room-wide design makes it easy to ventilate naturally and no doubt results in a significant amount of daylighting. Wide overhangs provide shade, and a grove of live oaks on the west side of the house provides protection from the late afternoon sun. The roof of the Texas White House was framed with preassembled roof trusses and covered with white galvanized tin to repel summer heat and reduce the cooling load.

Backup heating and cooling are provided by a ground-source heat pump. This system uses 75 percent less electricity than a conventional heating and cooling system and even heats the pool, included at the insistence of the Bush daughters.

Rainwater from the roof is collected in a 25,000-gallon cistern and used for irrigating the flower garden and newly planted trees. But that's not all. The Bushes' home is equipped with a wastewater recycling system. It collects and purifies blackwater from toilets and graywater from sinks, showers, and washing machines. The purified wastewater then empties into the cistern, providing an ample supply of irrigation water in this hot, dry landscape. The water is no doubt a valuable ally in Laura Bush's quest to restore native wildflowers and grasses on the property.

The house was also designed for accessibility. The layout is very simple and straightforward. There are no stairs and no thresholds to pose difficulties to aged parents or relatives.

Most rooms of this eight-room "eco-mansion" are small and simple and sparingly attired. Most of the square footage is concentrated in the living room, dining room, and kitchen—a strategy for making do with less space.

Designed by David Heymann, associate dean of the School of Architecture at the University of Texas, this modest home (by multimillionaire standards) stays warm in the winter and cool and comfortable in the summer, using a fraction of the energy of a similarly sized home. "The features are environment-friendly," notes Heymann, "but the reason for them was practical—to save money and to save water, which is scarce in this dry, hot part of Texas."

In closing, I like to think of green building as a way that humanity expresses its collective concern for the future and its willingness to take action that benefits all species now and in the long term. You can be a part of the important movement, even if on a small scale. Of course, the more you do to create a truly sustainable shelter, the brighter your—and our— future will be.

Resource Guide

This resource guide contains chapter-by-chapter listings of books, articles, videos, magazines, newsletters, organizations, and suppliers. Because addresses, phone numbers, and web sites change, I've tried wherever possible to provide multiple access points for each resource.

Chapter 1—Green Building and Beyond

Publications

Borer, Pat, and Cindy Harris. *The Whole House Book: Ecological Building Design and Materials.* Machynlleth, U.K.: Centre for Alternative Technology Publications, 1988. A detailed treatment of green building.

Chiras, Daniel D. *Lessons from Nature: Learning to Live Sustainably on the Earth.* Washington, D.C.: Island Press, 1992. For those interested in learning more about principles of ecological sustainability.

———. "Principles of Sustainable Development: A New Paradigm for the Twenty-First Century." *Environmental Carcinogenesis and Ecotoxicology Reviews* C13, no. 2 (1995): 143–78. A detailed exploration of the social, economic, and environmental principles of sustainability.

Dickinson, D. *Small Houses for the Next Century.* 2nd ed. New York: McGraw-Hill, 1995. One of several books on the subject. Full of good information.

Johnston, David. *Building Green in a Black and White World.* Washington, D.C.: Home Builder Press, 2000. A book for builders who want to learn more about the business side of green building.

Susanka, Sarah. *Creating the Not So Big House.* Newtown, Conn.: Taunton Press, 2000. A follow-up to Susanka's popular book listed below.

———. *The Not So Big House.* Newtown, Conn.: Taunton Press, 1997. A highly popular introduction to the art of building small houses, with floor plans and many exquisite photos.

Sustainable Buildings Industry Council. *Green Building Guidelines: Meeting the Demand for Low-Energy, Resource-Efficient Homes.* Washington, D.C.: SBIC, 2002. General guide to green building, covering many important topics.

Magazines and Newsletters

Environmental Building News. BuildingGreen, Inc., 122 Birge Street, Suite 30, Brattleboro, VT 05301. Tel: (802) 257-7300. Web site: www.buildinggreen.com. The nation's leading source of objective information on green building, including alternative energy and back-up heating systems. Archives containing all issues published from 1992 to 2001 are available on a CD-ROM.

Environmental Design and Construction. 81 Landers Street, San Francisco, CA 94114. Tel: (415) 863-2614. Web site: www.EDCmag.com. Publishes numerous articles on green building; geared more toward commercial buildings.

Mother Earth News. 1503 SW 42nd St., Topeka, KS 66609. Tel: (785) 274-4300. Web site: www.motherearthnews.com. Publishes a wide assortment of stories on green building, from natural building to solar and wind energy to natural swimming pools to green building materials.

Natural Home. 201 East Fourth St., Loveland, CO 80537. Tel: (800) 272-2193. Web site: www.naturalhomemagazine.com. Publishes numerous articles on green building, especially natural building and healthy building products, with lots of inspiring photographs.

Organizations

American Institute of Architects. 1735 New York Ave. NW, Washington, DC 20006. Tel: (800) 242-3837. Web site: www.aia.org. Their National and State Committees on the Environment are actively promoting green building practices, and have been for many years.

BuildingGreen, Inc. 122 Birge St., Suite 30, Brattleboro, VT 05301. Tel: (802) 257-7300. Web site: www.buildinggreen.com. Publishes *Environmental Building News, GreenSpec Directory* (a comprehensive listing of green building materials), *Green Building Advisor* (a CD-ROM that provides advice on incorporating green building materials and techniques in residential and commercial applications), and Premium Online Resources (a web site containing an electronic version of its newsletter).

Building Industry Professionals for Environmental Responsibility. 5245 College Ave., #225, Oakland, CA 94618. Web site: www.biperusa.biz. A national nonprofit organization that promotes environmentally sustainable building.

Center for Resourceful Building Technology. 127 N. Higgins, Suite 201, Missoula, MT 59802. Web site: www.crbt.org. A project of the National Center for Appropriate Technology. Promotes environmentally responsible construction.

Ecological Building Network. 209 Caledonia Street, Sausalito, CA 94965-1926. Tel: (415) 331-7630. Web site: www.ecobuildnetwork.org. Seeks ways to build environmentally sustainable shelter in both industrial and nonindustrial nations.

National Association of Home Builders Research Center. 400 Prince George's Blvd., Upper Marlboro, MD 20744. Tel: (301) 249-4000. Web site: www.nahbrc.org. A leader in green building, including energy efficiency. Sponsors important conferences, research, and publications. For a listing of their books contact www.builderbooks.com.

Chapter 2—Site Matters

Publications

Center for Resourceful Building Technology Staff. *Reducing Construction and Demolition Waste.* Missoula, Mont.: National Center for Appropriate Technology, 1995. A guide for builders and homeowners on job-site recycling.

Chiras, Daniel D. *The Natural House: A Complete Guide to Healthy, Energy-Efficient, Environmental Homes.* White River Junction, Vt.: Chelsea Green, 2000. Chapter 13 contains more detailed coverage of a number of aspects of site selection briefly mentioned in this book.

Clark, Sam. *The Real Goods Independent Builder: Designing and Building a House Your Own Way.* White River Junction, Vt.: Chelsea Green, 1996. Check out the chapters on choosing a site and site planning.

Rousseau, D., and J. Wasley. *Healthy by Design: Building and Remodeling Solutions for Creating Healthy Homes.* 2nd ed. Point Roberts, Wash.: Hartley and Marks, 1999. This book offers a great deal of advice on site selection.

Smith, Michael G. *The Cobber's Companion: How to Build Your Own Earthen Home.* 3rd ed. Cottage Grove, Ore.: The Cob Cottage, 1998. This book has an excellent section on siting a home.

On-Line Publications

Bernard, K. E., C. Dennis, and W. R. Jacobi. *Protecting Trees During Construction.* Available from Colorado State University Extension Service at www.ext.colostate.edu/pubs/garden/07420.html.

Johnson, Gary R. *Protecting Trees from Construction Damage: A Homeowner's Guide.* Available on-line from the University of Minnesota Extension Service at www.extension.umn.edu/distribution/housingandclothing/DK6135.html.

Organizations

The National Arbor Day Foundation, Building With Trees Program (in cooperation with the National Association of Home Builders). 100 Arbor Avenue, Nebraska City, NE 68410. Web site: www.arborday.org/programs/buildwtrees.html. For information on ways to protect trees during construction.

Chapter 3—The Healthy House

Publications

American Lung Association, U.S. Environmental Protection Agency, U.S. Consumer Product Safety Commission, and American Medical Association. *Indoor Air Pollution: An Introduction for Health Professionals.* Publication No. 1994-523-217/81322. Washington, D.C.: U.S. Government Printing Office, 1994. Dynamite reference for more detailed information on the health effects of the most common indoor air pollutants. Very valuable for diagnosing problems caused by indoor air pollution. Also contains an extensive bibliography of research papers on the subject. Available at www.epa.gov/iaq/pubs/hpguide.html.

Baker-Laporte, Paula, Erica Elliot, and John Banta. *Prescriptions for a Healthy House: A Practical Guide for Architects, Builders, and Homeowners.* 2nd ed. Gabriola Island, B.C.: New Society Publishers, 2001. Contains a great amount of useful information.

Borer, Pat, and Cindy Harris. *The Whole House Book: Ecological Building Design and Materials.* Machynlleth, U.K.: Centre for Alternative Technology Publications, 1998. Contains a wealth of information on building healthy, environmentally friendly homes.

Bower, John. *The Healthy House: How to Buy One, How to Build One, How to Cure a Sick One.* 3rd ed. Bloomington, Ind.: The Healthy House Institute, 1997. A detailed guide to all aspects of home construction.

Bower, John, and Lynn Marie Bower. *The Healthy House Answer Book: Answers to the 133 Most Commonly Asked Questions.* Bloomington, Ind.: The Healthy House Institute, 1997. Great resource for those who just want to learn the basics.

Davis, Andrew N., and Paul E. Schaffman. *The Home Environmental Sourcebook: 50 Environmental Hazards to Avoid When Buying, Selling, or Maintaining a Home.* New York: Henry Holt, 1997. Overview of the sources of health hazards in homes.

U.S. Consumer Product Safety Commission, U.S. Environmental Protection Agency, and the American Lung Association. *What You Should Know About Combustion Appliances and Indoor Air Pollution*. Washington, D.C.: EPA, undated. A great introduction to the effects of indoor air pollutants from combustion sources. Available at www.epa.gov/iaq/pubs/combust.html.

U.S. Environmental Protection Agency. *Model Standards and Techniques for Control of Radon in New Residential Buildings*. Washington, D.C.: EPA, 1994. This on-line document provides detailed and fairly technical information on ways to prevent radon from becoming a problem in new construction. Available at www.epa.gov/iaq/radon/pubs/newconst.html.

U.S. Environmental Protection Agency and the U.S. Consumer Product Safety Commission. *The Inside Story: A Guide to Indoor Air Quality*. EPA Document No. 402-K-93-007. Washington, D.C.: U.S. Government Printing Office, 1995. Very helpful on-line publication for those interested in learning more about indoor air quality issues and solutions. You can access it at www.epa.gov/iaq/pubs/images/the_inside_story.pdf.

U.S. Environmental Protection Agency, U.S. Department of Health and Human Services, and U.S. Public Health Service. *A Citizen's Guide to Radon: The Guide to Protecting Yourself and Your Family from Radon*. 4th ed. Washington, D.C.: EPA, 2002. A very basic on-line introduction to radon. Available at: www.epa.gov/iaq/radon/pubs/citguide.html.

Organizations

Air Conditioning and Refrigeration Institute (ARI). 4100 N. Fairfax Dr., Suite 200, Arlington, VA 22203. Tel: (703) 524-8800. Web site: www.ari.org. Information on in-duct air filtration/air cleaning devices.

American Academy of Environmental Medicine. 7701 East Kellogg, Suite 625, Wichita, KS 67207. Tel: (316) 684-5500. Web site: www.aaem.com.. Contact them for the name of a physician who is qualified to diagnose and treat multiple chemical sensitivity.

American Society of Heating, Refrigerating, and Air-Conditioning Engineers (ASHRAE). 1791 Tullie Circle, NE, Atlanta, GA 30329. Web site: www. ashrae.org. Provides information on air filters.

Association of Home Appliance Manufacturers (AHAM). 1111 19th St. NW, Suite 402, Washington, DC 20036. Tel: (202) 872-5955. Web site: www.aham.org. For information on standards for portable air cleaners.

The Healthy House Institute. 430 N. Sewell Road, Bloomington, IN 47408. Tel: (812) 332-5073. Web site: www.hhinst.com. Offers books and videos on healthy building.

Indoor Air Quality Information Clearinghouse. P.O. Box 37133, Washington, D.C. 20013-7133. Tel: (800) 438-4318. Distributes EPA publications, answers questions, and makes referrals to other nonprofit and government organizations.

Multiple Chemical Sensitivity Referral and Resources. 508 Westgate Road, Baltimore, MD 21229. Te: (410) 362-6400. Web site: http://www.mcsrr.org/. Professional outreach, patient support, and public advocacy devoted to the diagnosis, treatment, accommodation, and prevention of Multiple Chemical Sensitivity disorders.

The National Safety Council's Radon Hotline. Tel: (800) SOS-RADON. Web site: www.ncs.org/ehc/radon.htm. Calling this number or contacting the web site will give you access to local contacts who can answer radon questions.

U.S. Consumer Product Safety Commission. Washington, DC 20207-0001. Tel: (800) 638-CPSC. Web site: www.cpsc.gov. Contact them for information on potentially hazardous products or to report one yourself.

Suppliers

See list of green building material suppliers below.

Chapter 4—Green Building Materials

Publications

Demkin, Joseph A., ed., *The Environmental Resource Guide.* Washington, D.C.: American Institute of Architects, 1992. A massive publication that provides detailed life-cycle analyses of many construction materials. For a copy, contact the distributor at www.wiley.com.

Chappell, Steve K., ed. *The Alternative Building Sourcebook.* Brownfield, Maine: Fox Maple Press, 1998. Lists more than 900 products and professional services in the area of natural and sustainable building.

Chiras, Dan. "Green Remodeling: Keeping It Clean." *Solar Today* (May/June 2001): 24–27. Describes a strategy for remodeling a home to prevent indoor air pollution.

City of Austin's Green Building Program. *Sustainable Building Sourcebook.* Austin: City of Austin Green Builder Program. Excellent resource available on-line at www.greenbuilder.com/sourcebook.

Holmes, Dwight, Larry Strain, Alex Wilson, and Sandra Leibowitz. *Environmental Building News. GreenSpec Directory: Product Directory and Guideline Specifications.* Brattleboro, Vt.: BuildingGreen, Inc. 2003. 4th ed. Guideline specifications make this an extremely valuable resource for commercial builders and architects.

Hermannsson, John. *Green Building Resource Guide.* Newtown, Conn.: Taunton Press, 1997. A gold mine of information on environmentally friendly building materials. Reader beware: Not all building materials in books such as this pass the sustainability test.

Lawrence, Robyn Griggs. "Classy Trash." *Natural Home* (July/August 2002): 44–51. Great story about a home built from waste paperboard and plastic.

Pearson, David. *The Natural House Catalog: Everything You Need to Create An Environmentally Friendly Home.* New York: Simon and Schuster, 1996. Information on building and furnishing a home, including a list of environmentally friendly products and services.

Spiegel, Ross, and Dru Meadows. *Green Building Materials: A Guide to Product Selection and Specification.* New York: John Wiley and Sons, 1999. The newest entry into the green building materials books.

Manufacturers

Publications

Because there are many manufacturers of green building materials, please refer to *GreenSpec Directory, Green Building Resource Guide, Green Building Materials,* or *Sustainable Building Sourcebook* (listed above) for information on specific products and their manufacturers.

On-Line Sources

For on-line information on manufacturers, contact Austin's Green Building Program at www.greenbuilder.com/sourcebook.

You can also contact the Center for Resourceful Building Technology's e-Guide, which provides a searchable database on green building materials and their manufacturers at www.crbt.org.

Yet another on-line source is Oikos Green Building Product Information at www.oikos.com/products.

Wholesale and Retail Outlets

Building for Health Materials Center. P.O. Box 113, Carbondale, CO 81623.

Tel: (800) 292-4838. Web site: www.buildingforhealth.com. Offers a complete line of healthy, environmentally safe building materials and home appliances including straw bale construction products; natural plastering products; flooring; natural paints, oils, stains, and

finishes; sealants; and construction materials. Offers special pricing for owner-builders and
   contractors.
EcoBuild. P.O. Box 4655, Boulder, CO 80306. Tel: (303) 545-6255.
   Web site: www.eco-build.com. This company works specifically with builders and general
   contractors, providing consultation and green building materials at competitive prices.
Eco-Products, Inc. 3655 Frontier Ave., Boulder, CO 80301. Tel: (303) 449-1876. Web site:
   www.ecoproducts.com. Offers a variety of green building products including plastic lumber.
Eco-Wise. 110 W. Elizabeth, Austin, TX 78704. Tel: (512) 326-4474. Web site:
   www.ecowise.com. Carries a wide range of environmental building materials, including
   Livos and Auro nontoxic natural finishes and adhesives.
Environmental Building Supplies. 819 SE Taylor St., Portland, OR 97214.
   Tel: (503) 222-3881. Web site: www.ecohaus.com. Green building materials outlet for the
   Pacific Northwest.
Environmental Depot: Environmental Construction Outfitters of New York. 901 E. 134th
   St., Bronx, NY 10454. Tel: (800) 238-5008. Web site: www.environproducts.com. Sells an
   assortment of green building materials.
Environmental Home Center. 1724 4th Ave. South, Seattle, WA 98134. Tel: (800) 281-9785.
   Web site: www.built-e.com. Offers a variety of green building materials.
Planetary Solutions for the Built Environment. 2030 17th Street, Boulder, CO 80302. Tel:
   (303) 442-6228. Web site: www.planetearth.com. Long-time green building materials
   supplier. Offers paints, flooring, tile, and much more.
Real Goods. 13771 S. Highway 101, Hopland, CA 95449. Tel: (800) 762-7325. Web site:
   www.realgoods.com. Sells a wide range of environmentally responsible products for
   homes, from solar and wind energy equipment to water efficiency products to air filters
   and environmentally responsible furnishings.

Chapter 5—Wood-Wise Construction

Publications

Edminster, Ann, and Sami Yassa. *Efficient Wood Use in Residential Construction: A Practical Guide
   to Saving Wood, Money, and Forests.* New York: Natural Resources Defense Council, 1998.
   Describes how to reduce lumber use by 30 percent without compromising the structural
   integrity of a home. Available in print and on-line at
   www.nrdc.org/cities/building/rwoodus.asp.
Imhoff, Dan. *Building with Vision: Optimizing and Finding Alternatives to Wood.* Healdsburg,
   Calif.: Watershed Media, 2001. Covers many important ways to reduce wood use.
National Association of Home Builders. *Alternative Framing Materials in Residential Construction:
   Three Case Studies.* Upper Marlboro, Md.: NAHBRC, 1994.
Randall, Robert, ed., *Residential Structure and Framing: Practical Engineering and Advanced
   Framing Techniques for Builders.* Richmont, Vt.: Builderburg Group, 1999. A collection of
   articles from *The Journal of Light Construction* on a wide range of topics, including engi-
   neered lumber, advanced framing, and steel framing.

Organizations

Forest Stewardship Council. Provides information on FSC-certified lumber. Web site:
   www.fscoax.org.

Chapter 6—Energy-Efficient Design and Construction

<u>Creating an Energy-Efficient Building Envelope</u>

Publications

The Best of Fine Homebuilding: Energy-Efficient Building. Newtown, Conn.: Taunton Press, 1999. A collection of detailed, somewhat technical articles on a wide assortment of topics related to energy efficiency including insulation, energy-saving details, windows, house-wraps, skylights, and heating systems.

Carmody, John, Stephen Selkowitz, and Lisa Heschong. *Residential Windows: A Guide to New Technologies and Energy Performance*. 2nd ed. New York: Norton, 2000. Extremely important reading for all passive solar home designers.

Chiras, Dan. "Minimize the Digging: Frost-Protected Shallow Foundations." *The Last Straw* 38 (Summer 2002): 10. A brief overview of frost-protected shallow foundations.

Chiras, Dan. "Retrofitting a Foundation for Energy Efficiency." *The Last Straw* 38 (Summer 2002): 11. Describes ways to retrofit foundations to reduce heat loss.

Hurst-Wajszczuk, Joe. "Save Energy and Money—Now." *Mother Earth News* (October/November 2001): 24–33. Useful tips on saving energy in new and existing homes.

Loken, Steve. *ReCRAFT 90: The Construction of a Resource-Efficient House*. Missoula, Mont.: National Center for Appropriate Technology, Center for Resourceful Building Technology, 1997. Field notes, lessons learned, and other information obtained from experience building a demonstration home in Missoula, Montana.

Lstiburek, Joe, and Betsy Pettit. *EEBA Builder's Guide—Cold Climate*. Bloomington, Minn.: Energy and Environmental Building Association, 2002. Superb resource for advice on building in cold climates.

———. *EEBA Builder's Guide—Hot-Arid/Mixed-Dry Climate*. Bloomington, Minn.: Energy and Environmental Building Association, 2000. Superb resource for advice on building in hot, arid climates.

———. *EEBA Builder's Guide—Mixed-Humid Climate*. Bloomington, Minn.: Energy and Environmental Building Association, 2001. Superb resource for advice on this climate.

Magwood, Chris, ed. "Roofs and Foundation Issue." *The Last Straw* 38 (September 2002). An excellent resource for those who want to learn about energy- and material-efficient foundations.

Mumma, Tracy. *Guide to Resource Efficient Building Elements*. Missoula, Mont.: National Center for Appropriate Technology, Center for Resourceful Building Technology, 1997. A handy guide to materials that help improve the efficiency of homes and other buildings. Available in updated versions on-line and free at www.crbt.org.

National Association of Home Builders Research Center. *Design Guide for Frost-Protected Shallow Foundations*. Upper Marlboro, Md.: NAHB Research Center, 1996. Also available on-line from www.nahbrc.org.

Pahl, Greg. *Natural Home Heating*. White River Junction, Vt.: Chelsea Green, 2003. A useful overview of home heating.

Sikora, Jeannie L. *Profit from Building Green: Award Winning Tips to Build Energy Efficient Homes*. Washington, D.C.: BuilderBooks, 2002. A brief but informative overview of energy-conservation strategies.

Wilson, Alex. "Windows: Looking through the Options." *Solar Today* (May/June 2001): 36–39. A great overview of windows with a useful checklist for those in the market to buy new ones.

Wilson, Alex, Jennifer Thorne, and John Morrill. *Consumer Guide to Home Energy Savings*. 8th ed. Washington, D.C.: American Council for an Energy-Efficient Economy, 2003. Excellent book, full of information on energy-saving appliances.

Organizations

American Council for an Energy-Efficient Economy. 1001 Connecticut Avenue NW, Suite 801, Washington, DC 20036. Tel: (202) 429-0063. Web site: www.aceee.org. Numerous excellent publications on energy efficiency, including *Consumer Guide to Home Energy Savings*.

Building America Program. U.S. Department of Energy. Office of Building Systems, EE-41, 1000 Independence Avenue SW, Washington, DC 20585. Tel: (202) 586-9472. Leaders in promoting energy efficiency and renewable energy to achieve zero-energy buildings.

Cellulose Insulation Manufacturers Association. Your place to "shop" for information on cellulose insulation. 136 S. Keowee St., Dayton, OH 45402. Tel: (937) 222-2462. Web site: www.cellulose.org.

Consumers Union. Tel: (914) 378-2000. Web site: www.consumersunion.org. Publishes *Consumer Reports* (www.consumerreports.org) and *Consumer Reports Annual Buying Guide*, which rate appliances for reliability, convenience, and efficiency.

Energy and Environmental Building Association. 10740 Lyndale Ave. S, Suite 10W, Bloomington, MN 55420-5615. Tel: (952) 881-1098. Web site: www.eeba.org. Offers conferences, workshops, publications, and an on-line bookstore.

Energy Efficiency and Renewable Energy Clearinghouse. P.O. Box 3048, Merrifield, VA 22116. Tel: (800) 363-3732. Great source for a variety of useful information on energy efficiency.

Insulating Concrete Forms Association. 1730 Dewes St., Suite 2, Glenview, IL 60025. Tel: (847) 657-9730. Web site: www.forms.org. A great place to begin your research on ICFs.

National Fenestration Rating Council. 8484 Georgia Ave., Suite 320, Silver Spring, MD 20910. Tel: (301) 589-1776. Web site: www.nfrc.org. For information on energy efficiency of windows.

National Insulation Association. 99 Canal Center Plaza, Suite 222, Alexandria, VA 22314. Tel: (703) 683-6422. Web site: www.insulation.org. Offers a wide range of information on different types of insulation.

U.S. Department of Energy and Environmental Protection Agency's Energy Star program. Tel: (888) 782-7937. Web site: www.energystar.gov.

Energy-Efficient Heating Systems

Publications

The Best of Fine Homebuilding: Energy-Efficient Building. Newtown, Conn.: Taunton Press, 1999. Contains a collection of extremely useful articles on mechanical heating systems.

Fust, Art. "A Simple Warm Floor Heating System." *The Last Straw* 32 (Winter 2000): 25–26. Contains much useful information on radiant-floor heat.

Grahl, Christine L. "The Radiant Flooring Revolution." *Environmental Design and Construction* (January/February 2000): 38-40. Superb introduction to radiant-floor heating.

Hyatt, Rod. "Hydronic Heating on Renewable Energy." *Home Power* 79 (October/November 2000): 36–42. Provides a lot of practical advice on building your own radiant-floor heating system and powering it with photovoltaic panels.

Malin, Nadav, and Alex Wilson. "Ground-Source Heat Pumps: Are They Green?" *Environmental Building News* 9 (July/August 2000): 1, 16–22. Detailed overview of the operation and pros and cons of ground-source heat pumps.

National Renewable Energy Laboratory. "Geothermal Heat Pumps." Published on-line at www.eren.doe.gov/erec/factsheets/geo_heatpumps.html. Great overview of ground-source heat pumps.

O'Connell, John, and Bruce Harley. "Choosing Ductwork." *Fine Homebuilding* (June/July 1997): 98–101. Essential reading for anyone interested in installing a forced-air heating system.

Siegenthaler, John. "Hydronic Radiant-Floor Heating." *Fine Homebuilding*
   (October/November 1996): 58–63. Extremely useful reference. Well written, thorough,
   and well illustrated.

———. *Modern Hydronic Heating for Residential and Light Commercial Buildings.* 2nd ed.
Clifton Park, N.Y.: Thomson/Delmar Learning, 2003. Everything you would ever want to
   know about hydronic heating.

Wilson, Alex. "A Primer on Heating Systems." *Fine Homebuilding* (February/March 1997):
   50–55. Superb overview of furnaces, boilers, and heat systems.

———. "Radiant-Floor Heating: When It Does—and Doesn't—Make Sense." *Environmental
Building News* (January 2002): 1, 9–14. Valuable reading.

Organizations

U.S. Consumer Product Safety Commission. Office of Information and Public Affairs,
   CPSC, Washington, DC 20207-0001. Hotline: (800) 638-2772. Web site: www.cpsc.gov.
   Offers a wealth of information on space heaters, including safety precautions.

Geo-Heat Center, Oregon Institute of Technology, 3201 Campus Dr., Klamath Falls, OR
   97601. Tel: (541) 885-1750. Web site: geoheat.oit.edu. Technical information on heat pumps.

Geothermal Heat Pump Consortium, Inc. 6700 Alexander Bell Dr., Suite 120, Columbia,
   MD 21046. Tel: (410) 953-7150. Web site: www.geoexchange.org. General and technical
   information on heat pumps.

International Ground Source Heat Pump Association. Oklahoma State University, 499 Cordell
   South, Stillwater, OK 74078. Tel: (800) 626-4747. Web site: www.igshpa.okstate.edu.
   Provides a list of equipment manufacturers, installers by state, and numerous other
   resources for contractors, homeowners, students, and the general public.

Radiant Panel Association. P.O. Box 717, 1399 S. Garfield Ave., Loveland, CO 80537. Tel:
   (800) 660-7187. Web site: www.radiantpanelassociation.org. Professional organization
   consisting of radiant heating and cooling contractors, wholesalers, manufacturers, and
   professionals.

U.S. Department of Energy, Office of Geothermal Technologies. EE-12, 1000 Independence
   Avenue SW, Washington, DC 20585-0121. Tel: (202) 586-5340. Carries out research on
   ground-source heat pumps and works closely with industry to implement new ideas.

Efficient Wood-Burning Technologies

Publications

Barden, Albert A. *The AlbieCore Construction Manual.* Norridgewock, Maine.: Maine Wood
   Heat Company, 1995. Detailed construction manual.

———. *The Finnish Fireplace Construction Manual.* Norridgewock, Maine.: Maine Wood Heat
   Company, Inc., 1988. The only complete English-language primer on making masonry
   heaters. Available through the Maine Wood Heat Co., Inc., 254 Fr. Rasle Rd.,
   Norridgewock, ME 04957. Tel: (207) 696-5442. eb site: www.mainewoodheat.com.

Government of British Columbia, Ministry of Water, Land and Air Protection. "Reducing
   Wood Stove Smoke: A Burning Issue." On-line publication (Feb. 2002). Web site:
   wlapwww.gov.bc.ca/air/particulates/rwssabi.html.

Gulland, John. "Woodstove Buyer's Guide." *Mother Earth News* (December/January 2002):
   32–43. Superb overview of woodstoves, with a useful table to help you select a model
   that meets your needs.

Hyytiainen, Heikki, and Albert Barden. *Finnish Fireplaces: Heart of the Home.* Norridgewock,
   Maine: Maine Wood Heat Company, Inc., 1988. A valuable resource for anyone wanting
   to learn more about Finnish masonry stoves. Available through the Maine Wood Heat
   Company, listed above.

Johnson, Dave. *The Good Woodcutter's Guide: Chain Saws, Woodlots, and Portable Sawmills.* White River Junction, Vt.: Chelsea Green, 1998. A practical guide to felling trees and cutting firewood safely.

Lyle, David. *The Book of Masonry Stoves: Rediscovering an Old Way of Warming.* White River Junction, Vt.: Chelsea Green, 1998. This book contains a wealth of information on the history, function, design, and construction of masonry stoves.

Organizations

Hearth, Patio, and Barbecue Association (formerly the Hearth Products Association). 1601 North Kent Street, Suite 1001, Arlington, VA 22209. Web site: www.hpba.org. International trade association that promotes the interests of the hearth products industry. Offers lots of valuable information.

Masonry Heater Association of North America. 1252 Stock Farm Road, Randolph, VT 05060. Tel: (802) 728-5896. Web site: www.mha-net.org. Publishes a valuable newsletter and has a web site with links to dealers and masons who design and build masonry stoves.

Wood Heat Organization, Inc. 410 Bank Street, Suite 117, Ottawa, Ontario, Canada K2P 1Y8. Web site: www.woodheat.org. Promotes safe, responsible use of wood for heating.

Chapter 7—Accessibility, Ergonomics, and Adaptability

Publications

Altman, Adelaide. *Elderhouse: Planning Your Best Home Ever.* White River Junction, Vt.: Chelsea Green, 2002. A practical guide to help prevent accidents, ensure comfort, and maintain an independent, sustainable lifestyle in your own home as you age.

Boehland, Jessica. "Future-Proofing Your Building: Designing for Flexibility and Adaptive Reuse." *Environmental Building News* (February 2003): 1, 7–14. An overview of this important subject, geared primarily toward commercial buildings.

Clark, Sam. *The Real Goods Independent Builder: Designing and Building a House Your Own Way.* White River Junction, Vt.: Chelsea Green, 1996. A great guide to building your own home.

Diffrient, Niels, Alvin Tilley, and Joan Bardagjy. *Humanscale 1/2/3.* Boston: MIT Press, 1974. A technical reference on ergonomic design.

Friedman, Avi. *The Adaptable House: Designing Homes for Change.* New York: McGraw-Hill, 2002. Great resource for those who want to delve into this subject in much greater detail.

Inkeles, Gordon. *Ergonomic Living: How to Create a User-Friendly Home and Office.* New York: Simon and Schuster, 1994. Full of useful information on ergonomic design.

NAHB Research Center. *Directory of Accessible Building Products: Making Houses User-Friendly for Everyone.* Upper Marlboro, Md.: NAHB Research Center, 2003. An annual catalog of products for accessible design.

———. *Residential Remodeling and Universal Design: Making Homes More Comfortable and Accessible.* Upper Marlboro, Md.: NAHB Research Center, not dated. An informative guide for builders.

Peterson, M. J. *Universal Bathroom Planning: Design That Adapts to People.* Hackettstown, N.J.: National Kitchen and Bath Association, 1996. A useful guide for accessible design of bathrooms. Finding a copy may be challenging as the book is now out of print.

———. *Universal Kitchen Planning: Design That Adapts to People.* Hackettstown, N.J: National Kitchen and Bath Association, 1996. A useful guide for accessible design of kitchens. You can obtain a copy by calling (800) 843-6522. Web site: www.nkba.org.

Organizations

Lighthouse International. 111 East 59th Street, New York, NY 10022. Tel: (800) 829-0500.
Web site: www.lighthouse.org. Information for people suffering from visual impairment.

Paralyzed Veterans of America. 801 18th St. NW, Washington, DC 20006. Tel: (800) 424-
8200. Web site: www.pva.org. Produces information on accessibility, including the *Access
Information Bulletins*.

U.S. Department of Housing and Urban Development. P.O. Box 23268, Washington, DC
20026-3268. Tel: (800) 245-2691. Web site: www.huduser.org. General information and
guidelines on accessible design.

Chapter 8—Using Concrete and Steel to Build Green

Publications

Imhoff, Dan. *Building with Vision: Optimizing and Finding Alternatives to Wood*. Healdsburg,
Calif.: Watershed Media, 2001. Covers many topics, including the use of light-gauge steel
for framing.

Randall, Robert, ed. *Residential Structure and Framing: Practical Engineering and Advanced
Framing Techniques for Builders*. Richmond, Vt.: Builderburg Group, 1999. Contains a number of chapters on steel framing.

Organizations

Monolithic Dome Institute. 177 Dome Park Place, Italy, TX 76651. Web site:
www.monolithicdome.com. Log on to their web site to learn about their conferences,
workshops, books, CDs, building plans, and publications and to read feature articles.

Steel Framing Alliance (formerly North American Steel Framing Alliance). 1201 15th St.
NW, Suite 320, Washington, DC 20005. Tel: (202) 785-2022. Web site:
www.steelframingalliance.com. Offers a wealth of useful information about the use of
steel for framing homes, including many publications.

Chapter 9—Natural Building

General Publications

Chiras, Daniel D. *The Natural House: A Complete Guide to Healthy, Energy-Efficient,
Environmental Homes*. White River Junction, Vt.: Chelsea Green, 2000. Information on
natural building with discussions of the pros and cons of each type.

Elizabeth, Lynne, and Cassandra Adams, eds. *Alternative Construction: Contemporary Natural
Building Methods*. New York: Wiley, 2000. A compilation of articles on numerous natural
building techniques with a little information on natural plasters.

Kennedy, Joseph E., Michael G. Smith, and Catherine Wanek, eds. *The Art of Natural Building:
Design, Construction, and Resources*. Gabriola Island, B.C.: New Society Publishers, 2001.
Contains an assortment of articles on natural building.

Minke, Gernot. *Earth Construction Handbook: The Building Material Earth in Modern Architecture*.
Southampton, England: WIT Press/Computational Mechanics Publications, 2000.
Contains a great deal of technical information and a good section on earth plasters, called
loam plasters by the author.

Magazines

Natural Home. 201 E. 4th Street, Loveland, CO 80537. Tel: (800) 272-2193. Web site: www.nat-uralhomemagazine.com. Covers a wide range of topics vital to healthy, natural building.

Straw Bale and Straw-Clay Construction

Publications

King, Bruce. *Buildings of Earth and Straw: Structural Design for Rammed Earth and Straw Bale Architecture.* Sausalito, Calif.: Ecological Design Press, 1996. A great book for the technically minded reader. Contains information on tests run on straw bale structures.

Lacinski, Paul, and Michel Bergeron. *Serious Straw Bale: A Home Construction Guide for All Climates.* White River Junction, Vt.: Chelsea Green, 2000. Contains a great deal of information on building with straw bales and plastering in cold and wet climates. Detailed coverage of lime plaster.

Laporte, Robert. *MoosePrints: A Holistic Home Building Guide.* Fairfield, Iowa.: Natural House Building Center, 1993. The only published source on straw-clay construction. Contains some excellent illustrations, but only a fraction of the information you will need to learn this technique.

Magwood, Chris, and Peter Mack. *Straw Bale Building: How to Plan, Design, and Build Straw Bale.* Gabriola Island, B.C.: New Society Publishers, 2000. A wonderfully written book on building straw bale in a variety of climates, especially northern climates. Contains a fair amount of information on plastering.

Myhrman, Matts, and S. O. MacDonald. *Build It with Bales: A Step-by-Step Guide to Straw-Bale Construction, Version Two.* Tucson, Ariz.: Out on Bale, 1998. A superbly illustrated and recently updated manual on straw bale construction. Contains a fair amount of information on wall preparation, plasters, and plastering.

Steen, Athena S., Bill Steen, David Bainbridge, and David Eisenberg. *The Straw Bale House.* White River Junction, Vt.: Chelsea Green, 1994. The best-selling book that helped fuel interest in straw bale construction. Tons of information on straw-bale construction, wall preparation, and plasters.

Magazines and Newsletters

The Last Straw. P.O. Box 22706, Lincoln, NE 68542-2706. Tel: (402) 483-5135. Web site: www.thelaststraw.org. Quarterly journal containing the latest information on straw-bale construction. Annual resource issue contains a gold mine of information. Publishes articles on natural plasters. This is an absolute must for all straw bale enthusiasts!

Videos

Building with Straw, Vol. 1: A Straw-Bale Workshop. Black Range Films, 1994. Covers a weekend workshop in which volunteers helped build a two-story greenhouse addition on a lodge. To order, log on to www.strawbalecentral.com.

Building with Straw, Vol. 2: A Straw-Bale Home Tour. Black Range Films, 1994. A tour of ten straw bale structures in New Mexico and Arizona. To order, log on to www.strawbalecentral.com.

Building with Straw, Vol. 3: Straw-Bale Code Testing. Black Range Films, 1994. Takes you on a tour of ten straw bale structures in New Mexico and Arizona. Presents the insights of the owners/builders. To order, log on to www.strawbalecentral.com.

How to Build Your Elegant Home with Straw Bales. By Steve Kemble and Carol Escott. Covers the specifics of building a load-bearing straw bale home. Comes with a manual. To order, log on to www.strawbalecentral.com.

The Straw Bale Solution. Black Range Films. Narrated by Bill and Athena Steen and produced by Catherine Wanek. Features interviews with architects, engineers, and owner-builders. Covers the basics of straw bale construction and much more. To order, log on to www.strawbalecentral.com.

Organizations

Austrian Straw Bale Network. A-3720 Baierdorf 6, Austria. Web site: www.baubiologie.at. (In German.)

California Straw Building Association. P.O. Box 1293, Angels Camp, CA 95222-1293. Tel: (209) 785-7077. Web site: www.strawbuilding.org. This group is involved in testing straw bale structures. They also offer workshops and sponsor conferences.

The Canelo Project. HC1, Box 324, Elgin, AZ 85611. Web site: www.caneloproject. com. Founded and run by Athena and Bill Steen, contributing authors of *The Straw Bale House*. They offer workshops, videos, and books on straw bale construction as well as information on building codes and results of tests on straw bale homes.

Center for Maximum Potential Building Systems. 8604 FM 969, Austin, TX 78724. Tel: (512) 928-4786. Working at the cutting edge of building materials, systems, and methods. Led by Pliny Fisk III.

Development Center for Appropriate Technology. Contact them in care of David Eisenberg, P.O. Box 27513, Tucson, AZ 85726-7513. Tel: (520) 624-6628. Web site: www.dcat.net. Offers a variety of services including consulting, research, testing, assistance with code issues, project support, instruction, and workshops.

European Straw Bale Construction Network. To join the mailing list, e-mail strawbale-1@eyfa.org.

GreenFire Institute. 1509 Queen Anne Ave. N #606, Seattle, WA 98103. E-mail: greenfire@delphi.com. Offers straw bale workshops, design consultation, full design, building consultation, and full building options, all using straw or other sustainable materials.

Japan Straw Bale House Association. 8-9 Honcho, Utsunomiya, Tochigi, Japan 3200033. Web site: www.geocities.co.jp/NatureLand/1946/. (In Japanese.)

Norwegian Straw and Earth Building Organization. Wemhus, N-1540 Vestby, Norway. E-mail: arild.berg3@chello.no.

Straw Bale Association of Nebraska. 2110 S. 33rd St., Lincoln, NE 68506-6001. Tel: (805) 483-5135. Active in promoting straw bale construction and the MidAmerica Straw Bale Association.

Straw Bale Association of Texas. P.O. Box 49381, Austin, TX 78763. Tel: (512) 302-6766. Web site: www.greenbuilder.com/sbat. Sponsors monthly meetings, publishes a newsletter, and provides a host of other resources.

Straw Bale Building Association for Wales, Ireland, Scotland, and England. SBBA, Hollinroyd Farms, Butts Lane, Todmorden, OL14 8RJ, United Kingdom. Tel: 01442 825421. Exchanges information and experience in straw bale construction.

Straw Bale Building Association of Australia. Contact at sbaoa@yahoo.com.au.

Straw Bale Construction Association of New Mexico. Catherine Wanek, Route 2, Box 119, Kingston, NM 88042. Tel: (505) 895-5652. E-mail: blackrange@zianet.com.

Adobe and Cob Building

Publications

Bee, Becky. *The Cob Builder's Handbook: You Can Hand-Sculpt Your Own Home*. Murphy, Ore.: Groundworks, 1997. Amply illustrated and clearly written introduction to cob building with a brief section on plasters and plastering.

Bourgeois, Jean-Louis. *Spectacular Vernacular: The Adobe Tradition*. New York: Aperture Foundation, 1989. Superb and beautifully photographed overview of adobe building throughout the world.

Evans, Ianto, Michael Smith, and Linda Smiley. *The Hand-Sculpted House: A Practical and Philosophical Guide to Building a Cob Cottage*. White River Junction, Vt.: Chelsea Green, 2002. Superb resource! A must-read for anyone interested in cob building.

McHenry, Paul G. Jr. *Adobe and Rammed Earth Buildings: Design and Construction*. Tucson, Ariz.: University of Arizona Press, 1989. Excellent reference, covering history, soil selection, adobe brick manufacturing, adobe wall construction, and many more topics. Good coverage of earthen plastering.

———. *Adobe: Build It Yourself*. 2nd ed. Tucson, Ariz.: University of Arizona Press, 1985. Highly readable and surprisingly thorough introduction to many aspects of adobe construction. Focuses on cement and gypsum plaster.

Smith, Michael G. *The Cobber's Companion: How to Build Your Own Earthen Home*. 2nd ed. Cottage Grove, Ore.: The Cob Cottage Company, 1998. Well-written introduction to cob with many excellent and useful illustrations.

Stedman, Myrtle, and Wilfred Stedman. *Adobe Architecture*. Santa Fe, N.Mex.: Sunstone Press, 1987. Contains numerous drawings of houses and floor plans and well-illustrated basic information on making adobe bricks and laying up walls.

Magazines and Newsletters

The CobWeb. The only cob-focused periodical. Published twice yearly by The Cob Cottage Company, P.O. Box 123, Cottage Grove, OR 97424. Tel: (541) 942-2005. Web site: www.cobcottage.com.

Videos

Building with the Earth: Oregon's Cob Cottage Co. Great resource. Obtain from The Cob Cottage Company, P.O. Box 123, Cottage Grove, OR. 97424. Tel: (541) 942-2005. Web site: www.cobcottage.com.

Organizations

Center for Alternative Technology. Machynlleth, Powys, United Kingdom SY20 9AZ. Phone: 01654 705950. Web site: www.cat.org.uk. This educational group offers workshops on earth building and natural finishes, among other topics.

Rammed Earth, Rammed Earth Tires, and Earthbags

Publications

Easton, David. *The Rammed Earth House*. White River Junction, Vt.: Chelsea Green, 1996. An informative, highly readable book. A must for anyone considering this technology. No discussion of plaster or plastering.

Hunter, Kaki, and Doni Kiffmeyer. *Earthbag Building, The Tools, Tricks, and Techniques*. Gabriola Island, B.C.: New Society Publishers, 2004. Detailed book on earthbag construction with information on plastering. Informative and well organized.

King, Bruce. *Buildings of Earth and Straw: Structural Design for Rammed Earth and Straw Bale Architecture*. Sausalito, Calif.: Ecological Design Press, 1996. Another essential reading for anyone interested in building a rammed earth home.

Middleton, G. F. *Earth Wall Construction*. North Ryde, Australia: CSIRO Publishers, 1987. A manual on rammed earth showing a unique forming system. Appendices contain structural and insulation calculations.

Reynolds, Michael. *Earthship Volume 1: How to Build Your Own.* Taos, N.Mex.: Solar Survival Press, 1990. A must-read for those wanting to understand the basics of early Earthship design. This book contains some outdated information, however, so be careful. Be sure to read the more current volumes and check out the *Earthship Chronicles* for up-to-date information.

————. *Earthship Volume 2: Systems and Components.* Taos, N.Mex.: Solar Survival Press, 1990. Explains the various systems such as graywater, solar electric, and domestic hot water. Essential reading for all people interested in sustainable housing.

————. *Earthship Volume 3: Evolution Beyond Economics.* Taos, N.Mex.: Solar Survival Press, 1993. Presents many of the new developments. The latest information, however, is always to be learned in workshops, tours of new houses, and the *Earthship Chronicles*.

Wojciechowska, Paulina. *Building with Earth: A Guide to Flexible-Form Earthbag Construction.* White River Junction, Vt.: Chelsea Green, 2001. Describes earthbag construction and offers some details on plastering.

Magazines and Newsletters

Earthship Chronicles. Published by Earthship Global Operations, P.O. Box 2009, El Prado, NM 87529. Tel: (505) 751-0462. Pamphlets issued periodically to disseminate new information: graywater, catchwater, blackwater, mass vs. insulation, and equipment catalog.

Solar Survival Newsletter. Available from Solar Survival Architecture, P.O. Box 1041, Taos, NM 87571. E-mail: solarsurvival@earthship.org.

Videos

Building for the Future. This is a video about the building of my house. It explains how it was built and discusses many green building products. To order, contact me at (303) 674-9688 or via e-mail at danchiras@msn.com.

Dennis Weaver's Earthship. Shows construction of actor Dennis Weaver's Earthship. Well done and very informative. Helpful in securing building permits. Available from Earthship Biotecture at their on-line store at earthship.org.

The Earthship Documentary. Describes the history of Earthship construction, the underlying philosophy behind this unique structure, and building techniques. Available from Earthship Biotecture at their on-line store listed above.

Earthship Next Generation. A look at new Earthship designs. Available from Earthship Biotecture at their on-line store listed above.

From the Ground Up. Takes you through the process of building an Earthship. Available from Earthship Biotecture at their on-line store listed above.

Rammed Earth Construction. 1985. A 29-minute video produced by Hans-Ernst Weitzel. To order, call Bullfrog Films at (800) 543-3764 or visit their web site at www.bullfrogfilms.com.

The Rammed Earth Renaissance Video. Lycum Productions. This 31-minute video features David Easton and serves as an excellent introduction to the subject or a companion to *The Rammed Earth House.* Available from Chelsea Green (www.chelseagreen.com).

Organizations

CalEarth, California Institute of Earth Art and Architecture. CalEarth/Geltaftan Foundation, 10376 Shangri La Avenue, Hesperia, CA 92345. Tel: (760) 244-0614. Web site: www.calearth.org. Offers information on earthbag construction, including an on-line newsletter.

Chapter 10—Earth-Sheltered Architecture

Publications

Chiras, Dan. "The Down Earth Home," *Mother Earth News.* A brief overview of earth-sheltered building.

Oehler, Mike. *The $50 & Up Underground House Book: How to Design and Build Underground.* Bonners Ferry, Idaho: Mole Publishing Co.,1997. An interesting little book that presents an unusual way of building underground.

Reynolds, Michael. *Earthship Volume 1: How to Build Your Own.* Taos, N.Mex.: Solar Survival Press, 1990. A must-read for those wanting to understand the basics of early Earthship design. This book contains some outdated information, however, so be careful. Be sure to read the more current volumes and check out the *Earthship Chronicles* for up-to-date information.

———. *Earthship Volume 2: Systems and Components.* Taos, N.Mex.: Solar Survival Press, 1990. Explains the various systems such as graywater, solar electric, and domestic hot water. Essential reading for all people interested in sustainable housing.

———. *Earthship Volume 3: Evolution Beyond Economics.* Taos, N.Mex.: Solar Survival Press, 1993. Presents many of the new developments. The latest information, however, is always to be learned in workshops, tours of new houses, and the *Earthship Chronicles*.

Roy, Robert L. *The Complete Book of Underground Houses: How to Build a Low-Cost Home.* New York: Sterling Publishing Co., 1994.

Wells, Malcolm. *The Earth-Sheltered House: An Architect's Sketchbook.* White River Junction, Vt.: Chelsea Green, 1998. Great little book on earth-sheltered design.

———. *How to Build an Underground House.* Self-published, 1991. Overview of earth-sheltered building.

Organizations

American Underground Construction Association. 3001 Hennepin Ave. South, Suite D202, Minneapolis, MN 55408. Tel: (612) 825-8933. Web site: www.auca.org. Conferences and referrals to earth-sheltered professionals.

Chapters 11 and 12—Passive Solar Heating and Cooling

Chiras, Daniel D. "Build a Solar Home and Let the Sunshine in." *Mother Earth News* (August/September 2002): 74–81. A survey of passive solar design principles and a case study showing the economics of passive solar heating.

———. "Learning from Mistakes of the Past." *The Last Straw* 36 (Winter 2001): 15–16. Describes common errors in passive solar design.

Chiras, Daniel D., ed. "Solar Solutions." *The Last Straw* 36 (Winter 2001). A collection of over a dozen articles, many by the author, on passive solar heating, integrated design, thermal mass, and more.

Cole, Nancy, and P. J. Skerrett. *Renewables Are Ready: People Creating Renewable Energy Solutions.* White River Junction, Vt.: Chelsea Green, 1995. Contains numerous interesting case studies showing how people have applied various solar technologies, including passive solar.

Crosbie, Michael, ed. *The Passive Solar Design and Construction Handbook.* New York: John Wiley and Sons, 1997. A pricey and fairly technical manual on passive solar homes. Contains detailed drawings and case studies.

Crowther, Richard I. *Affordable Passive Solar Homes: Low-Cost Compact Designs*. Denver, Colo.: SciTech Publishing, 1984. Contains some valuable background information on passive solar design and numerous specific designs.

Energy Division, North Carolina Department of Commerce. *Solar Homes for North Carolina: A Guide to Building and Planning Solar Homes*. Raleigh, N.C.: North Carolina Solar Center, 1999. Available on-line at the North Carolina Solar Center's web site, www.ncsu.edu.

Givoni, Baruch. *Passive and Low Energy Cooling of Buildings*. New York: Van Nostrand Reinhold, 1994. A fairly technical book, but one of a few resources on the subject.

Kachadorian, James. *The Passive Solar House*. White River Junction, Vt.: Chelsea Green, 1997. Presents a lot of good information on passive solar heating and an interesting design that has reportedly been fairly successful in cold climates.

Kubsch, Erwin. *Home Owners Guide to Free Heat: Cutting Your Heating Bills Over 50%*. Sheridan, Wyo.: Sunstore Farms, 1991. A self-published book with lots of good basic information.

McIntyre, Maureen, ed. *Solar Energy: Today's Technologies for a Sustainable Future*. Boulder, Colo.: American Solar Energy Society, 1997. An extremely valuable resource with numerous case studies showing how passive solar heating can be used in different climates, even some fairly solar-deprived places.

Olson, Ken, and Joe Schwartz. "Home Sweet Solar Home: A Passive Solar Design Primer." *Home Power* (August/September 2002): 86–94. Superb introduction to passive solar design principles.

Passive Solar Industries Council. *Passive Solar Design Strategies: Guidelines for Home Building*. Washington, D.C.: PSIC, undated. Extremely useful book with worksheets for calculating a house's energy demand, the amount of backup heat required, the temperature swing you can expect given the amount of thermal mass you've installed, and the estimated cooling load. You can order a copy from the SBIC (www.sbicouncil.org) with detailed information for your state, so you can design a home to meet the requirements of your site.

Potts, Michael. *The New Independent Home: People and Houses That Harvest the Sun, Wind, and Water*. White River Junction, Vt.: Chelsea Green, 1999. Delightfully readable book with lots of good information.

Reynolds, Michael. *Comfort in Any Climate*. Taos, N.Mex.: Solar Survival Press, 1990. A brief but informative treatise on passive heating and cooling.

Sklar, Scott, and Kenneth Sheinkopf. *Consumer Guide to Solar Energy: New Ways to Lower Utility Costs and Take Control of Your Energy Needs*. Chicago: Bonus Books, 1995. Delightfully written introduction to many different solar applications, including passive solar heating.

Solar Survival Architecture. "Thermal Mass vs. Insulation." In *Earthship Chronicles*. Taos, N.M.: Solar Survival Architecture, 1998. Booklets provided by the SSA. Basic treatise on passive solar heating and cooling.

Sustainable Buildings Industry Council. *Designing Low-Energy Buildings: Passive Solar Strategies* and *Energy-10 Software*. SBIC, 1996. A superb resource! This book of design guidelines and the Energy-10 software that comes with it enable builders to analyze the energy and cost savings in building designs. Help permit region-specific design.

Taylor, John S. *A Shelter Sketchbook: Timeless Building Solutions*. White River Junction, Vt.: Chelsea Green, 1983. Pictorial history of building that will open your eyes to intriguing design solutions to achieve comfort, efficiency, convenience, and beauty.

Van Dresser, Peter. *Passive Solar House Basics*. Santa Fe, N.Mex.: Ancient City Press, 1996. This brief book provides basics on passive solar design and construction, primarily of adobe homes. Contains sample house plans, ideas for solar water heating, and much more.

Magazines and Newsletters

Backwoods Home Magazine. P.O. Box 712, Gold Beach, OR 97444. Tel: (800) 835-2418. Web site: www.backwoodshome.com. Publishes articles on all aspects of self-reliant living, including renewable energy strategies such as solar.

Buildings Inside and Out. Newsletter of the Sustainable Buildings Industries Council (formerly Passive Solar Industries Council). See their listing under "Organizations" on page 277.

CADDET InfoPoint. Web site: www.caddet-re.org. Quarterly magazine published by the Centre for the Analysis and Dissemination of Demonstrated Energy Technologies. Covers a wide range of renewable energy topics.

New Earth Times (formerly *Dry Country News*). Box 23-J, Radium Springs, NM 88054. Tel: (505) 526-1853. Web site: www.zianet.com/earth. A new magazine devoted to living close to, and in harmony with, nature. Covers all aspects of natural life including home building and renewable energy.

EERE Network News. Newsletter of the U.S. Department of Energy's Energy Efficiency and Renewable Energy Network. An electronic newsletter from the DOE. To subscribe, log on to www.cere.energy.gov.

Home Energy. 2124 Kittredge St., No. 95, Berkeley, CA 94704. Tel: (510) 524-5405. Web site: www.homeenergy.org. Great resource for those who want to learn more about ways to save energy in conventional home construction.

Home Power. P.O. Box 520, Ashland, OR 97520. Tel: (800) 707-6585. Web site: www.home-power.com. Publishes numerous articles on PVs, wind energy, and microhydroelectric and occasionally an article or two on passive solar heating and cooling.

The Last Straw. P.O. Box 22706, Lincoln, NE 68542-2706. Tel: (402) 483-5135. Web site: www.thelaststraw.org. This journal publishes articles on natural building and features articles on passive solar heating and cooling.

Solar Today. ASES, 2400 Central Ave., Suite A, Boulder, CO 80301. Tel: (303) 443-3130. Web site: www.solartoday.org. This magazine, published by the American Solar Energy Society, contains a wealth of information on passive solar, solar thermal, photovoltaics, hydrogen, and other topics. Also lists names of engineers, builders, and installers and lists workshops and conferences.

Videos

The Solar-Powered Home. With Rob Roy. An 84-minute video that examines basic principles, components, set-up, and system planning for an off-grid home featuring tips from America's leading experts in the field of home power. Can be purchased from the Earthwood Building School at 366 Murtagh Hill Road, West Chazy, NY 12992. Tel: (518) 493-7744. Web site: www.cordwoodmasonry.com.

Organizations

American Solar Energy Society. 2400 Central Avenue, Suite A, Boulder, CO 80301. Web site: www.ases.org. Publishes *Solar Today* magazine and sponsors an annual national meeting. Also publishes an on-line catalog of publications and sponsors the National Tour of Solar Homes. Contact this organization to find out about an ASES chapter in your area.

Center for Building Science. Web site: eetd.lbl.gov. Lawrence Berkeley National Laboratory's Center for Building Science works to develop and commercialize energy-efficient technologies and to document ways of improving energy efficiency of homes and other buildings while protecting air quality.

Renewable Energy Policy Project and CREST (Center for Renewable Energy and Sustainable Technologies). 1612 K St. NW, Suite 202, Washington, DC 20006. Tel: (202)

293-2898. Web site: solstice.crest.org. Nonprofit organization dedicated to renewable energy, energy efficiency, and sustainable living.

El Paso Solar Energy Association. P.O. Box 26384, El Paso, TX 79926. Web site: www.epsea.org/design.html. Active in solar energy, especially passive solar design and construction.

Energy Efficiency and Renewable Energy Clearinghouse. P.O. Box 3048, Merrifield, VA 22116. Tel: (800) 363-3732. Great source of a variety of useful information on renewable energy.

Florida Solar Energy Center (FSEC). 1679 Clearlake Road, Cocoa, FL 32922. Tel: (321) 638-1000. Web site: www.fsec.ucf.edu. A research institute of the University of Central Florida. Research and education on passive solar, cooling, and photovoltaics.

Midwest Renewable Energy Association. 7558 Deer Rd., Custer, WI 54423. Tel: (715) 592-6595. Web site: www.the-mrea.org. Actively promotes solar energy and offers valuable workshops.

National Renewable Energy Laboratory (NREL). Center for Buildings and Thermal Systems. 1617 Cole Blvd., Golden, CO 80401. Tel: (303) 275-3000. Web site: www.nrel.gov/buildings_thermal/. Key player in research and education on energy efficiency and passive solar heating and cooling.

North Carolina Solar Center. Box 7401, Raleigh, NC 27695. Tel: (919) 515-5666. Web site: www.ncsc.ncsu.edu. Offers workshops, tours, publications, and much more.

Renewable Energy Training and Education Center. 1679 Clearlake Road, Cocoa, FL 32922. Tel: (321) 638-1007. Offers hands-on training and courses in the U.S. and abroad for those interested in becoming certified in solar installation.

Solar Energy International. P.O. Box 715, Carbondale, CO 81623. Tel: (970) 963-8855. Web site: www.solarenergy.org. Offers a wide range of workshops on solar energy, wind energy, and natural building.

Sustainable Buildings Industries Council (SBIC). 1331 H Street NW, Suite 1000, Washington, DC 20005. Tel: (202) 628-7400. Web site: www.sbicouncil.org. This organization has a terrific web site with information on workshops, books and publications, and links to many other international, national, and state solar energy organizations. Publishes a newsletter, *Buildings Inside and Out*.

Chapter 13—Green Power

Publications

Butti, Ken, and John Perlin. *A Golden Thread: 2500 Years of Solar Architecture and Technology*. Palo Alto, Calif.: Cheshire Books, 1980. Delightful history of solar energy.

Davidson, Joel. *The New Solar Electric Home: The Photovoltaics How-To Handbook*. Ann Arbor, Mich.: Aatec Publications, 1987. Comprehensive and highly readable guide to photovoltaics, although it is a bit out of date.

Gipe, Paul. *Wind Power for Home and Business: Renewable Energy for the 1990s and Beyond*. White River Junction, Vt.: Chelsea Green, 1993. Comprehensive, technical coverage of home wind power.

Gourley, Colleen. "Production Builders Go Solar." *Solar Today* (January/February 2002): 24–27. An inspiring story about the incorporation of solar electricity into homes in California by large-scale production builders.

Hackleman, Michael, and Claire Anderson. "Harvest the Wind." *Mother Earth News* (June/July 2002): 70–81. A wonderful introduction to wind power.

Jeffrey, Kevin. *Independent Energy Guide: Electrical Power for Home, Boat, and RV.* Ashland, Mass.: Orwell Cove Press, 1995. Contains a wealth of information on solar electric systems and wind generators, and it is fairly easy to read.

Komp, Richard J. *Practical Photovoltaics: Electricity from Solar Cells.* 3rd ed. Ann Arbor, Mich.: Aatec Publications, 1999. Fairly popular book on PVs.

Linkous, Clovis A. "Solar Energy Hydrogen—Partners in a Clean Energy Economy." *Solar Today* 13 (1999): 22–25. A good, but detailed and somewhat technical, article on hydrogen production.

NREL. *The Borrower's Guide to Financing Solar Energy Systems.* Golden, Colo.: National Renewable Energy Laboratory, 1998. Provides information about nationwide financing programs for photovoltaics and passive solar heating. Contact NREL's Document Distribution Service at (303) 275-4363 for a free copy.

———. *The Colorado Consumer's Guide to Buying a Solar Electric System.* Golden, Colo.: National Renewable Energy Laboratory, 1998. Provides basic information about purchasing, financing, and installing photovoltaic systems in Colorado that is applicable to many other states and countries. Contact NREL's Document Distribution Service at (303) 275-4363 for a free copy.

Peavey, Michael A. *Fuel from Water: Energy Independence with Hydrogen.* 8th ed. Louisville, Ky.: Merit Products, 1998. Technical analysis for engineers and chemists.

Potts, Michael. *The New Independent Home: People and Houses That Harvest the Sun, Wind, and Water.* White River Junction, Vt.: Chelsea Green, 1999. Delightfully readable book with lots of good information.

Rastelli, Linda. "Energy Independence with All the Comforts." *Solar Today* (January/February 2002): 28–31. An inspiring story about a passive solar/solar electric home in the Washington, D.C., area.

Roberts, Simon. *Solar Electricity: A Practical Guide to Designing and Installing Small Photovoltaic Systems.* Saddle River, N.J.: Prentice-Hall, 1991. Good reference but a bit dated.

Sagrillo, Mick. "Apples and Oranges 2002: Choosing a Home-Sized Wind Generator." *Home Power* (August/September 2002): 50–66. An extremely useful comparison of popular wind generators with lots of good advice on choosing a wind machine that works best for you. A must-read for anyone interested in buying a wind generator.

Schaeffer, John, ed. *Solar Living Source Book.* 10th ed. Ukiah, Ca.: Gaiam Real Goods, 1999. Contains an enormous amount of background information on wind, solar, and microhydroelectric.

Seuss, Terri, and Cheryl Long. "Eliminate Your Electric Bill: Go Solar, Be Secure." *Mother Earth News* (February/March 2002): 72–82. An excellent discussion of solar roofing materials.

Solar Energy International. *Photovoltaic Design Handbook, Version 2.* Carbondale, Colo.: Solar Energy International. A manual on designing, installing, and maintaining a PV system. Used in SEI's PV design and installation workshops.

Strong, Steven J., and William G. Scheller. *The Solar Electric House: Energy for the Environmentally Responsive, Energy-Independent Home.* Still River, Mass.: Sustainability Press, 1993. A comprehensive and more technical guide to solar electricity.

Videos

An Introduction to Residential Microhydro Power with Don Harris. Available from Scott S. Andrews, P.O. Box 3027, Sausalito, CA 94965. Tel: (415) 332-5191. Outstanding video packed with lots of useful information.

An Introduction to Residential Solar Electricity with Johnny Weiss. Good basic introduction to solar electricity. Available from Scott S. Andrews at the address listed above.

An Introduction to Residential Wind Power with Mick Sagrillo. A very informative video, especially for those wishing to install a medium-sized system. Available from Scott S. Andrews at the address listed above.

An Introduction to Solar Water Pumping with Windy Dankoff. A very useful introduction to the subject. Available from Scott S. Andrews at the address listed above.

An Introduction to Storage Batteries for Renewable Energy Systems with Richard Perez. This is one of the best videos in the series. It's full of great information. Available from Scott S. Andrews at the address listed above.

Newsletters and Magazines

Home Power. P.O. Box 520, Ashland, OR 97520. Tel: (800) 707-6585. Web site: www.home-power.com. Publishes numerous articles on PVs, wind energy, and microhydroelectric and occasionally an article or two on passive solar heating and cooling.

Solar Today. ASES, 2400 Central Ave., Suite A, Boulder, CO 80301. Tel: (303) 443-3130. Web site: www.solartoday.org. This magazine, published by the American Solar Energy Society, contains a wealth of information on passive solar, solar thermal, photovoltaics, hydrogen, and other topics. Also lists names of engineers, builders, and installers and lists workshops and conferences.

Wind Energy Weekly. Newsletter published by the American Wind Energy Association, listed under "Organizations," below.

Organizations

American Solar Energy Society. 2400 Central Avenue, Suite A, Boulder, CO 80301. Web site: www.ases.org. Publishes *Solar Today* magazine and sponsors an annual national meeting. Also publishes an on-line catalog of publications and sponsors the National Tour of Solar Homes. Contact this organization to find out about an ASES chapter in your area.

American Wind Energy Association. 122 C Street, NW, Suite 380, Washington, DC 20001. Tel: (202) 383-2500. Web site: www.awea.org. This organization also sponsors an annual conference on wind energy. Check out their web site, which contains a list of publications, their on-line newsletter, frequently asked questions, news releases, and links to companies and organizations.

British Wind Energy Association. 1 Aztec Row, Berners Road, N1 0PW, United Kingdom. Tel: 020-7689-1960. Web site: www.bwea.com. Actively promotes wind energy in Great Britain. Check out their web site for fact sheets, answers to frequently asked questions, links, and a directory of companies.

Centre for Alternative Technology. Machynlleth, Powys, United Kingdom SY20 9AZ. Tel: 01654 702782. Web site: www.cat.org.uk. This educational group in the United Kingdom offers workshops on alternative energy, including wind, solar, and microhydroelectric.

Center for Renewable Energy and Sustainable Technologies. 1612 K St. NW, Suite 202, Washington, DC 20006. Tel: (202) 293-2898. Web site: solstice.crest.org. Nonprofit organization dedicated to renewable energy, energy efficiency, and sustainable living.

European Wind Energy Association. Rue du Trone 26, B-1000, Brussels, Belgium. Tel: 32-2-546-1940. Web site: www.ewea.org. Promotes wind energy in Europe. The organization publishes *Wind Directions*. Their web site contains information on wind energy in Europe and offers a list of publications and links to other sites.

National Wind Technology Center of the National Renewable Energy Laboratory. 18200 State Highway 128, Boulder, CO 80303. Tel: (303) 384-6900. Web site: www.nrel.gov/wind. Their web site provides a search mode, so you can explore a great deal of information on wind energy, including a wind resource database.

Solar Energy International. P.O. Box 715, Carbondale, CO 81623. Tel: (970) 963-8855. Web
    site: www.solarenergy.org. Offers a wide range of workshops on solar energy, wind
    energy, and natural building.

Solar Living Institute. P.O. Box 836, Hopland, CA 95449. Tel: (707) 744-2017. Web site:
    www.solarliving.org. A nonprofit organization that offers frequent hands-on workshops
    on solar and wind energy and many other topics.

Chapter 14—Water and Waste

Publications

Banks, Suzy, and Richard Heinichen. *Rainwater Collection for the Mechanically Challenged*.
    Dripping Springs, Tex.: Tank Town Publishing, 1997. Humorous and informative guide to
    aboveground rainwater catchment systems.

Campbell, Stu. *The Home Water Supply: How to Find, Filter, Store, and Conserve It*. Pownal, Vt.:
    Storey Communications, 1983. Good resource on water supply systems, although it is
    dated. Unfortunately, it has very little about catchwater or graywater systems.

Del Porto, David, and Carol Steinfeld. *The Composting Toilet System Book*. Concord, Mass.:
    Center for Ecological Pollution Prevention, 1999. Contains detailed information on com-
    posting toilets and graywater systems.

Harper, Peter. *Fertile Waste: Managing Your Domestic Sewage*. Machynlleth, Powys, U.K.: Centre
    for Alternative Technology, 1998. This brief book offers some useful information on com-
    posting toilets and handling urine.

Jenkins, Joseph. *The Humanure Handbook. A Guide to Composting Human Manure*. 2nd ed.
    Grove City, Pa.: Jenkins Publishing, 1999. Excellent resource, well worth your time.

Ludwig, Art. *Builder's Greywater Guide: Installation of Greywater Systems in New Construction and
    Remodeling*. Santa Barbara, Calif.: Oasis Design, 1995. Considerable information on graywa-
    ter systems, including important information on safety and chemical contents of detergents.

———. *Create an Oasis with Greywater: Your Complete Guide to Choosing, Building, and Using
    Greywater Systems*. Santa Barbara, Calif.: Oasis Design, 1994. Fairly detailed discussion of
    the various types of graywater systems.

Solar Survival Architecture. "Black Water." *Earthship Chronicles*. Taos, N.M.: Solar Survival Press,
    1998. Provides an introduction to the blackwater systems under development by SSA.

———. "Catchwater." *Earthship Chronicles*. Taos, N.M.: Solar Survival Press, 1998. Focuses pri-
    marily on catchwater systems for Earthships, but has ideas that are relevant to all homes.

———. "Greywater." *Earthship Chronicles*. Taos, N.M.: Solar Survival Press, 1998. Focuses pri-
    marily on graywater systems for Earthships, but has ideas that are relevant to all homes.

U.S. Environmental Protection Agency. *U.S. EPA Guidelines for Water Reuse*. Washington,
    D.C.: U.S. EPA, 1992. Publication USEPA/USAID EPA625/R-92/004. You can obtain a
    copy of this document at the U.S. EPA National Center for Environmental Publications,
    P.O. Box 42419, Cincinnati, OH 45242. Tel: (800) 489-9198. Web site:
    www.epa.gov/epahome/publications.htm.

Videos

Rainwater Collection Systems. Morris Media Associates, Inc., 1992. A brief, informative video
    that comes with a 50-page booklet that provides more details on systems and information
    on equipment and suppliers. Available from Garden-Ville Nursery, 8648 Old BeeCave
    Road, Austin, TX 78735. Tel: (512) 288-6113. Web site: www.garden-ville.com.

Organizations

American Water Works Association. 6666 W. Quincy Avenue, Denver, CO 80235. Tel: (303) 794-7711. Web site: www.awwa.org. Concerned with many aspects of water, including water reuse. They publish proceedings from their water reuse conferences that offer valuable information

Rocky Mountain Institute. 1739 Snowmass Creek Road, Snowmass, CO 81654. Tel: (970) 927-3851. Web site: www.rmi.org. Check out the catalog of this outstanding organization for publications on water efficiency and water reuse.

Chapter 15—Environmental Landscaping

Publications

Buege, Douglas, and Vicky Uhland. "Natural Swimming Pools." *Mother Earth News* (August/September 2002): 64–73. A fairly thorough discussion of natural swimming pools, describing construction details.

Dramstad, Wenche E., James D. Olson, and Richard T. Forman. *Landscape Ecology: Principles in Landscape Architecture and Land-Use Planning.* Washington, D.C.: Island Press, 1996. A useful textbook on the subject.

Groesbeck, Wesley A., and Jan Striefel. *The Resource Guide to Sustainable Landscape Gardens.* Salt Lake City, Utah.: Environmental Resources, Inc., 1995. Excellent resource.

Matson, Tim. *Earth Ponds Sourcebook: The Pond Owner's Manual and Resource Guide.* Woodstock, Vt.: Countryman Press, 1997. Useful guide for making ponds and natural swimming pools.

Moffat, Anne S., and Marc Schiler. *Energy-Efficient and Environmental Landscaping.* South Newfane, Vt.: Appropriate Solutions Press, 1994. An excellent reference with an abundance of information on landscaping strategies and plant varieties suitable for different climate zones. This book also lists solar tables that will help you determine the path of the sun at different times of the year in your area.

Mollison, Bill. *Permaculture Two: Practical Design for Town and Country in Permanent Agriculture.* Tyalgum, Australia: Tagari Publications, 1979. A seminal work in the field of permaculture.

NREL. *Landscaping for Energy Efficiency.* Washington, D.C.: DOE Office of Energy Efficiency and Renewable Energy, 1995. DOE/GO-10095-046. Provides a decent, though somewhat disorganized, overview on the topic.

Reynolds, Kimberly A. "Happiness Is a Suburban Homestead." *Mother Earth News* (June/July 2002): 109, 128. A tale of partial independence in an urban environment.

Taute, Michelle. "Make a Splash," *Natural Home* (July/August 2002): 56–59. A brief but delightfully well-illustrated story on natural swimming pools.

Vivian, John. "The Working Lawn: A Step Beyond an Expanse of Green." *Mother Earth News* (June/July 2001): 66–74. A guide to converting lawn to a productive landscape.

Magazines

The Permaculture Activist. P.O. Box 1209, Black Mountain, NC 28711. Tel: (828) 669-6336. Web site: www.permacultureactivist.com. Publishes articles on a variety of subjects related to permaculture and includes an updated list of permaculture design courses.

Permaculture Magazine: Solutions for Sustainable Living. Permanent Publications, Freepost (SCE 8120) Petersfield GU32 1HR, United Kingdom. Web site: www.permaculture.co.uk. A quarterly journal, published in cooperation with the Permaculture Association of Great Britain, containing articles, book reviews, and solutions from Britain and Europe.

Organizations

Appropriate Technology Transfer for Rural Areas. P.O. Box 3657, Fayetteville, AR 72702. Tel: (800) 346-9140. This organization is actively involved in the permaculture movement.

Biotop. HauptstraBe 285, A-3411 Weidling, Austria. Tel: 43-0-2243-30406-21. Web site: www.biotop-gmbh.at. The Austrian company that has pioneered the natural swimming pool.

International Permaculture Institute. P.O. Box 1, Tyalgum, NSW 2484, Australia. Tel: (066)-793-442. An international coordinating organization for permaculture activities such as accreditation.

National Wildlife Federation, Backyard Wildlife Habitat Program. 1400 16th Street NW, Washington, DC 20036-2266. Tel: (800) 822-9919. Web site: www.nwf.org/backyard-wildlifehabitat. Contact them for information on creating wildlife habitat in your back-yard or on your land.

Epilogue—Making Your Dream Come True

Bridges, James E. *Mortgage Loans: What's Right for You?* Cincinnati: Betterway, 1997. This book will help you understand mortgages and help you pick the one that's best for you.

Freeman, Mark. *The Solar Home: How to Design and Build a House You Heat with the Sun.* Mechanicsburg, Pa.: Stackpole Books, 1994. This book contains a wealth of information on building your own home, including many practical aspects.

Heldmann, Carl. *Be Your Own House Contractor: Save 25% without Lifting a Hammer.* Pownal, Vt.: Storey Books, 1995. Although it is geared toward conventional home building, the book will walk you through the steps of building a home, giving advice on many issues such as permits and working with subcontractors.

Roy, Rob. *Mortgage-FREE! Radical Strategies for Home Ownership.* White River Junction, Vt.: Chelsea Green, 1998. Wonderful book that should be on of the top of your list if you're looking to build a home but wish to avoid the tyranny of mortgage payments.

Wilson, Alex, Jenifer L. Uncapher, Lisa McManigal, L. Hunter Lovins, Maureen Cureton, and William D. Browning. *Green Development: Integrating Ecology and Real Estate.* New York: John Wiley and Sons, 1998. Contains an enormous amount of information for professional and nonprofessional readers.

Organizations

Resnet. Residential Energy Services Network. P.O. Box 4561, Oceanside, CA 92052. Tel: (760) 806-3448. Web site: www.natresnet.org. A nationwide network of mortgage companies, real estate brokerages, builders, appraisers, utilities, and other housing and energy professionals. This organization is dedicated to improving the energy efficiency of the nation's housing stock and expanding the national availability of mortgage financing.

Green Building Recommendations

Below is a list of all major suggestions offered in this book. You can use this list when building a home, remodeling, or considering a home for purchase. Note that a few items are repeated, as they pertain to several aspects of green building.

Selecting a Place to Live

1. Build in already developed areas within cities and towns.
2. Choose a site with good solar exposure.
3. If considering the use of wind energy, choose a site with good, reliable winds.
4. Choose a site suitable for earth sheltering.
5. Avoid building in frost pockets.
6. Be on the lookout for favorable microclimates.
7. Select a well-drained site.
8. Select a site with stable (nonexpansive) subsoils.
9. Avoid building in hazardous areas, such as floodplains, arroyos, and locations potentially in the path of mud slides and avalanches.
10. Avoid building on or near marshy areas.
11. Do not drain wetlands to build a house.
12. Select a site with rich, productive soils for growing food and fiber.
13. Select a site that could provide some or all of your building materials.
14. Select a site with a reliable, clean water supply.
15. Select a site that is easily accessible and does not require extensive grading for placement of the house or driveway construction.
16. Avoid noisy areas.
17. Check out environmental and community amenities such as recycling facilities, bike paths, parks, and recreation. Consider living in a cohousing community, ecovillage, or "new town."

Siting a Home

18. Orient a home for maximum solar gain.
19. Don't destroy the beauty of a site or build on its most picturesque part.
20. Nestle the house sensitively in the landscape, protecting views and delicate ecological areas.
21. Consider cluster development.
22. Build a house that is unobtrusive in the landscape.

23. Design a home that is compatible with the characteristics of the site.

Protecting a Site during Construction

24. Minimize land disturbance during construction.
25. Designate one access route to the site.
26. Designate a parking area or ask workers to park on the road.
27. Designate an area for delivery of building materials and other supplies.
28. Designate an area for stockpiling materials to be recycled.
29. Cordon off trees and areas you want to protect.
30. Stockpile topsoil for later reapplication, and protect it from erosion.
31. Do not stockpile soils around trees.
32. Recycle all waste from the building site.
33. Do not dump hazardous materials on the building site.
34. Hire a tree specialist to help develop a tree protection strategy.
35. Carefully site the home to reduce the need to remove trees.
36. Protect trees from physical damage by wrapping cardboard around their trunks. Protect roots from damage.
37. Communicate the essential elements of the tree protection strategy to the contractor, subcontractors, and workers.
38. Include fines in your contracts with the contractor to pay for damage to trees.
39. Fertilize nearby trees before construction and water every two weeks during construction.
40. Recycle trees that must be cut down.
41. Control on-site erosion.
42. Design driveways to minimize erosion.

Building a Healthy House

43. Eliminate or reduce the use of oriented strand board, plywood, and other engineered lumber containing formaldehyde; specify low- or no-formaldehyde products instead.
44. Air out all engineered lumber before installation.
45. Eliminate or reduce the use of standard carpeting; install "chemical-free" carpets instead.
46. Install energy-efficient, closed-combustion-chamber, forced-vent furnace or boiler and water heater.
47. Specify the use of nontoxic (low- or no-VOC) paints, stains, and finishes.
48. Install insulation that does not outgas or is not produced with ozone-depleting chemicals.
49. Do not build near potential sources of air pollution such as power plants, factories, and pig farms.
50. Do not build near potential sources of noise such as airports, major highways, hospitals, nightclubs, and police stations.
51. Do not build near potential sources of low-frequency electromagnetic waves such as radio and TV towers and electrical substations.

52. Seal oriented strand board and other engineered lumber to prevent out-gassing.

53. Install vapor barriers and building wraps to reduce air infiltration to prevent toxics from entering indoor air.

54. Seal cracks in the building envelope to prevent pollutants from seeping into the interior.

55. Install radon protection such as polyethylene sheeting over the ground in crawl spaces.

56. Install metal-shielded wires to protect against magnetic radiation.

57. Equip the home with natural ventilation.

58. Install exhaust fans in bathrooms, kitchen, and laundry room.

59. Install a whole-house ventilation system with heat-recovery ventilator.

60. Install an air filtration system as last resort.

Reducing Wood Use

61. Renovate or remodel existing buildings.

62. Build small.

63. Build simply.

64. Design the home to be built in 2-foot increments.

65. Build for durability.

66. Build for adaptability.

67. Design using optimum value engineering.

68. Use engineered lumber and wood products.

69. Use prefabricated trusses and other factory-assembled building components.

70. Reduce wood waste by recycling and reusing scrap lumber.

71. Use salvaged and/or reclaimed lumber.

72. Use alternative materials such as steel, straw bales, and earthen materials.

73. Use certified wood.

Renewable Energy and Energy Efficiency

74. Build using an integrated or whole-building approach.

75. Build a small, space-efficient home.

76. Build using optimum value engineering.

77. Insulate the foundation and building envelope well.

78. Install the insulation correctly.

79. Seal cracks to prevent infiltration and exfiltration.

80. Ensure proper ventilation.

81. Install energy-efficient windows and doors.

82. Install shades or shutters to cover windows at night.

83. Design and build for passive solar heating.

84. Design and build for passive cooling.

85. Install an energy-efficient backup heating system.

86. Install a quiet, energy-efficient cooling system.

87. Correctly size the heating and cooling systems.

88. Install programmable thermostats.
89. Install energy-efficient appliances and electronics.
90. Install an energy-efficient lighting system (including compact fluorescent light bulbs, measures to prevent overlighting, task lighting, individual controls, and motion sensors and timers for light switches).

Accessibility

91. Install a ramp for wheelchair access or design the entrance so that it can be easily retrofitted.
92. Install wider interior and exterior doors.
93. Build interior hallways wider.
94. Create turn-around zones.
95. Design bathrooms for wheelchair access (raise tubs 3 inches, install grab bars, install sinks at wheelchair height, and build showers larger).
96. Build laundry rooms large enough for wheelchair access.
97. Install front-loading washers and dryers.
98. Design kitchens for wheelchair access (install countertops at wheelchair height).
99. Include a ground-level bedroom or a room that can be converted to one in the floor plan.
100. Install special doorknobs and handles for ease of use.
101. Minimize or omit thresholds.
102. Prewire an intercom system.

Ergonomic Design

103. Locate countertops at convenient heights for multiple users.
104. Install adjustable countertops.
105. Install switches and outlets in optimal reach zone.
106. Install cabinets in optimal reach zone.
107. Be sure the kitchen work area is sufficient and well organized for efficient preparation of meals.
108. Design the home for smooth flow of internal traffic.

Adaptability

109. Design the house so spare bedrooms, extra garage space, and basements can be converted to rentable space when family size shrinks.
110. Design the house so apartments (existing or planned) have private entrances.
111. Insulate for sound.

Earth Sheltering

112. Consider building an earth-sheltered home.

113. When building an earth-sheltered home, choose a site with permeable soils and good drainage, install a good drainage system around the house, waterproof walls and roof, insulate walls and roof, check for radon, build above the water table, build on south-facing slopes and incorporate passive solar design, backfill carefully, and hire a professional designer/builder.

Passive Solar Heating

114. Choose a site with good solar access—exposure to the sun from 9 A.M. to 3 P.M.

115. Avoid wooded lots or be prepared to cut down some trees on the south side of the house.

116. Avoid building in the path of buildings or geologic formations that will obstruct the sun.

117. Concentrate windows on the south side of the house.

118. Include overhangs.

119. Provide additional shade for the east and west sides of a house.

120. Install the proper amount of thermal mass to accommodate solar gain through south-facing windows.

121. Build an energy-efficient structure. Be sure to insulate and seal the building well.

122. Protect insulation from moisture.

123. Design the house so rooms are heated directly by the sun. If this is not possible, be sure to locate rooms with lower heat requirements on the north side of the house.

124. Create sun-free zones inside the home.

125. Provide efficient, nonpolluting, and properly sized backup heating and cooling systems.

126. Subject the house design to analysis by energy software before you begin to build.

Passive Cooling

127. Replace incandescent light bulbs with compact fluorescent light bulbs and halogen lighting.

128. Install task lighting.

129. Position windows to take advantage of daylighting.

130. Install energy-efficient appliances.

131. Provide space to cook and dry clothes outside when weather permits.

132. Isolate water heaters and laundry rooms from the main living area.

133. Orient the home properly to minimize external heat gain.

134. Place windows properly to minimize external heat gain.

135. Use skylights sparingly or install solar tube skylights.

136. Avoid unshaded two-story glass walls.

137. Plant shade trees around the house, but do not shade south-facing windows.

138. Install overhangs and mechanical shading devices.

139. Paint the house a light color.
140. Install light-colored roofing material.
141. Install radiant barriers.
142. Install low-E windows.
143. Insulate the house well.
144. Create an airtight buidling envelope.
145. Site the home to take advantage of breezes for natural ventilation.
146. Locate openable windows to provide cross-ventilation and capitalize on stack effect.
147. Design your home with an open floor plan.
148. Landscape to funnel breezes toward the house.
149. Install window and ceiling fans.
150. Install an attic fan or a whole-house fan.
151. Install thermal mass for nighttime cooling using natural ventilation, evaporative cooling, or air conditioner.
152. Install an energy-efficient cooling system.

Solar and Wind Energy

153. Install a solar electric system.
154. Install a wind generator.
155. Purchase green power from a local utility.

Water and Waste

156. Install water-efficient fixtures, including showerheads, faucets, and toilets, and water-efficient appliances, such as washing machines.
157. Install a catchwater system to meet some or all of your water needs.
158. Install a graywater system.
159. Install a blackwater system or composting toilet.

Energy-Efficient Landscaping

160. Plant trees and other vegetation to provide shade and natural cooling.
161. Plant a windbreak to protect against winter winds.

Natural Drainage

162. Utilize existing natural drainage by properly siting the home and protecting natural drainage from damage.
163. Protect vegetation to reduce on-site erosion.
164. Minimize impervious surfaces such as driveways, patios, and sidewalks.
165. Plant turf between impervious surfaces to absorb runoff.
166. Revegetate denuded landscape in the vicinity of the house.

Water-Efficient Landscaping

167. Apply a thick, rich layer of topsoil over the yard before replanting.
168. Plant native vegetation.
169. Plant xeric vegetation in arid or semiarid areas.
170. Mulch gardens, trees, and shrubs to reduce water demand.
171. Place plants that need more water near driveways, sidewalks, and downspouts.
172. Reduce grassy areas or plant low-water grasses.
173. Install an efficient automatic watering system with a timer, and water early and late in the day.
174. Be sure water systems don't spray onto impervious surfaces.
175. Install root-zone and drip-irrigation systems whenever possible.
176. Build a natural swimming pool.

Landscaping for Wildlife

177. Devote a portion of your yard to wildlife.
178. Plant vegetation that supports wildlife, providing shelter, nesting sites, and food.

Planting Edible Landscape

179. Use part of your land to grow vegetables, herbs, fruit, meat, and fiber.

EarthCraft House℠ Checklist

The EarthCraft House Checklist is a detailed point system to assist builders and designers and was provided courtesy of the Greater Atlanta Home Builder's Association.

	POINTS	WORKSPACE	FINAL
SITE PLANNING			
Erosion control site plan	8		
Workshop on erosion and sediment control	2		
Topsoil preservation	5		
Grind stumps and limbs for mulch	3		
Mill cleared logs	2		
Building With Trees (NAHB program)	25		
Builder may choose to certify house meets Building With Trees program or earn points from individual tree protection and planting measures, outlined below.			
Tree Protection and Planting Measures			
Tree preservation plan	5		
No trenching through tree root zone (per tree)	1		
No soil compaction of tree root zone	2		
Undisturbed areas	1		
Tree planting	4		
Wildlife habitat	2		
SITE PLANNING SUBTOTAL			

ENERGY-EFFICIENT BUILDING ENVELOPE AND SYSTEMS

Energy Star 90

Builder may choose to certify house meets Energy Star requirements or earn a minimum of 75 points from energy measures outlined below.

Energy Measures
Points must total at least as 75, but cannot exceed 85. Houses must meet or exceed the Georgia Energy Code.

Air Leakage Test
Builder can provide documented proof of certified test to homeowner or earn points for individual air sealing measures.

	POINTS	WORKSPACE	FINAL
Certify maximum 0.35 air change per hour	35		

Air Sealing Measures
maximum 30

	POINTS	WORKSPACE	FINAL
Bottom plate of exterior walls	2		
Floor penetrations between unconditioned and conditioned space	2		
Bath tub and shower drain	2		
Cantilevered floors sealed above supporting wall	2		

	POINTS	WORKSPACE	FINAL
Drywall sealed to bottom plate of exterior walls	2		
Fireplace air sealing package (all units)	2		
Drywall penetrations in exterior walls	2		
Exterior wall sheathing sealed at plates, seams, and openings	5		
Housewrap (unsealed at seams and openings)	2		
Housewrap (sealed at plates, seams, and openings)	8		
Window rough openings	2		
Door rough openings	1		
Airtight IC recessed lights or no recessed lights in insulated ceilings	4		
Attic access opening (pulldown stairs/scuttle hole)	2		
Attic kneewall doors (weatherstripped with latch)	2		
Attic kneewall has sealed exterior sheathing	5		
Chases sealed and insulated	5		
Ceiling penetrations sealed between unconditioned and conditioned space	2		
Ceiling drywall sealed to top plate	2		
Band joist between conditioned floors sealed	3		
AIR SEALING MEASURES SUBTOTAL			

Insulation

Homes with multiple foundation types must use foundation type of greatest area for points.

	POINTS	WORKSPACE	FINAL
★Slab insulation	2		
★Basement walls (continuous floor to ceiling R-10)	3		
★Framed floor over unconditioned space (R-19)	1		
★Sealed, insulated crawl space walls (R-10)	1		
★Cantilevered floor (R-30)	2		
Insulate fireplace chase	1		
Spray-applied wall insulation	4		
Exterior wall stud cavities (R-15)	1		
Insulated headers	2		
Insulated corners	2		
Insulated T-walls (exterior/interior wall intersection)	2		
Insulated wall sheathing (R-2.5 or greater)	2		
Insulated wall sheathing (R-5 or greater)	3		
Band joist insulated (R-19)	2		
Loose-fill attic insulation card and rulers	1		
Energy heel trusses or raised top plate	2		
Flat ceilings (R-30)	1		
Flat ceilings (R-38)	2		
Vaulted and tray ceilings (R-25)	1		
Vaulted and tray ceilings (R-30)	2		
Ceiling radiant heat barrier	1		
Attic kneewall stud cavities (min. R-19)	3		
Attic kneewall with insulated sheathing (R-5)	5		
Attic kneewall doors (R-19)	2		
Attic access doors (R19)	2		
INSULATION SUBTOTAL			

	POINTS	WORKSPACE	FINAL

Windows

	POINTS	WORKSPACE	FINAL
NFRC-rated windows (max U–56)	3		
Low-emissivity glazing	5		
Gas-filled double-glazed units	3		
Solar heat gain coefficient (max 0.4)	3		
1½-foot overhangs on all sides	1		
Solar shade screens	3		
West-facing glazing less than 2% of floor area	2		
East-facing glazing less than 3% of floor area	2		
Certified passive solar design (25% load reduction)	10		
WINDOWS SUBTOTAL			

Heating and Cooling Equipment
Builder must provide documented proof.

	POINTS	WORKSPACE	FINAL
*Cooling equipment sized within 10% of Manual J (all units)	5		
*Heating equipment sized within 10% of Manual J (all units)	5		
*Measured airflow within 10% of manufacturer's specifications	3		
90% AFUE furnace (per unit)	3		
SEER 12 cooling equipment (per unit)	2		
SEER 14 cooling equipment (per unit)	3		
HSPF 7.8 heat pump	2		
HSPF 8.0 heat pump	3		
Geothermal heat pump	4		
*Sensible heat fraction (max 0.7, all units)	2		
Programmable thermostat	1		
Outdoor thermostat for heat pump	1		
*Cooling equipment has non-CFC or -HCFC refrigerant	3		
Zone control—one system services multiple zones	5		
HEATING AND COOLING SUBTOTAL			

Ductwork/Air Handler
Builder must provide documented proof of certification to homeowner.

	POINTS	WORKSPACE	FINAL
*Certify duct leakage less than 5%	20		
Air handler located within conditioned space (all units)	5		
Ducts located within conditioned space (min. 90%)	5		
Duct seams and air handler sealed with mastic	10		
*Duct design complies with Manual D	5		
*Airflow for each duct run measured and balanced	3		
No ducts in exterior walls	3		
Longitudinal supply tank	1		
Multiple return ducts	2		
Interior doors with 1-inch clearance to finish floor	2		
Duct trunk lines outside conditioned space insulated to R–8	2		
DUCTWORK/AIR HANDLER SUBTOTAL			
ENERGY-EFFICIENT BUILDING ENVELOPE SUBTOAL			

minimum of 75, maximum of 85

	POINTS	WORKSPACE	FINAL

ENERGY-EFFICIENT LIGHTING/APPLIANCES

	POINTS	WORKSPACE	FINAL
Indoor fluorescent fixtures	2		
Recessed light fixtures are compact fluorescents	2		
Outdoor lighting control	2		
High-efficiency exterior lighting	2		
Energy-efficient dishwasher	1		
Energy-efficient refrigerator	2		
No garbage disposal	1		
ENERGY-EFFICIENT LIGHTING/APPLIANCES SUBTOTAL			

RESOURCE-EFFICIENT DESIGN

	POINTS	WORKSPACE	FINAL
Floor plan adheres to 2-ft. dimensions	2		
Interior living spaces adhere to 2-ft. dimensions	1		
Floor joists @ 24-in. centers (per floor)	3		
Floor joists @ 19.2-in. centers (per floor)	2		
Non-load-bearing wall studs @ 24-in. centers	2		
All wall studs @ 24-in. centers	3		
Window rough openings eliminate jack studs	2		
Nonstructural headers in non-load-bearing walls	2		
Single top plate with stacked framing	3		
Two-stud corners with drywall clips or alternative framing	3		
T-walls with drywall clips or alternative framing	3		
RESOURCE-EFFICIENT DESIGN SUBTOTAL			

RESOURCE-EFFICIENT BUILDING MATERIALS

Recycled/Natural-Content Materials

	POINTS	WORKSPACE	FINAL
Concrete with fly ash	3		
Insulation	1		
Flooring	1		
Carpet	1		
Carpet pad	1		
Outdoor decking and porches	2		
Air conditioner condensing unit pad	1		
RECYCLED/NATURAL-CONTENT MATERIALS SUBTOTAL			

Advanced Products

	POINTS	WORKSPACE	FINAL
Engineered floor framing	2		
Engineered roof framing	2		
OSB roof decking	1		
Nonsolid sawn wood or steel beams	1		
Nonsolid sawn wood or steel headers	1		
Engineered wall framing	1		

	POINTS	WORKSPACE	FINAL
Engineered interior trim	1		
Engineered exterior trim including cornice	1		
Steel interior walls	1		
Structural insulated panels (exterior walls)	5		
Structural insulated panels (roof)	3		
Precast autoclaved aerated concrete	5		
Insulated concrete forms	5		
ADVANCED PRODUCTS SUBTOTAL			

Durability

	POINTS	WORKSPACE	FINAL
Roofing (min. 25-year warranty)	1		
Roofing (min. 30-year warranty)	2		
Roofing (min. 40-year warranty)	3		
Light roof color (asphalt or fiberglass shingles)	1		
Light roof color (tile or metal)	2		
Roof drip edge	1		
Exterior cladding (min. three sides with 40-year warranty or masonry)	1		
Walls covered with builder paper or housewrap (drainage plane)	1		
Siding with vented rain screen	1		
Back-primed siding and trim	1		
Insulated glazing (min. 10-year warranty)	1		
Window and door head flashing	1		
Continuous foundation termite shield	1		
Roof gutters direct water away from foundation	1		
Covered entryway	1		
DURABILITY SUBTOTAL			
RESOURCE-EFFICIENT BUILDING MATERIALS SUBTOTAL			

WASTE MANAGEMENT

Waste Management Practices
Builder must provide documentation (receipt) of donated materials.

	POINTS	WORKSPACE	FINAL
Job site framing plan and cut list	10		
Central cut area	3		
*Donation of excess materials or reuse (min. $500/job)	1		
WASTE MANAGEMENT PRACTICES SUBTOTAL			

Recycle Construction Waste
Builder can receive points for individual materials or additional points for waste management plan.

	POINTS	WORKSPACE	FINAL
Posted job site waste management plan—*recycle 75% of three materials*	5		
Wood	3		
Cardboard	1		
Metal	1		
Drywall (recycled or grind and spread on-site)	3		

	POINTS	WORKSPACE	FINAL
Plastics	1		
Shingles	1		
RECYCLE CONSTRUCTION WASTE SUBTOTAL			
WASTE MANAGEMENT SUBTOTAL			

INDOOR AIR QUALITY

Combustion Safety

	POINTS	WORKSPACE	FINAL
Detached garage	5		
Attached garage—seal bottom plate and penetrations to conditioned space	4		
Attached garage—exhaust fan controlled by motion sensor or timer	2		
Direct-vent, sealed-combustion fireplace	3		
Furnace combustion closet isolated from conditioned area	4		
Water heater combustion closet isolated or power vented	4		
Carbon monoxide detector	4		
House depressurization test	4		
COMBUSTION SAFETY SUBTOTAL			

Moisture Control

	POINTS	WORKSPACE	FINAL
Drainage tile on top of footing	1		
Drainage tile at outside perimeter edge of footing	2		
Drainage board for below-grade walls	4		
Gravel bed beneath slab-on-grade floors	3		
Vapor barrier beneath slab (above gravel) and in crawl space	2		
Capillary break between foundation and framing	1		
MOISTURE CONTROL SUBTOTAL			

Ventilation

	POINTS	WORKSPACE	FINAL
Radon/soil gas vent system	3		
Radon test of home prior to occupancy	2		
High-efficiency, low-noise bath fans	3		
Tub/shower room fan controls	1		
Kitchen range hood vented to exterior	3		
Ceiling fans (min. three fans)	1		
Whole-house fan	2		
Controlled house ventilation (0.35 ACH)	4		
Dehumidification system	3		
Vented garage storage room	1		
No power roof vents	1		
Dampered fresh air intake	2		
VENTILATION SUBTOTAL			

Materials

	POINTS	WORKSPACE	FINAL
No urea formaldehyde materials inside conditioned space	2		
Urea formaldehyde materials inside conditioned space sealed	1		

	POINTS	WORKSPACE	FINAL
Low-VOC paints, stains, finishes	1		
Low-VOC sealants and adhesives	1		
Low-VOC carpet	1		
Alternative termite treatment	2		
Central vacuum system	1		
Filter/air cleaner with minimum 30% dust spot efficiency	2		
Protect ducts during construction	2		
MATERIALS SUBTOTAL			
INDOOR AIR QUALITY SUBTOTAL			

WATER—INDOOR

	POINTS	WORKSPACE	FINAL
Water filter (NSF certified)	1		
High-efficiency clothes washer	2		
Pressure-reducing valve	1		
High-efficiency plumbing fixtures	2		
Hot water demand recirculater	1		
Shower drain heat-recovery device	1		
Water heater (Energy Star: gas 0.62, electric 0.92)	2		
Water heater tank insulation	1		
Pipe insulation	1		
Heat traps	1		
Heat-recovery water heating	1		
Solar domestic water heating	3		
Heat pump water heater	2		
WATER—INDOOR SUBTOTAL			

WATER—OUTDOOR

	POINTS	WORKSPACE	FINAL
HBA Water Smart program	5		
Xeriscape resource	1		
Xeriscape plan	4		
Xeriscape installed	15		
Timer on hose bibs or irrigation system	1		
Efficient irrigation system (min. 50% plantings with drip system)	2		
Graywater irrigation	3		
Rainwater harvest system	3		
Permeable pavement driveway/parking area	1		
WATER—OUTDOOR SUBTOTAL			

HOMEBUYER EDUCATION/OPPORTUNITIES

	POINTS	WORKSPACE	FINAL
Guaranteed energy bills	15		
Review energy operation with homeowner	4		
Review irrigation system operation manuals with homeowner	2		
Built-in recycling center	2		
Local recycling contact	1		

	POINTS	WORKSPACE	FINAL
Household hazardous waste resources	1		
Environmental features checklist for walk–through	1		
HOMEBUYER EDUCATION/OPPORTUNITIES SUBTOTAL			

BUILDER OPERATIONS

	POINTS	WORKSPACE	FINAL
Builds 10% of total houses to EarthCraft House standards	3		
or builds 80% of total houses to EarthCraft House standards	5		
Markets EarthCraft House program	2		
Environmental checklist provided to all subcontractors	1		
Certified Professional Home Builder	3		
Uses HBA Homeowner Handbook for warranty standards	2		
BUILDER OPERATIONS SUBTOTAL			

BONUS POINTS

	POINTS	WORKSPACE	FINAL
Site located within 1/4 mile of mass transit	5		
Sidewalk connects house to business district	5		
Brownfield site	5		
Solar electric system	25		
Alternative fuel vehicles: electric charging station or natural gas pump	5		
American Lung Association Health House	5		
Exceeds Energy Star (1 point for each 1%) for a maximum of 5			
Innovation points *Builder submits specification for innovative products or design features to qualify for additional points*			
BONUS POINTS SUBTOTAL			

EARTHCRAFT HOUSE TOTALS

SITE PLANNING			
ENERGY-EFFICIENT BUILDING ENVELOPE AND SYSTEMS			
ENERGY-EFFICIENT LIGHTING/APPLIANCES			
RESOURCE-EFFICIENT DESIGN			
RESOURCE-EFFICIENT BUILDING MATERIALS			
WASTE MANAGEMENT			
INDOOR AIR QUALITY			
WATER—INDOOR			
WATER—OUTDOOR			
HOMEBUYER EDUCATION/OPPORTUNITIES			
BUILDER OPERATIONS			
BONUS POINTS			
GRAND TOTAL			

Built Green Communities Checklist

Mission Statement and Objectives

The Built Green Communities program, through a partnership of planners, developers, builders, lenders, and government agencies, will promote voluntary land-use and community design guidelines that will minimize environmental impact, promoting the understanding and acceptance of responsible community design to benefit all citizens.

Specific objectives of the program include preserving natural resources; balancing open space and density; reducing infrastructure costs through efficient design; encouraging cost-effective, innovative ideas and technologies; and creating diverse housing options. Built Green provides a guide for further information.

Built Green Communities checklist with best-case scoring shown

Percent by category of overall checklist makeup	
1. Buildings	9%
2. Site selection	13.5%
3. Transportation	17%
4. Planning and design	28%
5. Preservation, conservation, and restoration	19%
6. Community	13.5%

1. Buildings

Feature	Assessment	Possible Points	Best-Case Score
All homes registered under Built Green Colorado	NA	NA	NA
All homes meet 83 points on the E-Star home energy rating point scale, 100% verified by E-Star Colorado		35	
All homes meet 83 points on the E-Star home energy rating point scale, 25% verified by E-Star Colorado		25	
All homes meet 86 points on the E-Star home energy rating point scale, 100% verified by E-Star Colorado		45	
All homes meet 86 points on the E-Star home energy rating point scale, 25% verified by E-Star Colorado		35	
All homes meet 90 points on the E-Star home energy rating point scale, 100% verified by E-Star Colorado		60	60
All homes meet 90 points on the E-Star home energy rating point scale, 25% verified by E-Star Colorado		50	
	Subtotal Buildings		60
	Maximum Possible	60	
	Minimum for Qualification	N/A	

2. Site Selection

Feature	Assessment	Possible Points	Best-Case Score
Locate new developments adjacent★ to existing population and business centers with existing community services and facilities, and with utility infrastructure within ¼ mile of property line *★Adjacent means the new development has a shared boundary with existing development. Boundary may be interrupted by a road or alleyway.*		14	14
Promote infill development by locating new developments surrounded by existing population and business centers with existing community services and facilities		28	28
Promote brownfield development by remediating and building on a site classified as a brownfield		28	28
Participate in EPA's Brownfields Economic Redevelopment Initiative		10	10
Development is located within walking distance of existing open space or park *(available to the public; assumes walking speed of 265 ft. per minute)*			
0–5 minutes		10	10
6–10 minutes		6	
11–15 minutes		2	
Subtotal Site Selection			90
Maximum Possible		90	
Minimum for Qualification			

3. Transportation

Feature	Assessment	Possible Points	Best-Case Score
Create a mixed-use development with a variety of land uses and neighborhood gathering places (see Guide) and/or village centers within close proximity to minimize automobile trips, encourage walking, and increase social and economic vitality			
Potential uses for the community: Single-family (for sale) residential; Multi-family (rental) residential; Retail; Business; Recreational; Educational; Cultural★ *★Examples of cultural uses include theater, band shell, historical trail, museum, preserved historic site, etc.*			
Community includes a mix of:	5 uses	17	17
Community includes a mix of:	4 uses	14	
Community includes a mix of:	3 uses	10	
Community includes a mix of:	2 uses	6	
Walking distance (assumes walking speed of 265 ft. per minute) from 75% of the homes to activity center★, on- or off-site, is: *★"Activity center" is defined as including two or more of the following: Retail, Business, Recreational, Educational, Cultural*			
0–5 minutes		15	15
6–10 minutes		11	
11–15 minutes		7	

Feature	Assessment	Possible Points	Best-Case Score
Provide incentives to increase public transit use and to reduce auto dependency			
Bike racks or lockers at public areas		1	1
Bike/pedestrian paths to transit stations/stops		2	2
Permanent signage for transit information and access		1	1
Covered transit stations/stops		3	3
Develop under- or overpass elements for bike or pedestrian paths		8	8
Public transit accessibility			
Walking distance (assumes walking speed of 265 ft. per minute) from 75% of the homes to public transit is:			
0–5 minutes		12	12
6–10 minutes		8	
11–15 minutes		5	
Encourage new public transportation systems where existing systems are not adequate			
Advanced coordination with RTD	Define	5	5
Internal bus system	Define	13	13
Reserve land for future mass transit use.		Up to 13	13
Create nonmotorized, continuous and connected open space linkages between neighborhoods and basic services			
★"Internal" is within the boundaries of the community; "external" refers to links with features and/or services outside the community boundary.			
Internal and external★		12	12
External only		5	
Internal only		5	
Develope under-or-over pass elements for open space linkages interior or exterior		8	8
	Subtotal Transportation		110
	Maximum Possible	110	
	Minimum for Qualification		

4. Planning and Design

Feature	Assessment	Possible Points	Best-Case Score
Create efficient land use through developing land in higher gross residential density. Average gross density (du/ac)★ is:			
★du/ac = dwelling units per acre of development not including the publicly accessible open space			
14 or more		30	30
10 to 13.9		24	
7 to 9.9		16	
4 to 6.9		8	

Feature	Assessment	Possible Points	Best-Case Score
The following percentages of gross development area are permanently set aside as publicly accessible open space that neither encloses development nor is enclosed by development (Guide needs to make clear that donut open space doesn't count):			
40% or more		30	30
35 to 39.9%		24	
30 to 34.9%		16	
20 to 29.9%		8	
10 to 19.9%		4	
Promote energy conservation by maximizing solar exposure through access and building orientation. The following percentage of the homes have their predominant solar access facing within 15 degrees of solar south, and have 75% wintertime solar access (structural, see notes)			
85%		18	18
75 to 84.9%		14	
65 to 74.9%		10	
55 to 64.9%		5	
Reduce areas of imperviousness by use of relaxed front setbacks, narrower frontages, and minimized driveway lengths			
Push homes forward so that 50% of homes sit less than 18 ft. from front property line		6	6
Neck down driveway to maximum of 10 ft. width @ street connection		6	6
Use shared driveways to reduce impervious areas		5	5
Efficient street design			
Construct residential streets to 24-foot back of curb to back of curb		10	10
Construct residential streets to 28-foot back of curb to back of curb		6	
Construct residential streets to 32-foot back of curb to back of curb		3	
Provide detached walks on local streets		3	3
Water conservation through appropriate landscaping and water use			
Nonpotable means any nondrinkable water and includes both untreated and recycled; water treated to some extent can still be nonpotable.			
Common areas:			
Common areas irrigated with nonpotable★ water		15	15
50% of common area is restored or undisturbed native plant community to support native wildlife habitat, including a minimum of 6" topsoil, forbs, grasses, sedges, and woody plants		3	3
In all common nonturf areas, use only climate-tolerant plantings		3	3
Individual residences:			
All private lots irrigated with nonpotable★ water		15	15
In all private lots, nonturf areas use only native plants to support native wildlife habitat by providing cover and food sources		3	3
In all private lots, nonturf areas use only climate-tolerant plantings		3	3
Plant two 2-inch caliper or greater climate-tolerant trees per house (may substitute 6' or taller evergreens)		2	

Feature	Assessment	Possible Points	Best-Case Score
Plant four 2-inch caliper or greater climate-tolerant trees per house (may substitute 6' or taller evergreens)		7	7

Utilize the principles of xeriscape in landscape design and construction

Definition of efficient irrigation: System is professionally designed for even application of water and includes separate zones for turf and nonturf planting areas, low-flow irrigation for nonturf areas (i.e., drip or microspray), and rain or moisture sensors. Provide written maintenance guidelines for appropriate irrigation system management practices, such as regular adjustments based on changing seasonal needs of the landscape.

Common areas:

Feature	Assessment	Possible Points	Best-Case Score
Improve the soil (mix in organic material at 3 cubic yards/1000 s.f.)		3	3
Select appropriate turf areas (limit cool-season turf to no more than 50% of landscaped area; use native/drought tolerant-grasses in turf areas exceeding 50%)		3	3
Use appropriate mulches in bedding areas to a depth of 3"		3	3
Install efficient irrigation★ as defined in the Guide		3	3
Group plants according to moisture and sunlight needs		2	2
Provide landscaping maintenance guidelines		1	1

Individual residences, covenant required:

Feature	Assessment	Possible Points	Best-Case Score
Improve the soil (mix in organic material at 3 cubic yards/1000 s.f.)		3	3
Select appropriate turf areas (limit cool-season turf to no more than 50% of landscaped area; use native/drought-tolerant grasses in turf areas exceeding 50%)		3	3
Use appropriate mulches in bedding areas to a depth of 3"		3	3
Install efficient irrigation★ as defined in the Guide		3	3
Group plants according to moisture and sunlight needs		2	2
Provide landscaping maintenance guidelines		1	1
Subtotal Planning and Design			187
Maximum Possible		187	
Minimum for Qualification			

5. Preservation, Conservation, and Restoration

Feature	Assessment	Possible Points	Best-Case Score
Perform natural resources inventory to identify important natural features, i.e., plant and animal habitat, agricultural uses, sensitive natural areas, etc.		15	15
Provide long-term protection for existing watercourses, riparian areas, and significant wetlands and native growth in significant open areas			
Provide a long-term maintenance plan for these areas		5	5
Provide permanent conservation easement or other permanent legal means to preserve these areas		12	12

Feature	Assessment	Possible Points	Best-Case Score
Save and reuse topsoil that has nutrient value based upon soils analysis; remove, store, then replace topsoil to a depth of 4" in all disturbed areas (house pad and street excluded from replacement requirement)		10	10
Include a tree expert on the development team; conduct a tree survey prior to planning and design, and create a tree conservation plan			
Participate in the tree conservation program as defined in the Building With Trees Program created in 1998 by The National Arbor Day Foundation in cooperation with the National Association of Home Builders		5	5
Protect 66-100% of trees greater than 4 inches in diameter		4	4
Protect 25-65% of trees greater than 4 inches in diameter		3	
Replace all destroyed trees per caliper with minimum 4" caliper trees (i.e., destroyed 12" caliper tree replaced with three new 4" caliper):			
Replace trees at 2-to-1 ratio		9	9
Replace trees at 1-to-1 ratio		4	4
Implement landscaping techniques for energy conservation (e.g., evergreen trees for wind- and snowbreaks to reduce heating costs; deciduous trees for summer shading to reduce cooling needs/costs; vegetation and tree planting that encourages channeling of breezes for summer cooling; and the use of insulating shrubs and vines)			
Streets have deciduous trees planted for canopy cooling (see Guide to reference types of trees and proximity to street), 2-inch caliper planted a minimum of every 35 feet or as required by local jurisdiction, required by developer. (trees planted as replacements as above can also qualify for these points)		4	4
Integrate wetlands and landscape areas to promote infiltration of stormwater runoff. *Mitigation of wetland impacts can include: avoiding the impact altogether by not taking certain action or parts of an action; minimizing impacts by limiting the degree or magnitude of an action and its implementation; rectifying the impact by repairing, rehabilitating, or restoring the affected environment; reducing or eliminating the impact over time by preservation and maintenance operations during the life of the action; compensating for the impact by replacing or providing substitute resources or environments. See Guide.*			
Create on-site mitigation* for disturbed/destroyed wetland @ 1-to-1 ratio		3	
Create on-site mitigation for disturbed/destroyed wetland @ 2-to-1 ratio		6	
Create on-site mitigation for disturbed/destroyed wetland @ 3-to-1 ratio or greater		10	10
Enhance/create nonrequired wetlands		7	7
Create natural alternative to conventional stormwater detention (see Guide)		7	7
Create water-quality ponds to filter water		3	3
Create natural conveyance drainage system (e.g., eliminate curb and gutter and underground storm piping)		7	7
Create retention and infiltration ponds to retain 50% of increased flow beyond historical rates		3	

Feature	Assessment	Possible Points	Best-Case Score
Create retention and infiltration ponds for full retention of increased flow beyond historical rates		8	8
Employ sensitive site grading practices		12	12
Create on-site environmentally appropriate wastewater treatment facility			
Constructed wetland, living machine, or other biological wastewater treatment facility (would need specific info for Guide)		6	6
	Subtotal Preservation, Conservation, and Restoration		124
	Maximum Possible	124	
	Minimum for Qualification		

6. Community

Feature	Assessment	Possible Points	Best-Case Score
Establish community involvement through education of residents about what the development has done and what is expected of the residents to ensure that the development retains its green focus and by creating social initiatives to support the project and its principles			
An explicit plan of resource-conserving goals with implementation plan for how the community will meet the goals and create a self-sustaining program		9	9
Full-time paid staff position dedicated to helping the community meet environmental goals		8	8
Part-time paid staff or volunteer position dedicated to helping the community meet environmental goals		4	4
Homeowner guidebook to efficient operation of home with list of additional resources		3	3
Seminar or seminar series for each new family explaining efficient operation of home and other programs to reduce resource use (minimum 4 hours)		3	3
Ongoing series of open-space habitat restoration programs		1	1
Community events/festivals/performances that help educate children and adults about the community's environmental mission		1	1
Ongoing environmental award program		1	1
Neighborhood information kiosk and signage to promote green initiatives (e.g., sharing exchange for appliances and cars; flea market/community garage sale; babysitter exchange, etc)		1	1
Produce and distribute a regularly scheduled community newsletter		1	1
Establish and sponsor community meetings		1	1

Feature	Assessment	Possible Points	Best-Case Score
Establish and maintain an Internet community communication/ education link		1	1
Establish ongoing education and awareness with community signage		1	1
Create civic infrastructure by providing places for programs that bring people together, providing parks and land uses for the benefit of the overall community, and providing community facilities that are the heart of neighborhoods			
Parking is separated and residents access homes through a common green		1	1
50% or more of residences look out to a common green/pedestrian area		1	1
Provide improved common areas or parks serving the community; distribute improved common areas and parks throughout the community			
10–15% of total site		4	
16–20% of total site		8	
21% or more of total site		12	12
Walking distance to improved parks within community from 75% or more of the homes (assumes walking speed of 265 ft. per minute)			
0–5 minutes		8	8
6–10 minutes		4	
Provide a community meeting center, clubhouse, common house or other community-owned space within the development with any of the following features:			
A place for community meetings within the development		3	3
A place for residents to share common meals		3	3
Guest rooms for community use		1	1
An indoor children's playroom		1	1
A common laundry room		1	1
An office for community use		1	1
A community library or media room		1	1
A community teen room		1	1
Provide land and an organized program for community gardens within the development		2	2
Provide community recreation facilities within the development with any of the following features:			
A community swimming pool		3	3
A community exercise room or fitness course		1	1
Community tennis courts		1	1
Playground equipment, basketball court, volleyball court, soccer field, or softball field		1	1
Provide neighborhood educational facilities within the development including any of the following:			
Land dedication for a public library		3	3
Land dedication for a school		4	4

Feature	Assessment	Possible Points	Best-Case Score
Land dedication for a day-care facility		3	3
Land dedication for a mailing facility		3	3
Provide commercial facilities and uses that promote community interaction within the development including any of the following:			
A neighborhood coffeehouse or small restaurant		2	2
Small office spaces for rent to resident businesses		2	2
	Subtotal Community		94
	Maximum Possible	94	
	Minimum for Qualification		
	Final Score		
	Maximum Possible	665	

Minimum Points Requirement:

The qualifying threshold is to be set at 70% of total possible points. If all checklist items are available as possible selections on a given development, the minimum number of points required will be 465. The developer will be asked to submit a "narrative" explaining the choices made, what choices were prohibited from being made by local regulation, what choices may have been made as alternatives to checklist items while still meeting the intent of the checklist, etc. Given that some options will be a "no go" in some jurisdictions or will be in the realm of "can't do," "site doesn't have," "not possible," etc., as relates to the particular site, or even a particular city or town (rural vs. urban), some possible points can justifiably be eliminated as unobtainable on a given project. Those point values (even an entire category) could be deducted from the total possible points of 665, with the remaining number representing 100%, of which 70% must be obtained. In its review, it will be the job of the review board to determine, using the narrative and submittal of the developer, where the intent of the program is being met, even if not through an identified point on the checklist. There is discretion on the part of the review board in that regard, as well as the possibility that we could award "innovation credits."

State Energy Offices

Alabama
Science, Technology & Energy Division
Department of Economic and Community Affairs
Phone: (334) 242-5292
Web site: www.adeca.alabama.gov/columns.aspx?m=
4&id=19&id2=106

Alaska
Alaska Energy Authority
Alaska Industrial Development and Export Authority
Phone: (907) 269-4625
Web site: www.aidea.org/aea.htm

Arizona
Arizona Department of Commerce
Phone: (602) 771-1100
Web site: www.azcommerce.com/Energy/

Arkansas
Arkansas Energy Office
Arkansas Department of Economic Development
Phone: (501) 682-7377
Web site: www.aedc.state.ar.us/Energy/

California
California Energy Commission
Phone: (916) 654-4287
Web site: www.energy.ca.gov

Colorado
Governor's Office of Energy Management and
Conservation
Phone: 303-894-2383
Web site: www.state.co.us/oemc

Connecticut
Energy Research & Policy Development Unit
Connecticut Office of Policy and Management
Phone: (860) 418-6374
Web site: www.opm.state.ct.us/pdpd2/energy/
enserv.htm

Delaware
Energy Office
Division of Facility Management
Phone: (302) 739-1530
Web site: www2.state.de.us/dfm/energy/index.asp

District of Columbia
D.C. Energy Office
Phone: (202) 673-6718
Web site: www.dcenergy.org

Florida
Florida Energy Office
Florida Department of Environmental Protection
Phone: (850) 245-2940
Web site: www.dep.state.fl.us/energy/

Georgia
Division of Energy Resources
Georgia Environmental Facilities Authority
Phone: (404) 656-5176
Web site: www.gefa.org/energy_program.html

Hawaii
Energy Branch, Strategic Industries Division
Department of Business, Economic Development and
Tourism
Phone: (808) 587-3807
Web site: www.hawaii.gov/dbedt/ert/energy.html

Idaho
Energy Division
Idaho Department of Water Resources
Phone: (208) 327-7900
Web site: www.idwr.state.id.us/energy/

Illinois
Energy & Recycling Bureau
Illinois Department of Commerce and Economic
Opportunity
Phone: (217) 782-7500
Web site: www.illinoisbiz.biz/ho_recycling_energy.html

Indiana
Energy Policy Division
Indiana Department of Commerce
Phone: (317) 232-8939
Web site: www.state.in.us/doc/energy/

Iowa
Energy & Waste Management Bureau
Iowa Department of Natural Resources
Phone: (515) 281-8681
Web site: www.state.ia.us/dnr/energy

Kansas
Kansas Energy
Kansas Corporation Commission
Phone: (785) 271-3349
Web site: www.kcc.state.ks.us/energy/

Kentucky
Kentucky Division of Energy
Phone: (502) 564-7192
Web site: www.energy.ky.gov

Louisiana
Technology Assessment Division
Department of Natural Resources
Phone: (225) 342-1399
Web site: www.dnr.state.la.us/SEC/EXECDIV/
   TECHASMT/

Maine
Office of the Governor
Phone: (207) 287-4315

Maryland
Maryland Energy Administration
Phone: (410) 260-7511
Web site: www.energy.state.md.us

Massachusetts
Division of Energy Resources
Department of Economic Development
Phone: (617) 727-4732
Web site: www.magnet.state.ma.us/doer

Michigan
Michigan Public Service Commission
Michigan Department of Consumer and Industry
   Services
Phone: (517) 241-6180
Web site: www.michigan.gov/mpsc

Minnesota
Energy Division
Minnesota Department of Commerce
Phone: (651) 297-2545
Web site: www.commerce.state.mn.us

Mississippi
Energy Division
Mississippi Development Authority
Phone: (601) 359-6600
Web site: www.mississippi.org/programs/energy/
   energy_overview.htm

Missouri
Department of Natural Resources
Phone: (573) 751-4000
Web site: www.dnr.state.mo.us/energy/homeec.htm

Montana
Department of Environmental Quality
Phone: (406) 841-5240
Web site: www.deq.state.mt.us/energy/

Nebraska
Nebraska State Energy Office
Phone: (402) 471-2867
Web site: www.nol.org/home/NEO/

Nevada
Nevada State Office of Energy
Phone: (775) 687-5975
Web site: energy.state.nv.us

New Hampshire
Governor's Office of Energy & Community Services
Phone: (603) 271-2155
Web site: www.state.nh.us/governor/energycomm/
   index.html

New Jersey
Office of Clean Energy
New Jersey Board of Public Utilities
Phone: (609) 777-3335
Web site: www.bpu.state.nj.us

New Mexico
Energy Conservation and Management Division
New Mexico Energy, Minerals and Natural Resources
   Department
Phone: (505) 476-3310
Web site: www.emnrd.state.nm.us/ecmd/

New York
New York State Energy Research and Development
   Authority
Phone: (518) 862-1090
Web site: www.nyserda.org

North Carolina
State Energy Office
North Carolina Department of Administration
Phone: (919) 733-2230
Web site: www.energync.net

North Dakota
Division of Community Services
North Dakota Department of Commerce
Phone: (701) 328-5300
Web site: www.state.nd.us/dcs/Energy/default.html

Ohio
Office of Energy Efficiency
Ohio Department of Development
Phone: (614) 466-6797
Web site: www.odod.state.oh.us/cdd/oee/

Oklahoma
Office of Community Development
Oklahoma Department of Commerce
Phone: (405) 815-6552
Web site: www.odoc.state.ok.us/

Oregon
Oregon Office of Energy
Phone: (503) 378-4131
Web site: www.energy.state.or.us

Pennsylvania
Pennsylvania Energy
Department of Environmental Protection
Phone: (717) 783-0542
Web site: www.paenergy.state.pa.us/

Rhode Island
Rhode Island State Energy Office
Phone: (401) 222-3370
Web site: www.riseo.state.ri.us/

South Carolina
South Carolina Energy Office
Phone: (803) 737-8030
Web site: www.state.sc.us/energy/

South Dakota
Governor's Office of Economic Development
Phone: (605) 773-5032
Web site: www.state.sd.us/state/executive/oed/

Tennessee
Energy Division
Department of Economics & Community Development
Phone: (615) 741-2994
Web site: www.state.tn.us/ecd/energy.htm

Texas
State Energy Conservation Office
Texas Comptroller of Public Accounts
Phone: (512) 463-1931
Web site: www.seco.cpa.state.tx.us/

Utah
Utah Energy Office
Phone: (801) 538-5428
Web site: www.energy.utah.gov

Vermont
Energy Efficiency Division
Vermont Department of Public Service
Phone: (802) 828-2811
Web site: www.state.vt.us/psd/ee/ee.htm

Virginia
Division of Energy
Virginia Department of Mines, Minerals & Energy
Phone: (804) 692-3200
Web site: www.mme.state.va.us/de

Washington
Washington State University Energy Program
Phone: (360) 956-2000
Web site: www.energy.wsu.edu

West Virginia
Energy Efficiency Program
West Virginia Development Office
Phone: (304) 558-0350
Web site: www.wvdo.org/community/eep.html

Wisconsin
Division of Energy
Department of Administration
Phone: (608) 266-8234
Web site: www.doa.state.wi.us/energy/

Wyoming
Minerals, Energy & Transportation
Wyoming Business Council
Phone: (307) 777-2800
Web site: www.wyomingbusiness.org/minerals/

Glossary

Alternating-current electricity—Electricity created by the rapid flow of electrons back and forth along a wire at nearly the speed of light.

Blackwater—Water from toilets and kitchen sinks.

Brownfield—A previously developed site that may or may not be contaminated. Cleanup may or may not be necessary.

Building-integrated photovoltaics—Roofing materials or glass containing amorphous silicon that produces electricity when struck by sunlight.

Catchwater system—A system that captures water from roofs and other impervious surfaces (driveways, for example) and stores it in tanks for household or outside use.

Certified lumber—Wood produced by companies whose operations have been certified by an independent organization as sustainable.

Change order—A request to the builder to change part of the original construction contract.

Compact fluorescent light bulbs—Fluorescent light bulbs that screw into conventional light fixtures. They're color-adjusted to produce light similar in quality to that of ordinary light bulbs, but because they're so efficient, they use only about one-fourth of the electricity to produce the same amount of light.

Compressive strength—The ability of a material to withstand compression resulting from weight placed on it.

Cooling load—A term engineers use to describe the amount of energy that is required to keep a home cool during the cooling season.

Cross-ventilation—Airflow from one side of a house to another, usually created by opening strategically placed windows.

Daylighting—Using natural light from windows and skylights to illuminate rooms.

Direct-current electricity—Electricity formed by the one-way flow of electrons through a wire. It is produced by solar electric cells and wind generators and can be stored in batteries. In most homes, DC electricity is converted to alternating-current (AC) electricity, sacrificing some efficiency but gaining in convenience since most appliances and fixtures utilize AC.

Earth sheltering—Building houses or other buildings so they are partially protected by a blanket of soil. In some instances, exterior walls may be partially buried or bermed, that is, covered up to 3 or 4 feet. In others, walls may be completely buried and dirt may also cover the roof. This technique takes advantage of the Earth's relatively constant temperature and reduces heat loss in the winter and heat gain in the summer.

Earthships—Self-contained housing vessels that provide heat, electricity, food, water, and wastewater treatment so their occupants can "sail" into the future with little environmental impact and as little dependence on the outside world as possible.

Embodied energy—All of the energy that is required to make a product. This includes the energy needed to harvest or mine raw materials, process them, and manufacture a product. It also includes the energy required to transport raw materials to production facilities and finished products to stores and end users.

Ergonomic design—Designing and arranging components of a house for efficiency, ease of use, and safety.

Exfiltration—The movement of air out of a home, which makes a home less comfortable and more costly to heat and cool.

Exterior sheathing—Usually plywood or oriented strand board applied to the framing of a home on the exterior walls.

External heat gain—Heat that enters a building from outside sources, such as warm air leaking in through cracks in the building envelope or sunlight penetrating windows. External heat gain poses a problem during the cooling season, usually the summer.

Finger-jointed lumber—Lumber consisting of several pieces glued together at interlocking joints, much like interdigitation of fingers.

Formaldehyde—An organic chemical with many uses in industrial societies from embalming fluid to tissue preservative to a component of binding agents in many building materials, including oriented strand board, plywood, and fiberglass insulation. Formaldehyde is an irritant and thought to be one cause of multiple chemical sensitivity. It is also thought to be a possible carcinogen.

Grayfield—A site that has been previously built on but is not contaminated. No cleanup is necessary.

Graywater—Water from bathroom sinks, showers, and washing machines.

Green building program—A state- or city-run program designed to encourage builders to incorporate green building materials, techniques, and technologies in commercial, municipal, and residential buildings.

Greenfield—A previously undeveloped site.

Heating load—A term engineers use to describe the amount of heat energy that is required to keep a home warm during the heating season.

Heating season—That period of the year in which heat is required to maintain comfort.

Infiltration—The movement of air into a home, which makes the living space less comfortable. In the winter, infiltration can increase heating bills; in the summer, it can increase cooling bills.

Internal heat gain—Heat in a house generated by internal sources, such as light bulbs, appliances, and stoves.

Life-cycle cost—All of the costs of a product over its life cycle, from the extraction to the manufacture to the sale and use of a product to its ultimate disposal.

Living roof—A waterproofed, well-fortified roof planted with wildflowers and grasses.

Multiple chemical sensitivity—An immune system disorder characterized by severe, sometimes debilitating reaction to chemicals released by building materials, paints, stains, finishes, and other products.

Optimum value engineering—An approach to designing buildings that seeks ways to minimize material use without sacrificing structural integrity or violating building code requirements.

Oriented strand board—Sheathing made by gluing together large, flat wood chips.

Outgassing—The release of volatile chemicals from paints, stains, finishes, adhesives, building materials, furniture, and furnishings.

Roof sheathing—Typically plywood or oriented strand board nailed onto roof framing. Roofing materials such as shingles are applied to the roof sheathing, usually over a layer of waterproof material known as roof felt.

Roof truss—The framing that underlies many roofs. Trusses are typically made from smaller pieces of wood and are precisely engineered to support the projected loads.

R-value—A measure of how well a material resists heat conduction. The higher the R-value, the greater the heat resistance.

Septic tank—Usually a concrete tank buried underground that receives raw sewage and wastewater from a house. Solids remain in the tank and undergo decay, while liquids drain out of the tank into a leach field, a network of porous pipes buried in the ground.

Task lighting—A lighting system providing more intense light to high-use areas, such as kitchen counters, reading areas, and desks, where important tasks are carried out, rather than lighting an entire room to one level.

Tensile strength—The ability of a material to withstand lateral pressure.

Thermal bridging—Conduction of heat into or out of a building through framing members.

Thermal mass—solid materials, such as tile and concrete inside a passive solar house, that absorbs solar heat and help stabilize internal temperatures.

U-value—A measure of heat transmission through a material. It is the inverse of R-value. The lower the U-value, the less the heat transmission.

Water table—The upper boundary of groundwater.

Index

324 INDEX